Greener Marketing

A Global Perspective on Greening Marketing Practice

Edited by Martin Charter and Michael Jay Polonsky

Greener Marketing

A GLOBAL PERSPECTIVE ON GREENING MARKETING PRACTICE

EDITED BY
MARTIN CHARTER AND MICHAEL JAY POLONSKY

Published by Greenleaf Publishing Limited
Aizlewood Business Centre
Aizlewood's Mill
Nursery Street
Sheffield S3 8GG
UK

Typeset by Greenleaf Publishing Limited and printed on environmentally friendly, acid-free paper from managed forests by Biddles Ltd, King's Lynn, Norfolk

British Library Cataloguing in Publication Data:
Greener marketing : a global perspective on greening marketing practice. - 2nd ed.
1.Green marketing
I. Charter, Martin II. Polonsky, Michael J.
658.8'02

ISBN 978-1-874719-14-4

Contents

Preface

Greener Marketing: A Global Perspective on Greener Marketing Practice was conceived as a second edition to the highly successful *Greener Marketing: A Responsible Approach to Business*, the first book published by Greenleaf Publishing in 1992, in the wake of the Rio Summit and an awakening of interest in greener products by consumers and companies alike. It presented case studies of best practice from a range of organisations and brought together what was at that time quite disparate material, while retaining a practical bias.

With the rapid growth in this area over the last six years, it was soon clear that a mere update of the first edition would not be appropriate. The issues have moved on and the importance of the global economy has become even more evident. The second *Greener Marketing*, therefore, is a completely new collection offering current perspectives on a field that has grown and become far more sophisticated. The contributors to this volume provide a truly international focus, offering viewpoints from developing economies as well as leading-edge strategies and cases on such key areas as service provision, stakeholder communication and consumer attitudes.

It is fascinating to observe the distance travelled between the original volume and this, the second *Greener Marketing*; which in turn provides food for thought on what issues might be under discussion in *Greener Marketing 3*. Time will tell.

Foreword

SUCCESSFUL LEADERSHIP in the 21st century, be it in corporate management, product planning, sales and marketing or a service organisation, will require a more complex mix of skills than in the past. Managers will need to be able to recognise and understand trends and to navigate roadblocks effectively. They must be capable of developing efficient and/or innovative responses to key challenges, and they must be ready to profit from emerging opportunities.

Moving toward sustainability requires innovation and change. Although sustainability is a relatively new concept, business and other social institutions have considerable experience in balancing economic, social and environmental factors into their decision-making processes. Such decisions become more complex as business is faced with increasing challenges in environmental protection and as demand for more socially responsible and environmentally friendly products and services increases.

More and more managers are beginning to ask themselves questions such as: Do our customers really need the product the way we have designed it, or do they need another service? Some postulate that society simply has to learn to consume less. Others point out that if the 'less in quantity' is produced, used and disposed of irresponsibly, the net effect could be far worse than the 'more' that incorporates eco-efficient practices. Eco-efficiency is driven by the vision of making the challenge of sustainability a business opportunity.

As our companies continuously adapt to the changing dynamics in the markets in which they operate, business is implementing the vision through process changes, innovative product design, and through providing new services and functional offerings to meet the needs of their customers. These improvements require co-operation and dialogue between business, consumers and policy-makers.

Governments have a major role to play in helping to spread the applications of eco-efficiency. They have the means to promote and accelerate the process of change. And it is their responsibility to ensure free and open markets that provide the necessary encouragement for business to adapt eco-efficient practices.

In the dialogue with our consumers, it is well understood that the key to securing a real change towards sustainable development lies in shifting consumer behaviour.

Through marketing and advertisement, business reaches to the consumers and informs them. Many sectors of industry have become very adept in successfully employing marketing techniques, but the fact remains that business cannot sell what the consumer does not want. Market signals are thus the most effective drivers for changing production patterns and industry's competitive position.

For marketing managers, the challenge is to incorporate environmental considerations into the increasingly complex mix. ***Greener Marketing: A Global Perspective on Greening Marketing Practice*** is a practical guide for the business executive, manager and business student who wants to learn what the winners are doing and how to be successful in communicating to the market. The international team of contributors assembled here provide an impressive 25 chapters on strategic development, 'greening' the marketing mix and case studies that demonstrate that innovation is the key to sustainability.

I express my sincere hope that this new book will make a significant impression on the international dialogue between policy-makers, experts and business students alike.

Lutz-Günther Scheidt
Director, Environmental Center Europe,
Sony International (Europe) GmbH, Stuttgart

April 1999

Introduction

Martin Charter and Michael Jay Polonsky

AS SUSTAINABILITY evolves and starts to take account of social considerations, a change is occurring in green marketing. There is a growing recognition that, if we are to become more sustainable, we need a significant reduction in energy and resource consumption. But these environmental improvements will come not just from designing greener products and services. There will need to be a greatly improved understanding of customers—which means a greater need to focus on changing behaviour and reducing the environmental impact in the 'use' phase. This second phase of green marketing—the subject of this book—will have to be considerably more sophisticated than the first phase.

Progressing the greening of marketing to this next level requires that firms integrate environmental issues into the range of marketing mix variables. However, there may also be a need to develop markets wherein firms not only educate consumers about their products, but also educate consumers about the environmental issues involved. They may need to identify why their specific products address these issues in a more effective fashion than competitors. Such a perspective assumes that firms are broadening the appeal of their green marketing to cover the gamut of corporate activities, rather than simply designing green products targeted at a narrow green niche. In this way, the greening of marketing is no longer a fringe issue but a mainstream marketing mix issue—an issue of growing importance for all organisations.

There are some difficulties with the greening of marketing. The process of putting information across to consumers assumes that not only is the relevant environmental information available, but that consumers are able accurately and effectively to interpret it. Such a task is extremely problematic, especially in the context of continually evolving environmental awareness and data. In order to be able to communicate effectively, marketers and their organisations need to update their knowledge base on an ongoing basis. There is currently a growing emphasis that, within the greening process, firms need to be 'learning organisations', able to interact with various external stakeholders and internalise new information such that it improves organisational processes. In the context of the greening of marketing, a learning philosophy also requires more internal communication to ensure that the environmental consequences of a firm's activities are considered in the design and development stages, rather than simply trying to develop 'end-of-pipe' types of solution.

The greening of marketing has to overcome a range of other hurdles as well. In many cases, the greening process requires a substantive change in consumer behaviour. While an educational process might assist in identifying why changes in behaviour should occur, individuals find making discontinuous change difficult. For example, moving from the concept of owning products to either leasing them or sharing them may be too big a change for some consumers to make in a short time-frame. Firms can facilitate this process by making the change easier for consumers, by persuading them to adopt newer types of need-satisfying bundles which may not be based on traditional products. This book presents many examples of how the product-versus-service debate is developing.

Firms can extend beyond the basic marketing philosophy—whereby they create exchanges that satisfy consumers' needs—towards assisting consumers in identifying and adopting new, less environmentally harmful ways of satisfying their needs. This process will enable them to solidify their relationship with consumers by demonstrating that they truly understand consumers' underlying needs.

As this book clearly demonstrates, there are ample opportunities to green marketing profitably, as green products are not only often less expensive to produce, but may enable firms to recover valuable resources that were traditionally considered waste—for example, reprocessing used products. The fact that green marketing can be used to achieve long-term competitive advantage is especially important, as firms seek to differentiate themselves in an increasingly competitive business environment.

This book is divided into three sections. The opening section outlines broad strategic developments in the greening of marketing and discusses the broad greening debate. Subhabrata Bobby Banerjee explores the greening of marketing from a strategic perspective and explains the need to take on a stakeholder focus; Walter Wehrmeyer provides an overview of corporate environmental strategy development and its links to marketing; and Ken Peattie questions the various ways in which marketers think about addressing consumers' needs.

The second section deals with 'greening the marketing mix' and examines how marketers can green the tactics and strategies associated with the marketing management process. Jacquelyn Ottman focuses on the need to move beyond the eco-redesign of existing products and to move forward to eco-innovation of new products. Frank Belz imagines a future scenario where different strategies have been developed to satisfy customer needs, introducing the key topic of services as a replacement for product ownership. This theme is continued by Kai Hockerts, who provides an analysis and strategies for eco-service development and innovation.

Devashish Pujari and Gillian Wright provide background thinking on green product development and present findings from a survey of UK companies, whereas Robin Roy focuses on a single company, describing how Hoover has applied eco-design to its product range. Colin Beard and Rainer Hartmann provide a range of examples of how companies are implementing eco-innovation in product/service development.

We then turn to issues of external communication, beginning with two chapters on the subject of partnerships. Cathy Hartman, Edwin Stafford and Michael Polonsky use an analysis of a German example to provide a discussion of how firms can form alliances with environmental groups and how these groups can serve as bridges to other stakeholders; Easwar Iyer and Sara Gooding-Williams take a more empirical look at the subject, examining and classifying a range of NGOs, identifying how environmental alliances can be better developed, based on a study conducted in the US. Katharina Zöller then examines various types of opportunity for opening stakeholder firm communication, based on examples from both Germany and the US, while Lassi Linnanen, Elina Markkanen and Leena Ilmola examine other environmental communication strategies with examples of how issues can develop.

Shifting the focus of the book to the consumer, Hannu Kuusela and Mark Spence examine motivations and behaviour in relation to the purchase of energy-efficient durable goods. Daniel Ackerstein and Katherine Lemon report on the findings of an experimental design to examine the impact of greening the brand on consumer preferences and then present strategies based on these findings. Graham Earl and Roland Clift, on the other hand, examine the role of green marketing when dealing with business-to-business consumers, who are, it is generally perceived, substantially different to final consumers.

The third section of the book focuses on case studies of the marketing practices of specific firms, industries or regions. In doing so, the authors draw out broader green marketing implications as well as discussing what individual firms have learned or could learn from the specific cases described. The section begins with Alan Neale, who examines the infamous Brent Spar incident, in which Shell believed it was doing everything right but became embroiled in a communication and PR nightmare; the author describes how events unfolded and discusses how the company could have better managed the situation. Kate Kearins and Babs Klÿn then discuss The Body Shop International's approach to marketing and describe the problems associated with 'sticking one's head above the parapet'.

Rebecca Winthrop offers a developing-country perspective, examining the strategy used to green agroindustry in Costa Rica using an independent environmental certification programme. Sonja Grabner-Kräuter and Alexander Musch examine how an Austrian retailer used the development of a green product line to gain a competitive advantage in the industry. Dallas Hanson, Rhett Walker and John Steen look at approaches to eco-tourism in Tasmania, arguing that any such initiative is closely linked with the concept of stakeholder responsibility. There then follows two chapters by John Butler and Suthisak Kraisornsuthasinee on South-East Asia. The first examines the complexities of using green marketing strategies in countries where the concept is less established; the second describes how one firm in Thailand failed to maintain their environmental image because of a poorly managed strategy.

Norbert Wohlgemuth, Michael Getzner and Jacob Park illustrate the complexities of pricing and service provision associated with green electricity using a range of examples, particularly from the US. How do you green the provision of electricity when many consumers are not even aware that they have a choice between different alternatives? The importance of the educational aspect of green marketing strategy is highlighted here. Finally, John Cox, Joseph Sarkis and Wayne Wells examine the complexities of developing industrial ecosystems based on corporate recycling activities.

Greener Marketing: A Global Perspective on Greening Marketing Practice has been written to provide practitioners with best-practice examples and actionable recommendations on how to implement green marketing activities. It has been designed to provide information and ideas for those involved in marketing on how to incorporate green considerations into the marketing mix, as well as providing new perspectives on marketing for environmental managers. We hope that this volume will prove to be not merely a sequel to the first, but a benchmark reality check on international progress towards the greening of marketing practice.

In completing this project, we would like to thank a number of people, including the staff of Greenleaf Publishing for ensuring that the project moved along smoothly, the staff at the University of Newcastle, Australia, and the business partners at The Centre for Sustainable Design, Surrey Institute of Art & Design, University College, who have provided encouragement and support, as well as our respective families who have put up with our long hours.

Martin Charter
Michael Jay Polonsky

March 1999

1 STRATEGIC DEVELOPMENTS

1

Corporate Environmentalism and the Greening of Strategic Marketing

Implications for Marketing Theory and Practice

Subhabrata Bobby Banerjee

ENVIRONMENTALISM is enjoying a resurgence in today's society and the 1990s can be arguably called the decade of the environment. The previous surge in public interest in the environment began in the 1970s, a decade that saw the enactment of important legislation such as the Clean Air Act in 1970, the Resource Conservation and Recovery Act in 1976, and the Clean Water Act in 1977 (Cairncross 1992). Public and media attention waned in the 1980s; however, environmental problems worsened. The late 1980s and early 1990s saw a re-emergence of environmental concerns: environmental problems such as ozone depletion, global warming and rainforest destruction were high-priority items on the public agenda. The Rio Earth Summit of 1992 was a defining event in the business–environment relationship and corporate environmentalism became a mainstream rather than a fringe activity.

Public awareness of environmental issues is now almost universal. A national survey conducted by the Roper Organisation in 1992 found that 20% of North Americans constituted the 'True-Blue Green' segment (Roper 1992). This segment reflects the highest level of concern for the environment in terms of consumers' attitudes and behaviour. Interestingly, this greenest segment almost doubled from 11% in 1990 (Roper 1990), despite rising public concerns about economic recession, unemployment and healthcare. In recent years, this segment has shrunk to 10% (Speer 1997), as has the overall proportion of consumers who say they 'care about the environment' (63% in 1996, down from 72% in 1990). While the environment may not be the top-priority issue it was in the early 1990s, people still care about it: the apparent apathy is due to consumer perceptions that products and services are 'greener' than they were in the past (Speer 1997). Several recent surveys indicate that the state of the environment continues to remain a high-priority issue in many countries all over the world (Bonner 1997; Shanoff 1996).

Environmental issues have been studied by researchers in such diverse disciplines as economics, sociology, education and psychology since the 1970s. Interest in the environment appears to be reviving in recent years among marketing scholars: many international marketing conferences in the last four years have included special session papers on environmental issues, and several marketing journals have brought out special issues on the topic. In 1991, the American Marketing Association developed an environmental policy statement that urged all marketers to integrate environmental concerns into the business decision-making process, improve the accuracy of environmental claims for products and services, reduce environmental impact of their products, and work with industry, government and the public to find meaningful solutions to environmental problems (*Marketing News* 1991). The role of the business corporation in environmental protection is considered to be a major strategic focus for the nineties and, according to Varadarajan, 'enviropreneurial marketing appears to be taking roots in a growing number of organisations' (1992: 342). Enviropreneurial marketing is defined as 'the process for formulating and implementing entrepreneurial and environmentally beneficial marketing activities with the goal of creating revenue by providing exchanges that satisfy a firm's economic and social performance objectives' (Menon and Menon 1997: 54).

This growing trend appears to reflect changes in the external environment[1] of market systems: increased regulatory forces and public environmental concern have the potential to influence marketing actions at both micro and macro levels. Governmental monitoring and control of the ecological impact of business activity is a process that is designed to minimise the negative consequences of environmental damage. These macro-level actions attempt to address societal concerns about environmental issues and have strategic implications for business firms that are manifested at the micro level.

This chapter will discuss the phenomenon of corporate environmentalism, i.e. the process by which business firms integrate environmental concerns into their decision-making process. Using theoretical perspectives from strategic management and marketing, the chapter will examine the ways in which environmental concerns can be integrated into the strategic marketing decision process. Some plausible driving forces of corporate environmentalism will be identified and linked to strategic business actions. Possible consequences of an environmental strategy will also be discussed and the chapter will conclude by discussing implications for marketing.

1 The term 'environment' has multiple meanings. Throughout this chapter, the term will be used to include two meanings: the conventional academic connotation referring to surrounding conditions of a firm that influence its activities; and the 'green' meaning, the biophysical environment. When a distinction needs to be made, the terms 'biophysical' or 'natural' environment will be used for the latter. Obviously, corporate environmentalism refers to influences stemming from the natural environment.

Marketing and the External Environment

Traditionally, marketing theory has largely ignored the influence of the biophysical environment in the formulation of strategy. Conventional approaches to environmentalism have focused on 'green marketing', i.e. the implementation of marketing programmes directed at the environmentally conscious market segment (Henion 1976). This view has been criticised as being unnecessarily restrictive, and researchers have recommended that an environmental marketing programme should emerge from broader issues arising from the relationship of a firm and its stakeholders (Coddington 1993). As we will see in the next section, this chapter takes a much more comprehensive view of environmentalism, involving all levels of strategy.

Conceptualisation of the external environment of marketing has, until recently, ignored the biophysical environment. 'Environmental scanning' is used widely in the literature as an important part of the strategic planning process (Zeithaml and Zeithaml 1984), but the biophysical environment plays virtually no role in the development of strategy. The external environment as defined in environmental scanning has, until now, been restricted to the social, political, cultural, technological, economic or legislative environments. What is often overlooked in the analysis of external environments is that all its aspects operate within and are constrained by the biophysical environment. An increase in environmental legislation in recent years has resulted in heightened environmental awareness among the business community and many corporations have been compelled to integrate environmental issues into their strategic planning process in order to meet stricter environmental standards. However, this process is not an easy one and, as many researchers have pointed out, it is important to examine ecological constraints on strategy formulation and these considerations should be applied at the broadest corporate strategy level (Gladwin *et al.* 1995; Jennings and Zandbergen 1995; Shrivastava 1995a). However, theory development and empirical studies examining environmental influences on corporate strategy continue to be scarce.

Theoretical attempts to link the biophysical environment with business organisations have resulted in two research streams. One area focuses on the theoretical and paradigmatic implications of integrating the environment into strategy. Researchers have called for the re-evaluation of existing neoclassical economic paradigms and have discussed the emergence of new paradigms such as the ecocentric paradigm and the sustainable development paradigm (Gladwin *et al.* 1995; Purser *et al.* 1995). The basic premise of these arguments is that attention to the natural environment is lacking in the literature and, in cases where environmental issues have been addressed, ecological principles are either subsumed or disassociated with the economic paradigm. These views challenge traditional assumptions of the neoclassical economic paradigm. The main criticism of the current economic paradigm is that it presents a distorted picture of the economic situation by ignoring environmental damage (Passel 1990). Macroeconomic indicators such as GNP and GDP do not reflect the costs of environmental damage. For example, one of the world's worst environmental disasters, the *Exxon Valdez* oil spill, actually showed up as a gain in the United

States' GNP because of the products and services involved in the clean-up (Reilly 1990). Traditionally, environmental costs have been treated as 'externalities' arising from economic activity, and these costs are typically not borne by the producer and are thus not included in the market transaction. Public policy actions frequently attempt to internalise these externalities by estimating the external cost of pollution and by applying pollution taxes (Petulla 1980). These normative actions take into account the needs of the external stakeholders of business firms.

Stakeholder theory, business ethics and corporate social responsibility are some promising avenues for internalising environmental costs from the perspective of a business firm (Gladwin *et al.* 1995). The traditional view that the social responsibility of a firm means maximising profits for stockholders is being challenged by the stakeholder perspective (Bowie 1991; Westley and Vredenburg 1996; Klonoski 1991). Apart from stockholders, organisational stakeholders include employees, customers, the local community, government agencies, public interest groups, trade associations and competitors. Stakeholder theory implies that, since all stakeholders are legitimate partners in a business, a business firm must consider the impact of its actions on all stakeholder groups. Including the planet as a stakeholder (arguably the ultimate stakeholder) in this framework implies that business firms need to be accountable for environmental damage. According to a recent survey, an increasing number of North American consumers believe that companies should take greater steps to deal with environmental issues: 35% in 1996, up from 29% in 1993 (Lawrence 1998).

One problem with the stakeholder approach is that these theories only focus on 'what should be done' by organisations to address environmental problems and how organisations *should* be socially responsible and take into account the needs of all their stakeholders. Stakeholder theory does not provide too much detail on how to translate moral decisions into business actions, and there are pragmatic difficulties in operationalising stakeholder theory in real-world business decisions (Sternberg 1997). Different groups can have differing interests, and balancing competing interests to satisfy the needs of all stakeholders can be a difficult task (Fineman and Clarke 1996). In fact, some firms that attempted to position their operating strategy as ethical, environmentally and socially responsible are coming under increasing scrutiny by the public, media and government. False advertising claims about environmental attributes of products and services and attempts to 'greenwash' the public are among the criticisms directed at these firms (Entine 1995; Rosen 1995). The rise in public environmental concern in the late 1980s and the emergence of a 'green' market segment led many firms to try to position their products (often with dubious claims) as 'environmentally friendly'. Since 1990, the Federal Trade Commission (FTC) has listed 32 cases of 'suspicious environmentally related product claims' (Speer 1997) and threatened prosecution if the companies in question did not stop making the claims (the companies did stop making the claims).

The second research stream focuses on the strategic implications of environmental issues for the firm and describes the range of strategic options facing a firm in dealing with environmental issues. How does the environment impact on manufacturing strategy? How does it impact on marketing strategy or on new product development? This research examines environmental issues that can influence the behaviour of decision-makers within firms (Hanna 1995). It examines how managerial behaviour can be modified to address environmental issues and exemplifies the position of this chapter: namely, the conceptualisation of corporate environmentalism as the inclusion of environmental considerations in the strategic planning process of the firm (Maxwell *et al.* 1997).

Examples of this phenomenon abound in the business press. McDonald's launched a major waste reduction effort in alliance with the Environmental Defense Fund aimed at reducing waste by 80% in five years (Allen 1991). Procter & Gamble has invested in projects ranging from new, less environmentally harmful technologies to packaging modifications and industrial composting: under pressure from environmental groups, the company pledged to spend $20 million annually to develop composting facilities for disposable diapers (Coddington 1993). Firms are also engaged in actively seeking and cultivating environmentally conscious consumers by launching products with a 'green' appeal. For instance, the number of new green product introductions in national markets rose from 60 in 1986 to 810 in 1991. Moreover, the share of green products of all new product introductions rose from 1.1% in 1986 to 13.4% in 1991 (Ottman 1998). However, this figure has dropped over the last few years mainly due to increased consumer scepticism about environmental claims and the FTC's attempts to regulate their use (Kangun *et al.* 1991; Scammon and Mayer 1995). Green product introduction, especially in consumer markets, has slowed in recent years as the debate on what is really 'green' continues. However, many green products, once in niche markets, are now part of the general merchandise found in all sorts of store. Dollar sales for these products (diapers, feminine hygiene products, paper towels, dishwashing liquids) have shown sustained growth over the last few years (Speer 1997).

It can be argued that the effects of the behaviours described above are negligible in relation to the actual ecological impact of these firms' business activities and hence these actions are not really 'environmental'. The above examples do not imply that the firms in question are 'environmentally concerned' or 'socially responsible' companies. However, changes in the external environment have had an effect on these firms' strategies and led to some degree of integration of environmental issues. Corporate environmentalism as conceptualised in this chapter refers to the degree of integration of environmental issues into company strategy and is not intended to distinguish 'good' from 'bad' companies.

Corporate Environmentalism and the Business Firm

Changes in the external environment of the firm can influence their internal strategies. Policy actions designed to promote corporate environmentalism comprise one way of internalising the externalities of ecological damage. Corporate environmentalism involves the recognition by firms that environmental problems arise from the development, manufacture, distribution and consumption of their products and services. Including the biophysical environment in the strategic planning process and being responsive to environmental issues are important themes of corporate environmentalism. Based on an examination of the literature, a working definition of corporate environmentalism is proposed:

> **Corporate environmentalism is the organisation-wide recognition of the legitimacy and importance of the biophysical environment in the formulation of organisation strategy, inclusion of environmental impact of business actions in the strategic planning process, organisation-wide communication of corporate environmental goals, and the organisation-wide responsiveness to environmental issues.**

This definition comprises the two themes of corporate environmentalism that are found in the literature: the stakeholder concept and the strategy formulation concept. This conceptualisation of corporate environmentalism may appear to be too broad in scope, especially if traditional definitions of the role of marketing are considered. However, as will be discussed later, marketing has a crucial role to play in the development and implementation of corporate environmentalism. Marketing's role in corporate environmentalism becomes even more comprehensive if we employ a broader perspective of marketing's contribution to the strategic planning process of the firm (Day 1992; Varadarajan 1992).

One of the problems associated with examining the relationship between a business firm and the biophysical environment is the complexity of issues that arise. Every business firm can be said to have a negative impact on the natural environment, some more than others. It can be argued that there is no such thing as a 'green company'.

Objective standards for measuring the actual environmental impact of firms are extremely difficult to develop. There is growing work on developing standards for corporate environmental performance, but there is much disagreement on what these standards should be and who sets these standards. For example, the ISO 14000 series of environmental standards are process standards and not performance standards and do not provide quantifiable performance measures that a company should aim for. In any case, following a standardised system is voluntary. There is also considerable disagreement among environmental scientists about the long-term effects of different forms of pollution. While there are governmental guidelines for emissions and waste, these remain limited to the manufacturing process, and other areas of environmental impact, such as post-consumer waste, are often ignored. Another potential problem is that most firms will attempt to portray a green image regardless of the scope of their environmental activities. While the 'greenness' of firms must be established by assessing their actual 'downstream' environmental impact, corporate environmentalism as defined above can highlight the degree of integration of environmental issues into a firm's 'upstream' strategic planning process. As will be discussed later, environmental issues can be integrated at different levels of strategy depending on the characteristics of the firm and the industry, regulatory forces and public concern. The level of importance placed on environmental issues is important: the above definition of corporate environmentalism suggests that not only should environmental considerations be included in strategy, but should be integrated at the highest level of strategy.

One method of integrating environmental concerns into strategic planning involves employing the Total Quality Environmental Management (TQEM) approach, as advocated by the Global Environmental Management Initiative (GEMI), a group consisting of 20 leading companies. TQEM draws on the 1980s change in business philosophy to 'total quality management', which received widespread acceptance as a tool for improving corporate performance. In the 1990s, GEMI proposed applying TQM tools to environmental management and corporate environmental strategies (GEMI 1992a). The meaning of 'quality' was expanded to include environmental quality as well as product and process quality. A major implication of this conceptualisation is the integration of efforts within all business functions to improve environmental performance. In many cases, TQEM has led to an enhancement of product quality and corporate performance as well (GEMI 1992a).

For example, in order to comply with German environmental legislation, the Polaroid Corporation needed a significant reduction in its packaging for its new camera. The reduction in packaging meant that the product needed to be more durable to withstand possible damage in shipment. Driven by both regulatory pressures and a marketing strategy emphasising the environmental friendliness of the new package, Polaroid scientists redesigned the camera to make it more rugged, thus enhancing product quality (McCrea 1993). Combining 'quality' and 'environmental' concerns can also lead to significant cost advantages for companies, and many firms have made waste reduction a part of their corporate quality efforts. Firms such as Xerox, Procter & Gamble, IBM, DuPont, Digital Equipment Corporation, AT&T, Allied Signal and 3M have reduced pollution and lowered costs in diverse industries (Smith 1991).

In the examples discussed above, business firms performed a variety of activities in order reduce the environmental impact of their products and processes. These activities involved different levels of investment, both financial and managerial, and had different strategic implications for the firm. However, aligning environmental strategies with corporate strategies is by no means an easy task. A survey of 220 senior executives conducted by Booz, Allen & Hamilton indicated that a vast majority of managers thought that environmental issues were 'extremely important' to their firm (Newman and Breeden 1992). However, only 7% of those surveyed claimed to understand fully the environmental problems faced by their company. The survey found that most companies restrict their environmental activities to the manufacturing process and do not address environmental issues at the strategic level where competitive advantage can be created. The survey also found that companies that effectively managed environmental issues, as indicated by integration of environmental concerns throughout the organisation, received support from the highest echelons of management, and publicly demonstrated their environmental efforts (Newman and Breeden 1992).

How firms can internalise environmental concerns into their strategic planning process, what functional areas of the business are influenced by environmental considerations and at what level the firm can respond to environmental issues are addressed in the next section. Theoretical perspectives from strategic marketing and management are useful when integrating corporate environmentalism into the strategic planning process.

Strategic Management and Marketing Strategy: A False Dichotomy?

The strategic planning process will be used as a framework for understanding the concept of corporate environmentalism. Although the concept of strategic planning may appear to be outside the scope of traditional marketing strategy, Day and Wensley (1983) argue that there is a need to include the relationships of the marketing function of the firm with both internal and external forces. Traditional paradigms of marketing strategy view marketing planning as a lower-level activity arising from a corporate strategic plan (Greenley 1989; Walker and Reukert 1987). The dominant subject in the discipline of marketing is marketing management, which deals with the design of the marketing mix. Marketing strategy, on the other hand, focuses on competitive advantage and, as some authors claim, should be viewed as a part of corporate or business strategy (Wind and Robertson 1983). Proponents of the marketing strategy perspective argue that current research within the marketing discipline is limited in its scope and that the present paradigm is restrictive and does not emphasise innovation and adaptability, areas that are of crucial importance to the survival of the firm (Anderson 1982; Day and Wensley 1983; Wind and Robertson 1983). Important limitations that have been discussed in past literature include marketing's lack of interdisciplinary interaction, its short-term orientation, and the lack of competitive analysis (Wind and Robertson 1983).

One advantage of studying corporate environmentalism from a strategic perspective is that it overcomes the limitations mentioned earlier. The potential influence of corporate environmentalism is multidisciplinary and multifunctional: accounting, purchasing and production methods can all be influenced by a firm's commitment to environmental protection. For firms that make this strategic shift to corporate environmentalism, a marketing perspective for strategic planning at the corporate level is appropriate. If, as Anderson (1982) maintains, the objective of marketing is to maintain long-term customer support through customer satisfaction, effective corporate strategies for firm survival and growth should employ a marketing perspective. This involves not only focusing on customer satisfaction and identifying long-term strategies that assure customer support, but also communicating the importance of these strategic choices to the various departments within the firm as well as to other stakeholders. Thus, the role of 'internal marketing' is crucial in the acceptance of corporate environmentalism within the organisation. Internal marketing refers to the efforts and strategies employed to promote environmental awareness within the firm. Many large corporations have created special 'green teams' that are responsible for marketing environmental initiatives to different departments within the company.

In the case of a strategic emphasis on corporate environmentalism, a firm must adapt to changing needs of the market and also attempt to 'market' environmentalism to its internal constituencies or departments. To pursue this strategy, the value of environmentalism needs to be shared and communicated across departments. Before launching an environmental programme externally, a firm would also have to incorporate environmentalism into all of its functions. The interdependency of environmental marketing and environmental management thus broadens the role of the marketing function of the firm, and consequently the firm becomes more market-oriented (Coddington 1993; Kohli and Jaworski 1990). The meaning of the term 'market' becomes expanded to include customers, suppliers, regulatory agencies, the public and stakeholders. Thus, marketing plays a dual role in the strategic process: it identifies strategies that can provide long-term competitive advantage, and it is also involved in the implementation process, including communication of strategies across departments.

Firms concentrate their strategic efforts in a limited part of the market where they can maintain their distinctive competitive advantages (Henderson 1983). Long-term competitive advantage can arise from the emergence of strategic segments for environmental goods and services as well as the greening of existing products and services. Strategic segmentation is a broader concept than market or product positioning and involves a deeper understanding of the firm and its total environment. For example, a strategic segment could emerge when high levels of concern for environmental degradation in a market are matched with availability of recycled raw materials or the development of technologies that minimise impact on the natural environment. These new market conditions can influence a firm's strategies.

The changing needs of the marketplace is a source of structural fluctuations in any given industry (Porter 1981). A recognition of structural fluctuations may force firms to find new ways of innovating in order to survive in the marketplace. This can lead to changes in strategic decisions for these firms. Spurred by structural fluctuations such as increasing legislation and public concern, firms can choose to adopt corporate environmentalism as a strategy for surviving in the marketplace. These strategies can have feedback effects on market structure (Porter 1981), providing the setting for yet another series of structural changes in the environment.

The energy industry is a good example of the structural fluctuations and strategic changes that firms make. In some states in the US, regulators offer incentives to utilities for investing in energy conservation. Most utilities have realised that energy conservation is profitable, leading to the somewhat incongruous scenario where a firm spends money on advertising that urges their customers to use less of their product. The market for energy, traditionally based on increasing consumption, is now changing to an energy savings market, where utility companies follow an energy conservation strategy by attracting more consumers without the necessary increase in production capacity. While this may appear to be a simple example of demarketing (company efforts in reducing consumption of their products and services as is common during times of shortage), a closer look reveals that environmental regulations have been a major influence in this process. There is no 'shortage' of energy: more power plants can be constructed to meet the increased demand for energy. However, given the extent of environmental regulations governing the construction of new power plants and the substantial investment required to meet environmental laws, many utility companies have chosen to, or feel compelled to, adopt a conservation strategy. One would expect this conservation strategy to meet the financial goals of the companies involved. The strategic changes, though, occurred mainly as a response to changes in the regulatory climate as well as rising public environmental concern.

In what different ways can firms respond to changes in the external environment and formulate strategies to ensure their survival and growth? The strategic management process plays an important role in determining the effectiveness of strategies that integrate external constraints with corporate objectives. The following section discusses the various levels of the strategic management process and how corporate environmentalism can influence this process.

Schendel and Hofer's Levels of Strategy

In an overview of the strategic management process, Schendel and Hofer (1979) propose four hierarchical levels of strategy. These are enterprise strategy at the top, followed by corporate, business and functional strategies.

Enterprise Strategy

This is the broadest level of strategy that integrates the total organisation into its environment. Schendel and Hofer (1979) maintain that the focus of enterprise strategy is to examine the role the firm plays and should play in society. Firm governance, function and its form are areas that are addressed in enterprise strategy, questions that arise from a re-examination of the mission of the firm. A typical firm would perceive responsibility to its shareholders and customer satisfaction to be an acceptable mission. Environmental concerns can force the company to re-examine its mission and include other stakeholders, such as the public and environmental protection agencies, into the enterprise strategy level. Some firms (Ben & Jerry's and The Body Shop are examples that readily come to mind) attempted to incorporate these objectives at the enterprise strategy level with initial success. These two 'progressive' firms targeted the same segment using almost identical (and very effective) cause-related marketing strategies in two very different markets: ice cream and cosmetics. It is still early to comment on the long-term viability of 'green-niche' companies, although anecdotal evidence seems to indicate that, as a company grows, it does face problems of maintaining its environmental objectives and being profitable at the same time—as competition increases and so does pressure on profit margins. That fact that some of the environmental claims made by both these firms were later found to be more a result of creative imagination than actual fact (Entine 1995; Rosen 1995) raises important ethical and legal issues in the implementation of the stakeholder concept in enterprise strategy.

Environmental protection was the main theme that was used to position both the companies and their products: this was the stated criteria of their legitimacy in society. Historically, the economic function of a firm was the only criterion of its legitimacy in society. A business firm produced goods and services needed by society, and its existence was legitimised based on its ability to fulfil this need. Focusing on a firm's economic function as its sole criterion for legitimacy is inappropriate from a societal perspective, since the social and environmental impact of a firm's activities are not taken into account (Ansoff 1987; Gray 1992). Environmental concerns can lead to the development of new societal norms that determine the legitimacy of the business enterprise, and thus integrating corporate environmentalism at the enterprise strategy level is one way in which firms can seek institutional legitimacy in society.

Corporate Strategy

Corporate strategy involves identifying the kind of businesses that the firm should be involved in to meet its enterprise strategy goals. Integrating different businesses into a portfolio is an essential part of corporate strategy. Sociopolitical and cultural factors need to be considered in this level of strategy, although at the corporate strategy level, the focus is on the influence of these factors at the industry level, rather than at the broader society level. Product-market decisions are made at this level of strategy, as are decisions on technology development and use. If a firm makes a commitment to environmental protection at the enterprise strategy level, possible corporate strategic options are to enter environmental protection businesses or develop minimum-environmental-impact technologies. Both these strategies can also result in first-mover advantages.

The 'green market' for consumer products is also growing. Although consumers are rarely willing to pay a price premium for green products, there has been sustained growth in several environmentally positioned product categories. For instance, organic foods are expected to account for 10% of total foods sales by 2001 (Speer 1997). The 'natural products' industry (including fresh and organic foods and beverages, dietary supplements and personal-care products) grew 25% in 1996 for a sales value of $11.5 billion. The organic products segment (pesticide-free and fertiliser-free fresh produce, groceries and packaged goods) grew 26% to $3.5 billion. This industry has been experiencing high growth rates since 1992 (Speer 1997).

The environmental protection market is one of the fastest-growing ones in the G7 countries (Porter 1991). Growth in this sector was partly due to a shift in thinking from traditional 'end-of-pipe' technologies (which were proving to be prohibitively expensive) to pollution prevention and waste minimisation technologies. The focus shifted to clean production, design for the environment and eco-efficiency. Several small and medium-scale enterprises specialising in environmental technology have emerged over the last decade. These firms offer technical expertise and equipment, pollution control systems, waste management systems and serve a number of different manufacturing industries. The corporate strategy focus of these firms is to provide technology and services relating to environmental management: this defines their market and choice of business.

Apart from environmental management technology providers, the 1980s also saw substantial growth in the environmental management consultancy and services sector. The pattern of environmental consultancy is similar to that seen in green marketing: an initial boom in the number of companies offering environmental consultancy services was followed by a slowdown period, and this market is now dominated by a few large consultancy companies and many smaller firms offering specialised environmental services. The overall demand for external consultancy on environmental management has reduced as most companies, especially large industrial firms, have now acquired the expertise to develop and operate their own environmental management systems.

Business Strategy

A firm's business strategy involves the optimum allocation of its resources in order to achieve a competitive advantage. In addition, this level of strategy focuses on integrating the functional areas such as accounting and marketing into the business. Influences of corporate environmentalism on business strategies can include seeking cost advantages by adopting conservation procedures or attempting a premium positioning strategy by product differentiation on the 'green' dimension.

Competitive strategies are developed at this level of strategy in order to create and sustain a competitive advantage in the marketplace. Cost advantages, differentiated offerings and niche marketing are common business strategies used by firms (Porter 1980). Environmental issues can influence a firm's business strategy by providing a means of obtaining competitive advantage. New markets for environmental goods and services have emerged in recent years, now reaching a projected $600 billion per year by 2000.[2] Business strategies that exploit these markets include developing 'green' products aimed at the environmentally concerned segment of consumers, employing TQEM in business operations to take advantage of cost savings and increases in efficiency and productivity, and following a differentiation strategy by positioning the product/service on the green dimension.

Functional Strategy

Co-ordination is the key focus in a functional strategy: for example, the advertising plan for a new product launch is co-ordinated with personal selling and sales promotions. Co-ordination is not limited to one department: in the above example, the marketing activities must be co-ordinated with purchasing, accounting and production activities in order to implement a functional strategy successfully. The launch of a 'green' product, a strategic decision taken at the business strategy level, involves co-ordinating advertising and promotion efforts to position the product accordingly. Environmental issues can influence some or all elements of the marketing mix. Apart from product development ('biodegradable' detergents is an example), environmental issues can affect pricing (most 'natural' brands of foods, cosmetics and detergents are premium brands), and promotion decisions (e.g. green advertising). Thus, the greening of the marketing mix would be the basis of an environmental functional strategy.

---

2 Environmental Industries Commission press release, London, May 1997.

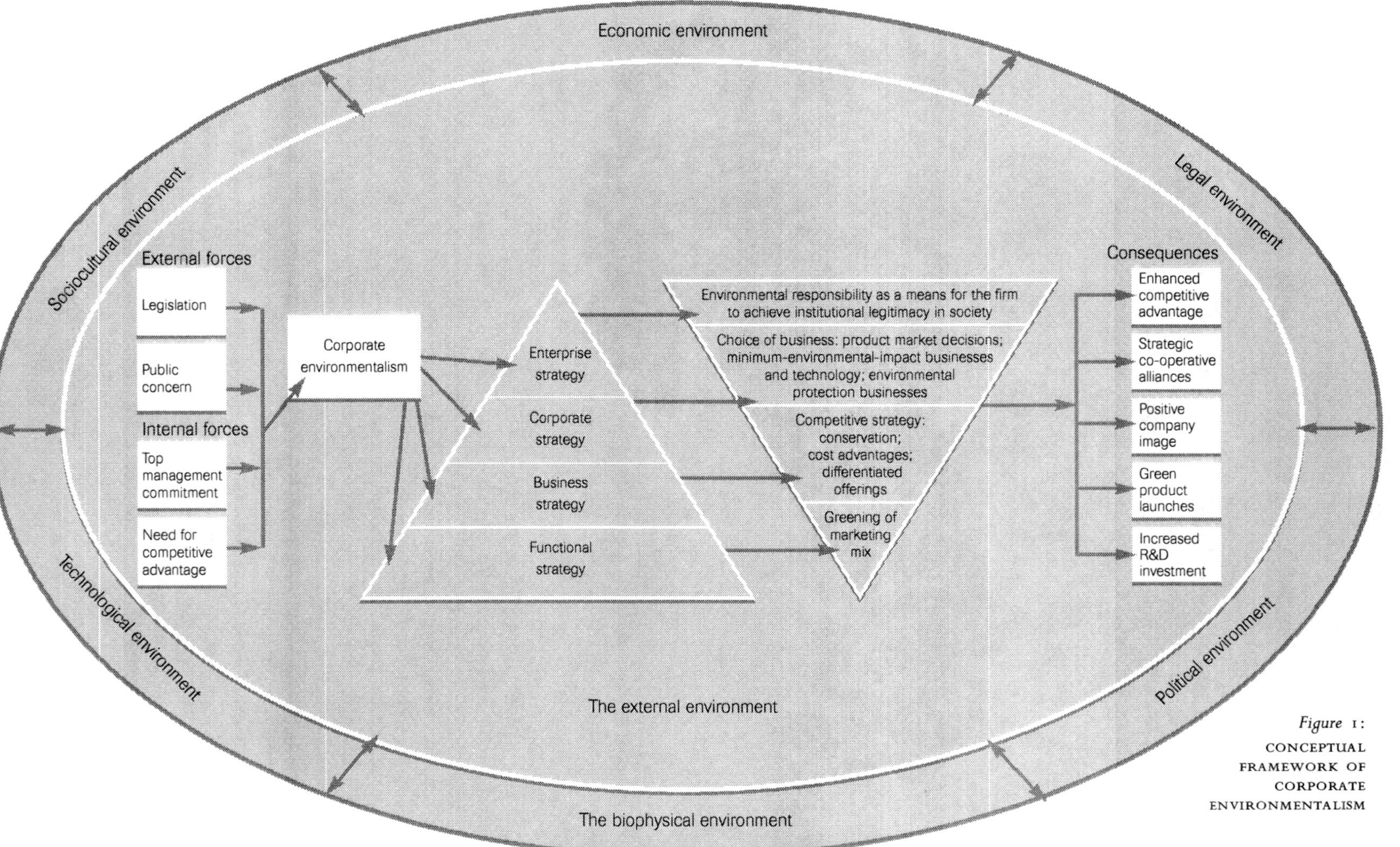

Figure 1: CONCEPTUAL FRAMEWORK OF CORPORATE ENVIRONMENTALISM

Implications of a Hierarchical Structure of Strategy

It is important to realise that these strategy levels are hierarchical and each level of strategy is constrained by the one above it. For instance, a functional strategy would be constrained by a firm's business strategy, which in turn is constrained by corporate and enterprise strategy. The different levels of strategy, influences from the external environment, and possible strategic responses by a firm are displayed in Figure 1.

From this hierarchical structure, it follows that, for a firm, the enterprise strategy is the most important level of strategy that drives the other levels. Enterprise strategy deals with the relationship of the firm and its environment and, consequently, conceptualisation of the organisational environment becomes crucial to the theory and practice of strategic management and marketing strategy (Throop 1993). Since the biophysical environment is largely ignored in conceptualising the organisational environment, most research on other levels of strategy has excluded the impact of the biophysical environment. This situation can be somewhat rectified by understanding that even the enterprise's strategic decisions are constrained by the biophysical environment. The relevance of the larger context of the biophysical environment would then influence strategic decisions at the corporate, business and functional levels.

However, it is important to note that the strategy process is not unidirectional, despite the hierarchical structure. Just as a firm can embody corporate environmentalism at the enterprise level, it is possible that the influence of the biophysical environment can occur at lower levels of strategy. For instance, at a purely functional level, packaging modifications to include more recycled content can be the basis of a 'green' appeal in the marketing function of a firm where advertising and promotion strategy highlight the environmental benefits of the package. The success of such a strategy can lead to the recognition of changing market needs, which in turn can influence business, corporate or enterprise strategy, thus providing a feedback loop.

In an attempt to develop a typology of environmental marketing strategies, Menon and Menon (1997) identified three approaches depending on the level of integration of environmental considerations. Environmental marketing strategies can be either strategic (formulated and implemented at the highest level of strategy), quasi-strategic (at the business strategy level) or tactical (functional). Decision-making on environmental issues can occur at various levels, from the top management team to strategic business unit managers or product and marketing managers.

The preceding discussion attempted to position corporate environmentalism in the strategic planning process of the firm. Having defined corporate environmentalism as the recognition by firms that the natural environment is crucial for their long-term viability, the question that now arises is: under what conditions do firms feel the need to respond to environmental issues and formulate strategies based on their impact on the natural environment? What external and internal forces lead to the adoption and implementation of corporate environmentalism?

For some firms, corporate environmentalism can be one strategy to pursue in its effort to seek a long-term competitive advantage in the market. There can be several

Determinant	Outcomes
Legislation	Higher levels of corporate environmentalism in firms operating in industries facing stricter legislation than firms in other industries. Corporate environmentalism integrated at higher levels of strategy (corporate or business strategy) leading to:
Corporate strategy	Greater levels of investment in environmental protection when corporate environmentalism is integrated at higher levels (corporate or business strategy)
	Green product positioning
	Cost advantages arising from environmental considerations, such as energy and resource conservation
Public concern	Higher levels of corporate environmentalism when organisational decision-makers perceive their customers to be environmentally conscious
	Negative public perceptions of a particular industry imply higher levels of corporate environmentalism among firms in these industries.
	Increased expenditure on green advertising emphasising the environmental benefits of products/services and promoting a green corporate image
Top management commitment	Higher levels of corporate environmentalism in firms where the top management team is supportive of environmental initiatives
	Members of the top management team will co-ordinate corporate environmental policies and programmes.
Competitive advantage	Higher levels of corporate environmentalism in firms that have experienced cost savings due to environmental initiatives
	New product development based on environmental considerations
	Developing new markets for environmental goods and services

Table 1: CORPORATE ENVIRONMENTALISM AND OUTCOMES

motivations for pursuing corporate environmentalism as a strategy: increased legislation, public concern, and rising costs of environmental protection are some factors. The following section will discuss some important determinants of corporate environmentalism and their strategic implications.

Determinants of Corporate Environmentalism

Two types of force or stakeholders can influence a firm's response to environmental issues. External forces such as the threat of legislation and public concern can lead to changes in a firm's strategies. The need for competitive advantage and top management commitment are internal forces that can influence a firm's strategies. The forces influencing corporate environmentalism along with potential outcomes are listed in Table 1.

How these forces influence the level of corporate environmentalism depends on a number of things: industry characteristics, company characteristics and market forces, to name a few. For instance, environmental legislation is more of a threat in certain industries, such as chemicals, engineering and utilities, than in others. Public concern for the environment may have a greater effect on strategic decision-making in certain visibly polluting industries, such as chemicals and oil—firms in these industries tend to have a poor public image on environmental issues. Many such industries currently use environmental responsibility as a theme in their corporate advertising campaigns (Banerjee *et al.* 1995). Thus, both internal and external forces influencing corporate environmentalism have varying effects on the level of corporate environmentalism, depending on the type of industry.

External Forces

Legislation. The threat of tougher legislation and the rising costs of complying with environmental regulations are possible motivating factors for firms to incorporate environmental concerns in their strategies (Banerjee 1998). The US Environmental Protection Agency (EPA) plays a major role in developing and enforcing environmental standards in different industries in the United States. Enforcement activities of the EPA have been steadily increasing since the 1970s, and their budget has increased from $205 million in 1970 to $7.4 billion in 1995.[3] Moreover, pollution standards are becoming stricter: the 1995 Clean Air Act has higher standards for air pollution and acid rain emissions which are expected to cost US business $21 billion annually.

Tougher legislation can affect a firm in two ways: first, the cost of compliance can become prohibitive. It is estimated that companies spend over $350 billion each year on environmental compliance. Given these legislative forces, many senior managers of firms perceive that a more effective strategy is to reduce emissions at the start instead of complying with clean-up and pollution control regulations.

---

3 EPA website, http://www.epa.gov, 1996.

Second, legislation can require substantial changes in product or package design or distribution channels. For example, several leading automobile manufacturers are incorporating the design for the environment process to include disassembly as a factor in their product design process (McCrea 1993). Another example is the reformulation of perfumes and colognes that is required to comply with new regulations proposed by the California Air Resources Board (Ottman 1998). In the international arena, a proposed German regulation involves manufacturers of electronics taking back used equipment (*Green Market Alert* 1993) and there are European Union directives regarding the recyclability of automobiles and electronic equipment. This can involve significant investments in channel structure as well as product design to include disassembly as a factor.

Another rising concern for firms is the increase in environmental liabilities and risks incurred. A recent survey indicated that, among the different kinds of environmental information required by investors, liabilities and litigation were ranked first and second respectively (Mastrandonas and Strife 1992). The US Securities and Exchange Commission has also mandated that corporations must disclose estimates of current and future environmental expenditures and liabilities. Companies are liable not only for any present damage to the environment but also for all future damage and they must disclose any environmental risks known to be potentially significant.

Thus, corporate environmentalism can stem from a perception that environmental legislation will increase and that the firm should not only respond to these pressures but also attempt to anticipate future legislation. Legislation can lead to different degrees of corporate environmentalism, depending on the level of strategy in which the firm includes environmental concerns. At a purely functional level, responses to legislative pressures can mean complying with existing regulations. At a higher corporate level of strategy, threat of environmental legislation and liability could influence decisions on new business opportunities.

Public concern. Another important reason for firms to develop an environmental orientation and strategy is the rise in public concern for the environment. There are literally hundreds of opinions polls on the environment conducted in Europe, Asia and the United States that indicate that environmental concern is a high-priority issue in both industrialised as well as developing countries. Research indicates that, besides the traditional reputational elements of a firm such as dependability, ethics and honesty, environmental attributes are becoming an integral part of corporate reputation. This includes corporate policies that demonstrate a concern for the environment, the degree of responsiveness to public concerns about the environment, environmental self-regulation and programmes to minimise generation of waste.

The results of many national polls tracking environmental concern among the general public indicate that environmental protection remains high on the agenda of the public worldwide, despite escalating economic woes in some areas. Three separate national surveys conducted in the US by the Yankelovich Organisation, the Roper Organisation and Simmons Market Research Bureau indicate that between 25% and 43% of the American population constitute the 'green' segment: consumers who are concerned about the environment (Earle 1993).

The need to maintain a good public image and respond to public concerns can lead to firms adopting corporate environmentalism. Many industries, such as the chemical industry or the oil industry, by the very nature of their products and processes, have a negative environmental image among the public. This probably explains why the most visible polluters such as the chemical industry and the oil industry are the ones that are publicly and privately paying the most attention to the environmental impact of their operations. A negative public image can influence firms in this industry to adopt corporate environmentalism as a strategy to survive and grow in the marketplace. All the large chemical corporations publicly affirm their commitment to environmental protection and have developed environmental mission statements or policy statements. Given the nature of their business, however, they continue to remain in one of the most polluting industries. Significant reductions in environmental impact will probably take some time, although the goal of zero emissions and declining levels of emissions in some cases is the direct result of integrating environmental concerns into business strategy.

In some cases, public perceptions of a company's environmental impact can lead to direct change in strategy. For years, McDonald's was the target of environmentalists as a large producer of solid waste and user of environmentally unsafe packaging. As environmental awareness grew, so did the negative public image of McDonald's. McDonald's abandoned their 'clamshell' polystyrene box after years of constant complaints from their customers and lobbying by environmental groups. Their initial response was to set up a recycling programme for polystyrene. However, due to operational problems and the public's negative perception of polystyrene, the recycling programme did not prove to be effective. In an effort to alleviate consumer concerns, McDonald's entered into an alliance with the Environmental Defense Fund hoping to influence public perception of their waste reduction activities. Despite investing in recycling programmes for polystyrene, McDonald's decided to replace the clamshell box with paper wrap, due to continued consumer pressure and the advice of its new environmental partner (Simon 1992).

Responding to public concerns goes beyond understanding and anticipating the needs of consumers and the market. Influences from suppliers, regulatory agencies, environmental organisations, employees and shareholders can play a major role in developing corporate environmentalism. Company environmental concerns are influencing purchasing procedures in business-to-business marketing. Major industrial customers of computer manufacturers are insisting on CFC-free products and other pro-environmental initiatives. This sets up a chain reaction wherein computer manufacturers now have to change their raw material requirements and insist on environmentally friendly products from their suppliers. This greening of the supply chain results in environmental concerns being addressed not only by the manufacturer but by their suppliers (and, in turn, influences their suppliers as well). Business-to-business purchase decisions are being increasingly influenced by environmental quality of the product, as corporate customers find their environmental performance is closely linked to that of their suppliers (Banerjee 1998).

Internal Forces

Top management commitment. The importance of top management in building a shared vision has been discussed in the organisation studies literature, especially in the area of organisational learning. Environmental problems associated with a firm's business activities can provide the stimulus for changing existing organisational patterns of behaviour (Banerjee 1998). Researchers have conceptualised organisational learning in a variety of ways: adaptation to changes in the environment, information-processing patterns within the firm, the development of new insights, new knowledge and new systems, and the institutionalisation of experience in organisations are some themes of organisational learning that have been discussed (Argyris and Schon 1978; Dechant and Marsick 1991; Levitt and March 1988; Shrivastava 1983). Research on organisational learning suggests that top management factors and intuitive judgement by leaders are necessary for high-level organisational learning to occur (Shrivastava 1983; Bennis and Nanus 1985). Environmental issues can challenge existing ways of thinking within a firm, and corporate environmentalism requires the support of top management for promoting an environmental ethic throughout the organisation. The development of new ways of thinking and efforts to develop an environmental orientation throughout the firm are possible when all organisational members share the vision of top management. Top management commitment is necessary for creating an organisational climate that is responsive to environmental issues facing the firm (Newman and Breeden 1992). The TQEM approach discussed earlier, for example, requires the support of top management in implementing environmental strategies (GEMI 1992a).

One way in which top management can manifest its public commitment to environmental protection is by developing an environmental mission or vision statement. Many corporations already have environmental mission statements and are making efforts to align environmental strategy with corporate strategy (Coddington 1993). Eastman Kodak Company has a 'vision of environmental responsibility' that includes a formal programme to guide managers in integrating environmental responsibility into their business operations (Poduska *et al.* 1992). Kodak's CEO chairs a special Management Committee on Environmental Responsibility which sets environmental goals and periodically reviews the company's progress on achieving these goals. Activities performed by this committee include educational and training programmes, communication of environmental practices and goals within the organisation, and environmental assessment programmes.

Other research that has examined the role of senior management is the literature on market orientation (Kohli and Jaworski 1990) and socially responsible organisational buying (Drumwright 1994). Top management commitment was found to be an important antecedent to a market orientation of a firm. Senior management factors were important in fostering a sense of customer focus, co-ordinated marketing and profitability, the three themes of a market orientation. Similarly, socially responsible organisational buying attempts to take into account the public consequences of organisational buying and often uses non-economic buying criteria such as utilising vendors from minority groups or environmental organisations. Corporate leaders, CEOs or owners of small firms were key decision-makers in this process.

Need for competitive advantage. Competitive strategy, involving the business and corporate strategy levels, has been the major focus in strategic management research. A long-term competitive advantage can be obtained by having lower costs than competitors or having differentiated offerings and being able to command a price premium. Corporate environmentalism can provide both these sources of advantage.

For instance, there are numerous cases where the installation of new environmentally friendly technologies has reduced costs for firms. New production processes and manufacturing changes in several firms (AT&T, Carrier, 3M, to name a few) have resulted in unexpected cost savings while meeting environmental protection goals (Naj 1990). A strategy of source reduction instead of pollution prevention has proven to be more advantageous for firms. Procter & Gamble used a dual source reduction strategy for the product (Downy softener) and the package (refillable pouch). The refillable package was found to be more cost-effective than the recycling option and led to a 95% reduction in waste (Simon 1992).

Preventing environmental damage is less expensive than cleaning up and can be cost-effective in the long run. The 3M company, which has a corporate-wide programme called 'Pollution Prevention Pays' (now called '3P Plus') realised the cost advantages of source reduction over conventional pollution control measures such as smokestack scrubbers. Its strategy was to reduce emissions before they were created with a view to cut costs and develop a stronger competitive position. The 3P Plus programme saved 3M over $1 billion between 1975 and 1993. Several such case studies have shown that environmental costs were transformed into cost savings leading to competitive advantage for the firm (Porter and van der Linde 1995).

Thus, purely financial reasons can motivate firms to focus on environmental activities. For instance, in the chemical industry, spending on environmental compliance is between 3% and 4% of sales per year (or about $10 billion per year). A strategic focus by firms in this industry to reduce costs of environmental compliance is manifested by their change in approach to environmental issues. Environmental management has evolved from mere compliance during the 1970s to the development of pollution reduction programmes and pollution prevention programmes in the 1980s and is now poised to become an integral part of company strategy (Buchholz 1993; Coddington 1993).

Constant innovation in cleaner technologies and products can enable firms to differentiate their products based on their environmental friendliness. This may allow the firm to command a price premium for their products and thus provide the competitive advantage. New marketing opportunities have emerged and will continue to emerge due to recognition of environmental problems. For instance, the US firm Pacific Gas and Electric dropped its proposed expansion plans for new nuclear power plants and decided that energy conservation was a more profitable investment. Environmental concerns were key issues driving this strategy. Also, DuPont used its technical knowledge derived from its in-house pollution prevention programme to set up a consulting operation. This new business is expected to yield annual revenues of $1 billion by 2000. New strategic markets that have emerged due to environmental concerns are growing rapidly. The environmental services sector has been growing at more than 20% annually over the past five years.

Consequences of Corporate Environmentalism

The preceding section has discussed conditions that can lead to a corporate environmental strategy. Corporate environmentalism has far-reaching consequences affecting the firm, customers, suppliers, employees, as well as society in general. This is a relatively new area of research, and studies examining the consequences of corporate environmentalism are virtually non-existent. This section will briefly discuss possible consequences of a strategic focus on environmental issues facing a firm.

Customer Satisfaction

If one motivation for firms is to respond to changing needs of its customers, a corporate environmental strategic focus should lead to enhanced customer satisfaction. The concept of customer satisfaction needs to be broadened beyond traditional criteria such as quality, service and value to include environmental quality. Communication of company environmental activities to customer groups, and company efforts to educate consumers about environmental issues are some important activities for effective implementation of an environmental strategy (Poduska *et al.* 1992). Environmentally concerned consumers are a significant niche market, and firms targeting this segment can increase customer satisfaction by offering minimum environmental impact products and services.

Positive Company Image

Organisational commitment to environmental protection can also lead to an enhanced corporate image. A majority of 'green' advertisements focus on corporate environmental activities, and many firms are clearly attempting to portray an image of environmental responsibility (Banerjee *et al.* 1995). Underlying themes in these advertisements include social responsibility, consumer and employee health and safety, and support for environmental causes. There is a danger of the positive image of a 'green' firm not living up to its environmental performance, as has been the case with Ben & Jerry's and The Body Shop. This might lead to a consumer backlash. Thus, it is important for a firm to appear credible in their green image campaigns.

Co-operative Alliances

In an attempt to enhance the effectiveness of their environmental strategies, firms may develop co-operative alliances with various groups: policy-makers, public environmental groups, and even with competitors. Some firms that have recognised the importance of environmental issues have already begun this process. The Irvine Company has an ongoing alliance with the Nature Conservancy to manage 17,000 acres of habitat in Southern California (O'Donnell 1993). The Dow Chemical Company has formed alliances with the Nature Conservancy and the National Fish and Wildlife Foundation to protect wetlands. The 'Buy Recycled Business Alliance' is a consortium of 25 major corporations, many of them direct competitors, involved in increasing demand for recycled products (O'Donnell 1993). These alliances go beyond conventional cause-related marketing and include specific activities, such as waste reduction and wetland preservation, as well as financial contributions. Consumer education is also a major theme of these alliances. Household recycling and composting, promoting use of recycled products and energy conservation are some programmes that have been conducted as a result of the alliances.

These types of alliance result in new markets for member firms as well as marketing opportunities for new firms capable of meeting the new demand. As firms increasingly integrate environmental issues into their strategic planning process, such co-operative alliances with other stakeholders will become more common.

Green Product Launches

An environmental strategic focus can influence the new product development process and lead to the launching of environmentally positioned products such as phosphate-free detergents, recyclable and recycled packages, catalytic converters and 'natural' pesticides. Retailers are also attempting to position their outlets environmentally: Eco-Mart, the first Wal-Mart 'green' prototype store carrying environmentally positioned products, opened in 1993. Environmental influences are not limited to merchandise: the store itself has been constructed using environmentally friendly building materials and uses less energy than conventional stores (Schlossberg 1993). Some retailers have even launched their own brand of environmentally positioned products.

Research and Development

Corporate environmentalism represents an ongoing process, and not a one-shot strategy. Firms who are attempting to link environmentalism with company strategies are investing substantial resources in research and development of environmentally friendly technologies, products and processes (Coddington 1993). Increased investment in research to minimise environmental impact of products and processes is one consequence of a corporate environmental strategy.

Enhanced Competitive Advantage

Traditional strategic performance measures such as profitability and market share are also possible consequences of corporate environmentalism. Although it may take some years before environmental investments yield the expected payoffs, many firms have already realised significant cost savings as a result of implementing environmental programmes (Porter and van der Linde 1995). Apart from cost advantages, environmental product positioning can also lead to product differentiation and gains in market share (Earle 1993). Source reduction strategies, the emergence of new markets for environmentally friendly products and technologies, and environmental influences on buyer preferences are some sources of competitive advantage.

Implications for Marketing Theory and Practice

Corporate environmentalism, as discussed in this chapter, has some important implications for marketers. Influences arising from changes in the external environment, such as increased regulatory pressures and public environmental concern, can result in the internalisation of environmental issues by a firm. Consequently, actions by the firm to minimise its environmental impact can have a positive effect on the natural environment (Harvey 1995). Integration of corporate environmentalism at higher levels of strategy implies that business firms can play a more proactive role in environmental protection and expend more resources in reducing the environmental impact of their actions. Other aspects of the external environment can also be impacted by marketing actions arising from corporate environmentalism. For example, corporate environmentalism can anticipate and even influence future environmental legislation, as did DuPont in the late 1980s. As the world's largest manufacturer of chlorofluorocarbons (CFCs), DuPont was faced with legislation regulating their use because of the damage they caused to the ozone layer. DuPont's first reaction (supported by all other CFC manufacturers) was to oppose the regulation claiming that the scientific evidence linking CFCs with ozone depletion was inconclusive. However, during this time, DuPont began investing about $4 million annually in research for CFC substitutes (Buchholz 1993). As evidence against CFCs mounted, and a large ozone hole was discovered over Antarctica, DuPont changed their position in 1987 at about the same time the Montreal Protocol limiting CFC production worldwide was signed. In a strategic move, DuPont supported the Montreal Protocol and called for international action curbing CFC production. In the new regulatory climate, DuPont is poised to become a world leader in CFC substitutes because of their investment in research and development for new products and their ability to influence legislation.

Corporate environmentalism can also provide the impetus for paradigm development in marketing. Explicit recognition of the importance of the natural environment challenges dominant economic paradigms. Research on environmentalism in diverse disciplines indicates that environmental destruction and its consequent implications for human survival is an anomaly in the current economic paradigm (Catton and Dunlap 1980). One of the main causes of environmental damage is the long-standing anthropocentric tradition of modern Western culture, wherein humans are viewed as being superior to the rest of nature (White 1967). This dualistic tradition is also evident in the development of scientific knowledge and the domination of nature through scientific and technological advancements. The major assumption was that the planet is an unlimited source of resources and opportunities; resource depletion was seen as a 'management' problem and something that could be overcome by advances in science and technology. In contrast, the ecological paradigm recognises that there are limits to growth and stresses the interdependence of the natural environment and human societies (Catton and Dunlap 1980). The notion of sustainable development attempts to seek a compromise between the two opposing paradigms of 'ecocentrism' and 'technocentrism' (Gladwin *et al.* 1995) and is becoming a significant field of study in economics, anthropology, sociology, science and technology studies, environmental science, and the different disciplines in management.

Implications for marketing theory arise from these emerging paradigms or worldviews. Stakeholder theory attempts to broaden the role and responsibility of business firms by including various groups in society that are impacted by business actions. However, the dilemma between social performance and economic performance of a firm is yet to be resolved. While management and marketing theories do account for social responsibility in varying degrees, it tends to be subsumed under traditional economic criteria. Most theoretical discussions of organisational stakeholders are inherently anthropocentric. Including the planet as a stakeholder adds a new dimension that can redefine several legitimacy challenges facing the firm. For instance, traditional accounting and valuation systems are concerned with private property rights and asset stewardship for shareholder interests (Rubenstein 1992). An ecological perspective implies accounting for common property resources and stakeholder rights. If the planet is a stakeholder, then asset stewardship has a deeper meaning: it involves stewardship of natural assets and must employ a longer-term perspective of costs and benefits. The social contract between a firm and society traditionally did not include the natural environment, and its inclusion in this contract has the potential to influence a range of strategic decisions for a business firm.

Corporate environmentalism can also influence public policy decisions that are designed to affect societal and environmental consequences of business actions. This chapter has discussed how the regulatory climate can influence the level of corporate environmentalism. Apart from enforcement of regulations, public policy research on regulation and consumer involvement can determine interventions that can raise the level of corporate environmentalism. This can involve developing and setting standards of environmental performance, developing consumer education programmes, providing incentives for environmental protection activities and developing joint environmental strategies with business. Several federal and state agencies have ongoing environmental programmes with business firms (Earle 1993). TQEM is a process that is designed to employ resources with minimum environmental impact in order to increase productivity and competitiveness. The International Organisation for Standardisation ISO 14000 series focuses on developing environmental standards and employs environmental criteria in their certification process. In the near future, business firms will be required to be certified under this series in order to compete in world markets.

Conclusion

This chapter has attempted to position the phenomenon of corporate environmentalism within the broader context of strategic marketing. Marketing has the potential to influence corporate environmentalism at all levels of strategy. Marketing's role becomes especially critical if one employs a broader perspective of its contribution to corporate strategy (Day 1992; Varadarajan 1992). For instance, marketing can provide a more comprehensive view of the firm's stakeholders by expanding traditional notions of the market to include competitors, suppliers, public agencies, customers and regulatory agencies. This perspective allows corporate environmentalism to be integrated at the higher levels of enterprise and corporate strategy. Market information would imply monitoring all external and internal stakeholders of the firm. As a result, marketing can influence product market and diversification decisions based on their environmental impact. For instance, market knowledge based on environmental criteria can help firms identify environmental protection businesses as a growth sector. Entry strategies and core competences required for this business can also be identified. Thus, employing a broader view of the marketing function can influence a firm's strategic competency.

At the business strategy level, marketing can influence product differentiation decisions based on the 'green' dimension. Cost advantages stemming from an environmental strategy can also provide a source of competitive advantage. At the functional level, corporate environmentalism can result in the greening of the marketing mix wherein product (green products and packaging), pricing (premium pricing), promotion (green advertising) and distribution (recycling as reverse channels of distribution) decisions can reflect an environmental strategy.

The link between quality enhancement and environmental quality is an important one: TQEM has emerged as the most popular method by which a firm addresses environmental concerns. Managers at all levels of a firm need to understand and implement TQEM policies that regularly assess their firms' environmental impact and set standards for improvement in environmental performance.

Green product positioning is another issue that warrants a closer look. Research on premium pricing for green products indicates that few consumers would pay higher prices for green products. Green product positioning might be more beneficial if it is a part of an overall environmental strategy of waste reduction, quality improvements, energy conservation and cost advantages. Approaches to environmental improvements need not be focused only on costs but can be viewed as opportunities to leverage and maintain competitive advantage. Managers of firms that are in the process of integrating environmental concerns need to enlist the support of top management in developing a comprehensive environmental strategy. Such a strategy can arise when there is a shift in the managerial mind-set from a compliance-based approach to one that focuses on environmental issues as a business opportunity.

Varadarajan (1992) identified corporate response to environmental problems as a strategic imperative of the nineties. Marketing is well suited to be the vehicle for this process of change. The business enterprise has been the major cause of environmental destruction on a global scale. The responsibility to reverse this process also rests on the business enterprise.

2

Reviewing Corporate Environmental Strategy

Patterns, Positions and Predicaments for an Uncertain Future

Walter Wehrmeyer

SINCE THE LATE 1980s, the way management has integrated environmental issues into business has shown remarkably little diversity. In fact, it is surprising to see that only a few methodologically distinct approaches have emerged: Western consumer societies that produce more than 15,000 product lines for a supermarket have not been able to produce more than a dozen corporate approaches to the environment that are distinct from each other. This may be unexpected, given that businesses are otherwise very well adapted to a great diversity of marketing circumstances. Inasmuch as the approach to marketing, market characteristics and demand patterns differ between, say, nuclear power stations or fish and chip shops, one would expect a much greater diversity of approaches between firms. As there are differences in the types of environmental problem, the relative importance of environment for the organisation and its main stakeholders and the potential (economic, technological and organisational) for the firm to address environmental issues, so one could expect a great variety of models and ideas, especially since firms are (still) frequently at the beginning of the environmental learning curve.

In addition, the dominant business rationale states that, through the conversion of physical and non-physical inputs to products, companies serve an economic function in meeting demand, serve a socio-psychological function in allowing consumers to express value preferences through product choice, serve a socio-fiscal function in providing employment and taxes, and serve an environmental function by promoting best resource use and, by implication, drive other, less environmentally efficient, producers out of the market. Yet there is very little diversity in the way organisations meet these heterogeneous demands. It seems that firms can have any approach to environmental management as long as it conforms to ISO 14001 or EMAS—a slight improvement on Henry Ford who is reputed to have said that customers could have any colour for their car as long as it was black. Moreover, apart from a small number of radical approaches, the business paradigm—that businesses are and remain generally beneficial socio-technical systems to the advantage of their diverse stakeholders—remains unchallenged.

The majority of approaches that describe or analyse business strategy towards the environment still have a predominantly technocratic definition of the environment. The interaction between the firm and its environment has a long tradition within corporate strategy, with contingency theorists arguing for a bespoke match between internal factors of the firm and characteristics of the external environment (Lawrence and Lorsch 1967). This argument is, if the definition of 'environment' is to include the natural environment, surprisingly modern, if industrial ecology approaches are to be considered. Being largely independent of, or at least *de facto* being ignorant of, the social definition of environmental problems, the majority of corporate environmental strategy (CES) approaches see environment as a technical problem: pollution that needs to be reduced, waste that needs to be avoided, risks that need to be minimised or insured against. In this perspective, conflicts over environmental uses, emotions about environmental risks or simply distrust of organisations' ability to preserve environmental qualities are interferences that should be avoided if the technologically superior decision about resource uses and product mixes are mediated correctly towards the various stakeholders. Reducing and de-emotionalising environmental issues to their technological dimension has made it easier to reinforce the business legitimacy. However, Brent Spar has demonstrated the essential fallacy of this: it ignores the often regional and frequently emotional and conflict-causing nature of environmental problems and largely—and potentially dangerously—ignores sustainable development, here seen as an umbrella that combines environmental protection and social care into an overall framework. Accordingly, a firm's marketing strategy has to address the social definition of environment to a much greater extent.

It is the purpose of this chapter to present and review the main CES approaches, and outline how environmental management has affected corporate strategy. The main motivation behind this chapter is to present the relatively poor diversity of approaches that have developed to integrate environmental management ideas and activities into organisational realities and to relate these to the marketing function. This will be done using Mintzberg's categorisation of strategy into five distinct definitions (Mintzberg and Quinn 1991).

Strategic Modes towards Environmental Business Management

A variety of factors have put increasing pressure on firms to incorporate environmental protection into business practice, which can be classified into economic and non-economic, with the former generally being of greater significance. On the economic side, incentives include the ability to respond to changing or changed consumer demand, the promise of cost savings from increased resource efficiency, or increased market share from a (qualitatively and/or environmentally) superior product. Economic disincentives include rising waste disposal costs, rising insurance costs[1] or, generally, rising costs of environmental compliance. Among the non-economic incentives, increasingly stringent legislation, improved environmental image,[2] better relations with outside stakeholders (particularly regulatory authorities, local neighbours and consumers), an improved health and safety record and better staff motivation and morale are frequent arguments (Peters 1991; Winter 1995; Groenewegen *et al.* 1996; Elkington 1997; Welford 1995)—all of which are essential for marketing and stakeholder management.

Given this rather impressive list of factors, it is hard to believe that the vast majority of firms, mostly small and medium-sized, still find it easier and presumably cheaper not to embark on an environmental change programme (Geiser and Crul 1996). However, the particular types of pressure also show that the main rationale for change is not the improvement of environmental quality but the maintained competitiveness of the organisation in the marketplace (Welford 1997). It will be argued below that failure to explain why businesses have not incorporated environment in a more comprehensive and more diverse way may be an indication of the state of environment-related organisation theory (Gladwin 1993) or, worse still, the failure of environment-related business analysts to understand the inner logic of organisations and change (Wehrmeyer 1995).

To better understand the corporate strategy implications of environmental management (EM), it is first necessary to define the concept of strategy: a term that probably has even more definitions and greater differences within these than sustainable development. In reviewing the concept, Mintzberg identifies five different definitions of strategy, labelled **plan**, **ploy**, **pattern**, **perspective** and **position** (Mintzberg and Quinn 1991). Strategy as *plan* refers to a consciously intended course of action, inevitably made in advance of the actual events occurring. A particular *plan* is the *ploy*—designed to outwit or outmanoeuvre an opponent, a competitor or a particular stakeholder. Strategy as ploy will be covered under the section 'Strategy as Plan' below.

Strategy as *pattern* is more complex, as it refers to the pattern that can be inferred from the behaviour of an observed organisation. In this, 'strategy is consistency in behaviour, whether or not intended' (Mintzberg and Quinn 1991: 13). This definition of strategy implies an observer and it can be the emergent result of unconscious or unplanned behaviour.[3] This part of the definition of strategy is thus in stark contrast

---

1 Buying insurance for the management of environmental risk is an economically sound but ethically odd way of reducing risk exposure for a firm. In this, the greatest benefit of the frequently cited 'reduced insurance cost' for environmental management should be seen in the reduced risk rather than the reduced payments to underwriters.

2 It should be noted that the strength of the 'image factor' largely depends on the relative importance of corporate or product image in the overall marketing strategy. For certain product categories, such as newspapers, tobacco or culture, product demand is determined by other factors and is largely inelastic to changes in the environmental image of the firm or its product.

3 The way Shell's behaviour in Brent Spar was seen as 'typical' is an example of stakeholders who saw a pattern of a company that was environmentally and socially uninterested—a pattern that certainly was neither planned nor intended by Shell (Löfstedt and Renn 1997).

to more classical definitions that presume purpose for all results of strategy deliberations or to definitions that presume strategic analysis and intellectual conceptualisation as the core determinants of corporate strategy (Porter 1985). Implementation of a deliberate strategy (*plan*) often yields different results; the emergent strategy then becomes a *pattern*.

In fact, the pattern that has been inferred may be radically different to the strategy that was intended. For example, the strategic pattern of The Body Shop is at times perceived to be radically different to its stated plan. A similar argument has been made regarding the oil industry (Ketola 1993). Strategy as *position* refers to the particular place of an organisation or its products in the environment—which implies that the match between organisation and its specific environment lies at the heart of corporate strategy and the organisation sees itself or its products within a wider context. This places the need for integration of the marketing function within the wider pattern of interaction between an organisation and its environment. Conversely, strategy as *perspective* refers to a particular way of perceiving the world—thus going further and having a vantage point that is different from *position*. In this, *perspective* is, like *pattern*, an abstraction, a concept, a way of understanding and making sense of the world.

This is also one of the reasons why—among many other contenders—this broad definition(s) is chosen here. The main reason is that the definition(s) put forward here largely allows the capture of the diversity of strategic modes that firms have adopted towards environmental protection and management.

Environmental Strategies as Plan

This category describes strategy approaches that offer intentional, planned strategies towards environmental improvement. In the majority of cases, it refers to improvements of aspects of an organisation or its products, usually in the form of the setting of an environmental policy[4] and its operationalised targets, the development of an environmental improvement programme and the process of equipping staff made responsible for the environment with the necessary resources (Sadgrove 1992; Elkington *et al.* 1991). In this approach, environment frequently starts as an addition to current business practice, often as an adjunct to health and safety, or, more frequently, as an extension of an existing quality programme. Yet it may transform the organisation more than initially anticipated (Elkington 1997) and, in the majority of cases, market demand for environmentally more responsible behaviour or products is the starting point. Greener marketing here is not the logical consequence but the starting point for the greening of the firm.

The defining criterion of this group of strategies, which is predominant within small and medium-sized firms, is that the initial business rationale—the perspective and the position—remains initially unaltered. Here, environment is an adjunct that

4 Here, (environmental) policy refers to the written statements of intent within an organised context.

is pursued for reasons indicated in the introduction. However, *plan* comprises a number of distinct approaches, including individual approaches to corporate strategy for environmental management as a plan, environmental management systems (and environmental performance indicator approaches in particular), and product-based approaches. These will be discussed briefly.

Individual approaches prescribe the range of activities organisations should pursue to harness the expressed benefits of environmental management in a consistent form and to the benefit of the firm. The main features of this approach include (a) the proposal of specific and frequently prescriptive sets of activities in the absence of arguments that fundamentally alter or change current business practices in the short run; (b) the focus on measures that improve the financial viability of the firm and a managerial approach that is based on existing planning procedures; and (c) continuous improvement as a guiding principle for further environmental action. The main starting point is the need to improve environmentally due to economic reasoning; the outcome is the reinforcement of existing business philosophy (Fiksel 1997). In fact, the threat of losing business against competitors is a dominant argument here.[5] Marketing lies at the heart of this approach: as firms move towards environmental incentives and avoid environmental disincentives, it is the promise of increased sales at higher unit prices that stimulates change. The idea is that greener products can be sold at a higher price—the **green premium** that often allows the recovery of environmental investments. Alternatively, it is argued that the development of environmentally enhanced product avoids the loss of market share or customer base.

Within this strategic approach, the use of checklists is common (Winter 1995; Callenbach *et al.* 1993) and indifferent to the particular circumstance of the firm. Often, a functional approach is adopted to classify the range of actions recommended, which then prescribes activities to be undertaken for various departments in an organisation (Sadgrove 1992; Elkington *et al.* 1991). The great advantage (and the reason for its popularity within the business community) lies in its simple applicability, the immediacy in which activities can be implemented and the implied certainty with which the prescribed measures promote or even deliver success (cf. Peters 1991). In addition, *planning* for environmental improvement provides practical help for busy environmental managers in their constant quest for arguments to convince superiors of the need for more environmental management or to develop sound elements of EM, such as environmental reporting (Deloitte *et al.* 1993). Some authors, such as Elkington (1997), present such a prescriptive plan within the wider argument that these changes, designed to improve the competitiveness of the firm, are in due course likely to alter radically the way the organisation is run and managed.

The principal disadvantage to using *plans* is the limited way in which they can be used. What happens to firms that have worked down the checklist: how can they 'go beyond' the prescribed plan? The need to develop individual approaches to specific environmental and organisational issues is stifled, and the capacity of the firm to

5 This is a curious point, because the environmental management debate started with concerns to save the planet (or parts thereof). The argument brought forward here is inverted—environment is saving (your) business.

cope with complexity and environmental change is not enhanced. Arguably, **individual approaches** are 'quick fixes'—with all their inherent problems, including lack of emphasis on organisational learning and creativity and implied reliance on consultants (or other externals) to guide the 'greening' process.

The **environmental management systems approach**, which intends to maximise the benefits of pursuing the environmental management agenda by developing a system—which may or may not be certified under ISO 14001 (BSI 1996) or the Eco-Management and Auditing Scheme (EMAS) of the European Union (Moxen and Strachan 1995). The great advantage of certified environmental management systems is that it provides firms with a clear framework as to how the process of environmental improvement can be organised. This allows for comparability, ease of implementation, cost savings in the design of the future system and some reassurance that what one does cannot be all that wrong.

However, as neither ISO 14001 nor EMAS comment on the environmental performance but on the process of its management, compliance is possible for any firm and it does not preclude prosecution for breaches in environmental compliance. In fact, several UK firms certified under ISO 14001 have already been prosecuted. Even though the popularity of certification is strong and growing, criticism of the standard and its beneficial effect on organisational change and long-term competitiveness is growing gradually (Gleckman and Krut 1996; Strachan 1997).

A number of product-based approaches towards CES have been developed, mainly in product markets where environmental image is arguably a factor in the purchasing decision. These plans are aimed at improving the environmental features or product quality, be this by changes in product design (Böttcher and Hartman 1997; Charter 1997), driving product and process innovation to include environmental improvements (Gouldson 1993), targeted environmental marketing (Polonsky and Mintu-Wimsatt 1995) or by integrating product design with the general strategic perspective of the firm (Oakley 1994). Here, it is not the organisation but a product that is to be environmentally enhanced. In fact, the focus on the product often precludes wider organisational change. Again, the extent to which environmental issues affect the firm are deliberately contained within a small segment of the organisation, although the changes that are being proposed can be justified with relative ease by the anticipated sales projections and the ethical or environmental legitimacy of the product is usually outside the decision-making parameters.

As part of the promotion of planned environmental strategies, specific efforts to develop a system of environmental performance indicators (EPIs) deserve a special place. Here, the strategic challenges lie in identifying significant environmental effects, developing appropriate environmental indicators (Wehrmeyer and Tyteca 1998); in integrating the environmental management system (EMS) with the overall management information system (MIS) (Sheldon 1997)—frequently enhanced by IT support—and the interface between environmental auditing and the EMS (Coulter 1994). Generally, the aim is to develop a system that, on one hand, remains flexible enough to account for changes over time and between sites or regions and does not stifle organisational learning and, on the other hand, provides stability and guidance for EPIs to be used as a measuring device over time.

Environmental Strategies as Pattern

The definition of CES as a *pattern* infers consistency—intended or otherwise—of action that provides the possibility of interpreting and analysing corporate behaviour. Whenever particular behaviour of a firm is attributed as 'typical for firm *x*', a comparison is made between the mental model of that firm's 'normal' behaviour and its particular actions at that time. *Pattern* thus refers to the strategy analysts' capacity to recognise aspects of corporate behaviour that are recurring, 'typical' or, ideally, 'characteristic'. As a result, organisations that comment on or habitually interpret corporate behaviour—such as academics, management consultancies, environmental pressure groups or industry federations and groupings—are frequently found among the authors of this definition of CES. In addition, the pattern is often based on the particular paradigm of the interpreter: Greenpeace's interpretation of industry's environmental management is quite different to that of, say, the CBI. However, it was Greenpeace's interpretation that was unexpectedly reinforced through the events of Brent Spar.

Equally, firms may try to influence the way their behaviour is seen by using pattern: Exxon's efforts to demonstrate that *Exxon Valdez* really was an aberration from the norm, or Shell's recent efforts to show that it has taken community participation seriously since Brent Spar, are examples of this. However, the latter example also shows that firms may try to change the pattern, and the conceptual mode by which it is evaluated. This is a difficult and time-consuming process and often implies cultural change as well as different policies. It also depends on how stable the interpreter's paradigm is: rather than changing its view on industry generally, Greenpeace has acknowledged Shell's recent Brent Spar initiatives, but has maintained its campaign against fossil-fuel-based firms.

Strategy as *pattern* requires the development and maintenance of such patterns, and they can be classified coarsely into those that are predominantly empirical and those that follow some form of theoretical or paradigmatic justification, which often refer to developmental models or historical changes. Using *patterns* is a process of pigeonholing that reduces the complexity of organisational action and which allows attribution and comparison of behaviour, yet may result in inappropriate conclusions if the pigeonholes are applied uncritically or in too simplified a manner.

One of the earliest and most popular CES *pattern* is that of Hunt and Auster (1991),[6] which is described below as an archetype of such approaches to CES.[7] However, they

6 Curiously, they also provide a 14-point checklist *plan* as to how environmental improvement programmes can be facilitated most competently.

7 A similar approach has been developed by Colby (1991), who focuses on paradigmatic differences between the patterns and applies them to economic and social development. He also ends up with five main categories, namely 'frontier economics', 'environmental protection', 'resource management', 'eco-development' and 'deep ecology'. Interestingly, the philosophical basis of his *pattern* is strong, as reference is made to the anthropocentricity of the first three and the eco- as well as biocentricity of the latter two. It also attributes different economic and socioeconomic theories to these stages and advantages, as well as disadvantages, for each stage are given. It therefore is also suited to be applied to wider productive social systems, such as countries.

use an unspecified number of company surveys and largely ignore the majority of organisations that know very little and do very little about their environmental impact. They conclude that five stages exist in the genesis of environmental management in firms. Acknowledging that different firms may be at various stages and that there is diversity in the categorisation of sectors within these stages, they name the stages as:

1. **The Beginner**. Firms at this stage see environmental management as essentially unnecessary and only minimal resources are allocated; no senior staff positions are allocated responsibility for environment, and general knowledge of the company's environmental effects is, accordingly, very poor and exposure to potential environmental accidents and risks is high. The majority of small and medium-sized firms are at this stage.

2. **The Fire-Fighter**. Firms at this stage perceive environmental issues as an inconvenience, a problem that should be addressed only as and when it arises. It is remarkably reactive position and offers little protection. Seeing environment as a 'hygiene factor' (Wehrmeyer 1998)—something that in itself is not a contributor to competitiveness—is also a rather cost-inefficient way of 'managing' environment, as risk containment and emergency response costs can easily outweigh the costs of a 'proactive' CES. Union Carbide before Bhopal could be cited as an example.

3. **The Concerned Citizen**. Firms at this stage perceive EM as worthwhile and the firm is, at least in principle, committed to pursue it. Mainly driven by ethical concerns or out of corporate social responsibility, the EM programme is limited to moderate efforts, which ensure compliance but are insufficient for the integration of EM with the rest of the organisation (see above), and company reporting mainly results in glossy advertising for the outside world and few internal reports.

4. **The Pragmatist**. This curiously labelled stage is characterised by comprehensive protection that sees EM as an important contributor to the organisation and its success. This requires staff involvement, usually sufficient funding, limited integration with other sections in the firm and, as a result, comprehensive protection from environmental risks. Firms such as Hewlett Packard, ICI, BP and NatWest may suitably fit into this stage.

5. **The Proactivist**. This stage offers maximum environmental performance and thus minimal risk by seeing environmental management as a priority area that requires active involvement of all staff, significant (and available) funding, and full integration of environmental work with all other parts of the organisation.

Another example of developing a *pattern* can be found in Western Mining's *Environmental Report*, which develops—again—a five-stage model of non-compliance, which is then used as a way of assessing its own performance. Here, the scale ranges from Level 1 (non-compliance with technical issues involving environmental laws and regulation, such as the late submission of a report) to Level 5 (non-compliance with a major issue with potentially serious environmental consequences and long-term impact, such as the breach of a major tailings retention system into surface water). Readers can then see progress towards its self-set development pattern: avoiding high-level non-compliances. A similar system can be found for the UK Business in the Environment report on the FTSE 100 companies' environmental reporting performance. Further, and curiously staying in the mode of providing five alternatives, the Global Environmental Management Initiative (GEMI) has developed a system that applies the Valdez Principles to a self-assessment system allowing firms to track their own progress towards these principles (GEMI 1992b).

Common to the above *patterns* is the relative ease with which managers or academics can attribute firms to these categories. They thus provide additional meaning and allow observers to 'make sense' of corporate behaviour. They allow progress to be tracked over time and the analysis of directional changes in CES, mainly to aspects internal to an organisation. However, even though most authors stress that it is not inevitable for a firm to progress to 'Level 5', the implicit assumption is that it is desirable to do so, be this desirability for the firm or for society as a whole.

Using a historical perspective, Schot and Fischer offer three stages, namely: 'Resistant Adaptation, 1970–1985'; 'Embracing Environmental Issues without Innovation, 1985–1992'; and 'Innovation, 1992–' (Fischer and Schot 1993). They also use Petulla's patterns to describe their first historical stage as 'Crisis-Oriented EM, Cost-Oriented EM and Enlightened EM' (Petulla 1987).[8]

Following this line, Hall and Roome also offer a three-category stage through which EM literature has transcended, again with the implied direction towards their latest stage, called *Environmental Strategy* (Hall and Roome 1996). Gladwin offers an approach to organisation theory that classifies the greening of industry into five patterns, which in turn refer to approaches in organisation analysis (Gladwin 1993). It should be noted that this *pattern* does not share the directional assumption of other approaches within this definition.

Environmental Strategies as Position

In this definition of CES, the core focus of the strategist is the particular place or position in which the organisation or its products are or should be. *Position* refers to a particular place in the outside world, and the match between that environment (or certain characteristics of it) and the organisation becomes the strategic goal of management development. Strategy as *position* requires market analysis and is, accordingly, a much more product- or market-driven strategy definition. In fact, the majority of green marketing strategies are about positioning products or image on markets. However, the paradigm sees the organisation within its specific socio-economic environment—the main difference between *position*, *pattern* and *perspec-*

8 It is interesting to note the similarity between Petulla's first category, which matches with Hunt and Auster's 'Fire-Fighter'; or Petulla's last category and Hunt and Auster's Stage 3.

tive. Pattern has no particular need to place the firm into its environment as it merely tries to make sense of its behaviour. For *pattern*, the core test is consistency. Equally, *perspective*, as it will be shown, is not about where the organisation (or its products) is in the world, but how the organisation sees the world, whereas *position* tries to place the organisation or its product(s) into the environment.

The main strategist within mainstream corporate strategy promoting *position* is Michael Porter of Harvard Business School, who argues that the essence of corporate strategy is to improve competitiveness and to cope with market competitors. Arguing that each market's competitive agenda depends on the bargaining power of suppliers and buyers and possible threat from new entrants or product substitutes, a firm's first step in developing strategy is to analyse the industry structure and to 'match the company's strength and weaknesses to it' (Porter 1979). Strategy development then often takes the form of differentiation and niche marketing where possible, or cost leadership through appropriate pricing in more homogenous product markets. *Positioning*, as in Porter's view, becomes the core of a greener marketing strategy.

The environmental implications of this strategy has been outlined by Porter in the famous 'Porter Hypothesis', which states that stringent environmental legislation leads to corporate competitiveness and international strategic advantages for those firms with tight domestic legislation (Esty and Porter 1998). The underlying rationale states that tight legislation leads to resource efficiency, to the innovation and implementation of cleaner production technologies and, accordingly, to lower unit production costs, an environmentally superior product and, subsequently, a more competitive product. In this, the **double dividend** is created through environmental legislation, and encompasses better environmental quality and improved firm competitiveness. This is at first hand a radical departure from traditional environmental management, which has consistently argued that tighter legislation leads to higher production costs as compliance costs rise, which in turn leads to less competitiveness. The reality probably lies somewhere between the two positions, in that technology modernisation will support a rise in productivity per input, but that the extra costs for this modernisation do have an effect on marginal unit costs (Xepapadeas and de Zeeuw 1998).

The principal idea of *position*—developing a (marketing) strategy from the match between an organisation and its specific environment—is not new but part of the well-established ground of management consultants and organisation analysts. The novelty is in the definition of what this environment constitutes, and a variety of *positioning* strategies emerge as variations on this theme. For instance, Steger's (1993) model is very similar to the original product mix strategy, developed at the time by the Boston Consulting Group. He relates an internal factor to an external one to arrive at a 2 × 2 model for the positioning of the firm in relation to environmental matters (Table 1).

In Table 1, *indifference* results from the little relevance environment has both with respect to market opportunity and with regard to the relative and perceived importance environmental issues and factors have on the organisation. Firms in this *position* have little to gain from, and little to worry about, environment, making a deliberate

		Environmental exposure	
		Low	High
Environmental market opportunity	Low	Indifference	Defensive
	High	Offensive	Innovative

Table 1: MODEL OF POSITIONING REGARDING THE ENVIRONMENT

Source: Steger 1993

or expansive environmental strategy highly unlikely. Accordingly, environmentally 'indifferent' firms are likely to have an environmentally indifferent marketing strategy.

Perhaps the most uncomfortable position for firms is where environmental implications are significant but potential gains from environmental improvements are unlikely to yield market opportunities. The tobacco industry may be an example of this. A firm that is unlikely to recover its environmental expenditures but is in the 'environmental spotlight' can, accordingly, be expected to adopt a minimalist position to try to reduce environmental costs wherever feasible—the typical *defensive* position. The appropriate marketing strategy would ignore environmental issues but focus on other, product- or image-related points.

Perhaps the most favourable position is that of *offensive*, where the firm has little environmental exposure, but significant environmental opportunities. The Body Shop may be an example. Here, small environmental investments to cover the environmental issues at hand are likely to yield substantial market opportunities, which means that the firm can quickly recover its environmental costs. A firm in this position can easily put environment at the heart of its operation with a marketing strategy that aggressively promotes environmental achievements. The 'high-risk–high-return' position is that of *innovation*, where the stakes, but also the potential benefits, are both high. Here, it depends on the particular way in which environment is being approached and success is not guaranteed. However, the marketing strategy is likely to promote the achievements of management in developing good environmental performance.

Other *positioning* models exist (cf. Lee and Green 1994) and even more can be developed with ease, provided the pigeonholing concept is maintained by providing a permutation between an internal and an external factor of choice. These can be relative growth of environmental market segment, perceived relevance of environmental issues (by firm or by customer), internal competence on environment, the rate of technological change or the degree of environmental regulation on the market.

Others are possible and each allows the formulation of a specific marketing strategy, based on the desired position of the product or its producer.

The beauty of such models is that they allow a categorisation without normative judgement as to where the organisation should go. It describes and, by doing so, somewhat legitimises the place a firm has adopted. Such models see firms or their products as a whole and they remain static as they do not prescribe how a firm turns itself from one position to another. In fact, if external forces remain unchanged, one optimal place exists for the firm and a transition from one position to another is unlikely and not even recommended. Here, strategy is really about being at the right place at the right time. However, as with *plan*, little evidence is provided as to how best to *position* a product, how to defend that *position* and how the existence of a firm in a *position* can be made more comfortable or competitive.

A particular marketing example of *position* is risk-hedging through product diversification. For example, during the late 1980s, washing powder producers responded to environmental demand by launching an environmentally friendlier product alongside its branded main product. At the time, it was uncertain whether environment would be a soon-receding fashion or a long-term trend. In the case of the former, the additional product could respond to a temporarily changed product demand, where the initial product development costs were to be recovered from higher product prices. Total market share of the combined product was unlikely to change. In the latter case, the additional product would establish a branding opportunity when the market was (still) young, which allowed for the possibility of phasing out the traditional product without change or the gradual inclusion of environmental criteria into the main product and its design—given that its development costs have been recovered from the environmentally friendlier product introduced recently. Again, loss of market share could be avoided. It was a safe strategy that crucially depended on the relative youth of the market concerned.

A particular permutation of this positioning approach lies in the burgeoning field of industrial ecology.[9] However, industrial ecology as a strategy (Lowe 1994), which includes siting firms in order to accommodate particular regional availability of resources is not new; what is new is that the range of resources now includes waste-streams. The main organisational challenges lie in managing supply as a strategic issue, the development of new partnerships across industry sectors that are unfamiliar with each other, and the need to stabilise supply at varying outputs[10] to meet the

---

9 Industrial ecology refers to an analytical and largely conceptual notion that focuses on the interaction of an industry with its specific physical environment with the aim of reducing environmental impacts across the process chain and converting material flows from its existing linear design into loops. It does so by perceiving the interactions of the organisation in question using the metaphor of ecology; hence material flows and the physical metabolism of the firm are under scrutiny, which nicely combines a life-cycle perspective with a systems approach (Allenby and Richards 1994). It also results in a paradigm that favours recycling over waste minimisation and which begins to develop material webs, but is as yet void of epistemology and an ethical base beyond broad affirmations about the need for greater environmental protection.

10 Given the often contractually defined need of one firm to produce sufficient waste that is an input to its waste customer, production down-times become more difficult to handle, as down-

requirements of a larger number of suppliers. The main marketing challenge of industrial ecology lies in developing long-term and reliable relationships with new customers for products that to date have been seen as worthless (or even costly) waste. This requires new competences in the product and its essential qualities, but also creates new dependencies as waste production may have delivery schedules equally as tight as the main product. In essence, this also calls for a better integration of production with the marketing function.

Environmental Strategies as Perspective

Environmental strategies as *perspective* are those that imply or apply the different outlook an organisation has (or should have) on its specific environment. This applies specifically to the marketing perspective. Whereas *position* viewed the firm or its product, *perspective* views the environment from the vantage point of the organisation. From this vantage point stems a specific approach to managing environmental issues. In this, *perspective* is the most paradigmatic of the definitions of CES and, accordingly, the approach to environmental issues depends on the corporate culture or, more generally, the organisational paradigm being adopted. A famous example of this is The Body Shop, where the core ethical values of one of its founders, Anita Roddick, have shaped an extremely successful marketing strategy with espoused ethical and some environmental beliefs at its heart. Another less known but probably more comprehensive example is that of Ecover, the Belgian detergent company which has developed a manufacturing strategy that successfully integrates environmental and economic criteria. There are a number of such approaches, mainly shaped by the founders' or the current managing board's beliefs and can be found in the co-operative movement or some paternalistic organisations, such as Marks & Spencer. The origin of these approaches lies within the business philosophy or the corporate culture of the organisation, rather than in the economic potential of the environment to produce benefits or cause harm to the firm. In this, many of the approaches arrive at environmental issues from the corporate social responsibility debate.

However, there are two *perspectives* that are beyond these individual approaches which originate more in the corporate social responsibility debate rather than in that of environmental management, one being that of 'total quality environmental management', the other is that of 'living systems'. In the former, the business philosophy of total quality management (TQM) is similar to the way businesses are managing (or should manage) environment (James 1994; Oakley 1994). In this, the core elements of TQM are applied to environmental improvement, namely the

---

times can affect an entire ecological web. This is a problem typical to the process industries, which have a variety of customers requiring different products from various stages of its processes. However, few firms in the manufacturing industry are prepared for this additional complexity, especially if existing contracts rely on Just-in-Time. In extreme cases, the environmental costs that may be produced to meet waste delivery deadlines can outweigh the benefits gained during normal operations.

commitment to continuous improvement (Coulter 1994) (which, not surprisingly, reappears in ISO 14001), the efforts to integrate environment into all levels and aspects of the organisation, a clear focus on the need and desire to measure environmental success (McGee and Bhushan 1993), and the belief that environment is an area that contributes to, rather than interferes with, the success and competitiveness of the firm (Blake 1994).

While this paradigm is comprehensive about the way in which environment is to be applied in the firm, it does not question the basic business rationale (Welford 1993). In this, the approach is very similar to those of strategies as *plans*. In fact, if the procedural principles of TQEM predominate over the underlying ethical principles, the prescriptions offered here can easily fall into the category of *plan*. It should be emphasised that the marketing strategy employed should follow the corporate perspective.

The other main *perspective* is that of self-organisation, which views organisations as 'living systems':

> Living systems are integrated wholes embedded in larger wholes on which they depend. The nature of any living system derives from its relationship among its component parts **and** from the relationship of the whole system to its environment. The basic principle of organisation that is characteristic of all life is the principle of self-organisation. This means that the order and functioning of a living system is not imposed on the environment but is established by the system itself (Callenbach *et al.* 1993: 91-92).

This means that the main (if not the only) purpose of an organisation (the relationship between its functional units) is to survive by avoiding undesirable circumstances or events and it does so in a continuous process of adapting its structure (the current manifestation of the organisation) to its specific environment. A couple of important conclusions follow from the acceptance of this as yet contentious notion. First, it implies that organisations have their own logic and are largely autonomous entities, which inevitably reduces the scope of managers to control and direct it. This means that marketing inevitably must be—in shape and extent—a direct outcome of the organisation's logic. Second, it provides for the first time a theory-based explanation as to why environment is so reluctantly taken up by organisations. Third, it provides a range of behavioural alternatives that can foster environmentally more acceptable behaviour in firms. Fourth, it renders much of our current approach to controlling environmental behaviour of organisations less efficient: instead of criminal prosecution of employees, more emphasis should be given towards the withdrawal or freezing of assets of the organisation (Probst 1987; Morgan 1997; Wheatley 1994).

This theory of self-organisation raises the interesting question of co-evolution between the structure and its specific (self-enacted) environment to maintain the organisation and how this co-evolution can best be driven towards greater environmental activities. From this perspective, the main elements within this environment that can trigger structural change of the firm—namely customers, non-governmental organisations (NGOs) and regulators—may need to become more involved within a collaborative network to drive the greening of the company. Here, participation and not regulation is the most likely driver towards better EM, which implies a radically different approach to marketing. Such a new approach can be characterised by a collaborative approach to marketing, a greater fostering of customer relations (similar to the way in which Japanese multinationals align themselves with long-term suppliers), closer proximity between market and production, and market research that aims to understand and adapt from the organisation's environment rather than to impose production output to an essentially saturated market.

It is interesting to note that both major new *perspectives*, industrial ecology and self-organisation, apply a systems perspective, focus on the interaction between the firm and its environment and suggest contractual and negotiation skills in the development of new partnership agreements as the core managerial skill of the future.

Conclusions

Given the remarkable variety of products, customers and markets as well as environmental problems, it is perhaps surprising to see how few distinct approaches have been developed to formulate ways in which strategic approaches towards environmental management can be categorised. Moreover, two features of the current environmental debate within organisations and within society seem to be fixed: on the organisational side, the profit motive—and its longer-term brother, competitiveness—continues to be the dominant driver for environmental action within firms. For better or worse, this means that organisations do not primarily pursue environment because it is an ethically right thing to do, or because not doing so would generate an unacceptable environmental stigma, but because it helps to make the firm more profitable and green marketing is subsumed under this. This, in a sense, has introduced an artificial ceiling onto the range of environmental activities that we can expect firms to undertake and the question remains whether we can successfully address the environmental problems at hand given this ceiling.[11] If anything, the legitimacy of organisations has increased as a result of corporate environmental management efforts.

On the societal side, it is highly unlikely that significant environmental problems will be addressed successfully, especially not in the short term. Global issues, such as the hole in the ozone layer, global climatic change or the environmental legacy of contaminated land, and the environmental impact of Western industrial metabolism, will continue to produce significant and often seemingly insurmountable challenges to human organisation and social welfare, although they are likely to take on different guises. The effects of environmental change are inadequately represented in the official statistics: nobody will directly suffer harm from global warming, but rather indirectly from consequent increased flooding, dryer (or wetter) regions or more

11 The question also arises of what can be done (by whom?) if environmental problems cannot be addressed successfully within this ceiling—and what 'success' means here anyway.

extreme temperature. This hides the costs of environmental damage and prolongs the debate about the significance of environmental issues.

So, where are the managerial and organisational solutions to the pressing environmental problems if businesses are recognised both as core polluters and core yet reluctant contributors to the development of appropriate technologies? The answer lies in the emergence of two distinct paradigms on corporate strategy, both of which impinge significantly on marketing. The first, here labelled the **environment-as-resource** perspective, continues to see environmental issues as a (now not so new anymore) challenge for businesses to respond to. The type of (market) response that is being triggered depends here on the predominant business *plan*, the past strategic *patterns*, or its market *position*. However, the core set of business values is maintained, the *raison d'être* of organisations is untouched if not reinforced and more solidly legitimised. The success in finding environmental improvements depends on finding consumers willing to pay for such products, the appropriate mix of environmental legislation and on the hope that per-unit environmental improvements are not over-compensated by higher levels of consumption.

The second perspective implies that the environmental debate has changed (and continues to do so) the way in which we perceive organisations. It is not only that the organisational environment has changed, and that consumers now want different products, which results in a different marketing mix and a slightly modified product. It is also that we have changed our view on organisations: for instance, 'stakeholders' (in themselves a relatively new breed) remain sceptical about the environmental announcements of an organisation—they want to be convinced and require increasingly complex reports. This reporting is not limited to environmental pollution; it also covers ethical issues, such as animal testing, workers' rights and the—conveniently ambiguous—sustainable development agenda.

In this, perhaps the most likely aspect of environmental management that can revolutionise organisations—if that is indeed what is required to 'solve' environmental problems at the speed we are able to produce them—is not the inclusion of ethical considerations, but openness. As Shell in Camisea has shown, the development of a 'transparent organisation' is possible, but it has wider implications for the firm than usually anticipated (Jones 1997). It means going beyond communication to include participation, to transcend the reporting function into a process of stakeholder dialogue, and it means that conflicting resource uses are resolved in a mutually acceptable way. This quest for openness and participation may address the very agenda that the technocratic approach to environment tried for many years to exclude (Kimmel 1989).

The latter point turns stakeholder management from a process where the organisation sits in the middle and has many stakeholders to address into a process where resources are at the core with many interested parties having a stake in it—and firms being only one. It implies that firms have a leased 'licence to operate', which requires periodic if not constant renewal and reaffirmation by the local communities, consumers, pressure groups and other stakeholders. In this, the second perspective is significantly wider, less clearly defined and more 'muddled'. It is also the one perspective that promises to change significantly the perspective society has on businesses.

3

Rethinking Marketing

Shifting to a Greener Paradigm

Ken Peattie

THE INITIAL REACTION of marketers, both practitioners and academics, to the green challenge has been to try to integrate it within the existing marketing world-view. When analysing the competitive environment, and seeking opportunities for generating competitive advantage, strong eco-performance is now recognised as a potential source of differentiation. When considering the structure of an industry, relationships with suppliers are being reconsidered to take into account the environmental impacts accounted for by the supply chain, as well as the issues relating to power balances and their impact on economic performance. When considering the macro-environment, conventional auditing frameworks such as 'PEST' analysis (which subdivides the external environment into political, economic, social and technological components) are being adjusted to take into account ecologically related shifts in social values, political agendas and technology. When it comes to considering customers, marketers have tried to identify the 'green consumer' and to work out what motivates green consumers to buy environmentally improved products, and how much extra they would pay for the privilege.

All of this activity appears perfectly sensible. However, if one steps back and tries to comprehend the enormity of the challenge that making substantive progress towards sustainability entails, then much of it seems not only insufficient, but largely misconceived. The difficulty stems from the fact that the existing world-view (or 'paradigm', to use the academic jargon) has brought us to the position whereby the earth's resources and systems are being used unsustainably. The pressure for this to continue is driven by continued population growth, by the material expectations of the one-fifth of the world's population living within the consumer society, and by the aspirations of the four-fifths that are not (Durning 1992). However, as the Brundtland Report (WCED 1987) made clear, following the conventional paths of 'development' will impoverish future generations and is increasingly likely to compromise our current quality of environment and life, in pursuit of economic growth. Real change will require business, and all its component disciplines, to go beyond trying to absorb the environment within its existing frame of reference, and instead to develop a new paradigm (Shrivastava 1994). What makes shifting to a greener marketing paradigm so difficult is that many of the fundamental and widely accepted assumptions and concepts of conventional marketing are rooted in unsustainable production and consumption. This chapter tries to identify some of the problems that will be posed by trying to green marketing practice without adjusting the conventional marketing mind-set.

Thinking Differently about Customers

The marketing philosophy revolves around doing business profitably by identifying and meeting the needs and wants of customers. However, the conventional marketing view of customers suffers from limitations that make it difficult for marketing to embrace sustainability. One limitation is that the needs and wants of customers are dealt with in the present tense only. If one accepts that the environment has limits, then it becomes clear that meeting the needs of today's customers unsustainably will reduce our ability to meet the needs of future generations of customers. 'Futurity' is one of the fundamental components of sustainability, but it has little place within conventional marketing thought.

The pioneering management theorist Henri Fayol once quipped about sending out for workers, but human beings turned up instead. Marketing also risks dehumanising customers into 'units of consumption' whose purpose is to consume, by trying to reduce their individuality and complexity into a small number of purchase-related 'want' factors. People are complex, individual, inconsistent and possessed of many and often conflicting wants and needs. In many cases, explicit material wants conflict with our inherent needs as human beings for a healthy diet, an active lifestyle, a manageable level of stress or a sustainable environment. Mulhern (1992) proposes a greater orientation of marketing towards consumer welfare, as well as wants, because the classical marketing concept is 'blind to the well-being of consumers'.

Rethinking Consumer Sovereignty

In marketing, 'the customer is king' and 'the customer is always right'. The concept of consumer sovereignty is fundamental both to marketing and to free-market economics, to the extent that it is often equated with the sanctity of individual liberty within democratic societies. Hutt (1936) expounded that 'the only grounds on which consumer sovereignty may be attacked, is by a refutation of our conception of liberty'. This approach plays into Adam Smith's 'invisible hands' in assuming that consumer freedom creates the optimal level of both satisfaction and social welfare. However, as the response to the passive smoking issue in the USA has demonstrated, liberty and satisfaction are also partly about freedom from the consequences (individual or collective) of others' consumption.

Consumer sovereignty is a curious concept, in that, by ascribing to consumers such a supposedly powerful role, it equates them with an institution that, in the modern age, only wields symbolic power, and which in the past often wielded real power in ways that were characterised by cruelty, ignorance, the pursuit of whims and manipulation by advisors.

Consumers, of course, lack the unified will of an individual ruler and, instead, resemble a democratic electorate who can only influence marketing strategy by combined and coherent action (for example, when mobilised by a consumer boycott). 'Dethroning' customers to view them as a constituency, combined with a focus on customer welfare as well as wants, demands that marketers no longer devolve responsibility for their decisions and actions by portraying themselves as passive servants of the customer. Although marketers may recoil from this as misguided and paternalistic, it is simply an extension of the kind of marketing that characterises professional services, in which the company takes some responsibility for the customer and for informing and guiding their actions. At present, there is a danger that the concept of consumer sovereignty is being used by companies as a shield and as an excuse. Volvo and McDonald's are both environmentally responsive companies, but each has fallen back on 'It's what the customer wants' as an argument to defend environmentally suboptimal strategies with the introduction of, respectively, the large-engined 850 car (Rothenberg *et al.* 1992) and the paper and plastic burger 'wrap' in the USA.

The faith placed in consumer choice as a means of guiding companies' strategies and activities assumes that consumers are faced with a full and free choice, and that they have the information with which to make appropriate decisions. In reality, consumer freedom of choice may be more illusory than real because choices are often carefully constructed by others (Tomlinson 1990). Sometimes choices are constructed by policy-makers: for example, in the UK economy, car ownership and use has been strongly promoted by reductions in public transport and the trend towards large 'out-of-town' retailing. To reach a supermarket shelf, a new green product usually has to overcome the preconceptions of corporate marketing managers and the retail buyer 'gatekeepers'. Einsmann (1992) describes how the product managers developing the Lenor refill pack at Procter & Gamble had to battle against conventional company wisdom about consumers' willingness to use refills, just to get the product as far as market testing. Even when customers have the opportunity to exercise choices in favour of the environment, do they have sufficient understanding and information to do so? Most market research exercises reveal consumers to be confused and lacking in confidence in relation to environmental issues and their relationship with purchase decisions.

So, although in the long term consumers are increasingly likely to use their purchasing power to discriminate in favour of the environment, to believe that this alone can make substantial progress towards sustainability appears hopelessly optimistic. The marketing discipline is strongly oriented towards the consumer, and the environmental marketing debate so far has mirrored this. However, New *et al.*

(1995) highlight the fact that the majority of economic activity lies in exchanges between companies rather than with the final consumer. Although the final consumer theoretically acts as the catalyst of, and focus for, events throughout the supply chain, in reality many companies are seeking to gain improved eco-performance from suppliers for reasons not related directly to consumers.

Rethinking Satisfaction

The fundamental aim of marketing is to create consumer satisfaction at a profit. This is a laudable aim, but it is one that has perhaps led to the marketing philosophy being accepted somewhat uncritically. Marketing is founded on the assumption that satisfaction is created by material consumption, even though there is evidence from fields such as psychology that the two are not that closely interconnected (Scitovsky 1976). What we mean by 'satisfaction' is also something that needs to be reconsidered in relation to sustainability. Although marketing is presented as a means of satisfying consumers' wants and needs, marketers' attention has usually been focused on the satisfaction of explicit material wants, even when these are at the expense of the customers' more fundamental needs. The marketing of cigarettes is a clear-cut example of this, as is the controversy over car marketing in the wake of the Ford Pinto scandal. Investigations into the development and marketing of this car, whose fuel tank tended to rupture to deadly effect if it was hit from behind, showed that safety compromises were deliberately made because of cost, a perceived lack of consumer concern about safety and, as one Ford engineer at the time expressed it, because safety modifications would mean 'You could only get one set of golf clubs in the trunk.' Lee Iacocca many times remarked that 'safety doesn't sell' (Trevino and Nelson 1995), and the obsession with speed and style over safety has dogged the car industry. Sustainability will require a greater focus not only on customer welfare, but on the satisfaction of needs as well as customer wants (Mulhern 1992).

Another problem with the existing marketing paradigm, when it talks about creating satisfaction, is that it takes a very narrow view of satisfaction entirely oriented around customers. The impact of marketing activities on non-consumers and the implications for their levels of satisfaction are rarely considered. This is simply seen as a set of 'externalities' which are left aside with the assumption that the public policy-makers will pick them up and deal with them if necessary. For example, in the way markets are defined, any person who has the potential to desire a product, but lacks the ability to pay for it, is not counted as part of the potential market and is screened out of the marketing thought process. These people may find themselves exposed to the marketing communications that promote the product, and may be encouraged to desire it, but their dissatisfaction at being unable to make a purchase exists outside the marketing frame of reference. Similarly, when the issue of car safety is discussed, it almost always equates to driver safety. Car safety ratings always favour the largest and heaviest cars, which are only 'safe' in an accident for the people who happen to be inside them. For cyclists, pedestrians and those in lighter vehicles, the reverse is true.

Rethinking Products

Although the marketing philosophy encourages companies to think about customer needs before they think about products, the product itself is still central to marketing. Marketing theory has helped to broaden marketers' concept of 'the product' beyond the core product to encompass all the tangible dimensions of the product (such as packaging) and also the augmented dimensions (consisting of supporting services and products). Environmentally concerned customers have shown that their purchase decisions can encompass factors that go beyond these conventional dimensions, to take account of the 'externalities' involved in production. This means that they may reject a technically excellent product because they become aware of the environmental harm caused in production or disposal, or because they disapprove of some activities of a producer, its suppliers or investors. Therefore customers may differentiate in favour of rod-and-line-caught tuna, or Café Direct's fair trade coffee; or may differentiate against CFC-driven aerosols, or against Nescafé because they disapprove of Nestlé's marketing of infant formula milk.

If customer satisfaction depends increasingly on the acceptability of the production process and all the other activities of the producer, we are approaching the situation where the company itself is becoming the product consumed (or the 'total' product; Peattie 1995). Drucker's (1973) famous concept that 'Marketing is the whole business seen from its final result, that is from the customers' point of view' will become a reality within a more sustainable economy, because green concern means that customers (or those who influence them) will be actively looking at all aspects of a company. As Bernstein (1992) comments, 'The consumer wants to know about the company. Companies won't be able to hide behind their brands. Who makes it will be as important as what goes into it since the former may reassure the customer about the latter.'

The conventional marketing vision placed the focus on the product as a market offering, and the marketing mix as the arena of interaction between the company and the customer. This left the production system in the shadows as a black box into which the customer was not encouraged to peer. In industrial markets, this attitude was changed with the introduction of total quality management, but a similar need to open up the black box is occurring in consumer markets and is being accelerated by laws such as the USA's 'Right to Know' legislation.

Rethinking Pricing

Markets are driven by prices. The price of a product usually covers the direct and indirect internal costs of production plus a profit margin to reward investors and for reinvestment. The socio-environmental costs of production, product use and disposal are largely excluded from the price that consumers pay. These costs are treated as 'externalities' which, in theory, are met by governments through the taxes that the company, its investors and employees pay. In practice, governments throughout the world are seeking to reduce tax burdens, and the costs of sustaining the environment are not being met. There is a danger that marketers and companies see this as 'not their problem', but companies are not politically neutral and in many cases lobby hard against the imposition of taxes proposed by democratically elected governments to meet the costs of environmental 'externalities'. In California, business spent $35 million on lobbying to prevent the state's 'Big Green' environment protection bill from becoming law.

The debate about the environment and marketing has been underpinned by an assumption that greener products will necessarily cost more, and by discussion about how much green premium the customer will be willing to pay. In many instances, this may be misleading because the additional cost comes from the 'end-of-pipe' philosophy which has dominated green technology development in key markets. Adding a catalytic converter to a car must add cost. However, evidence suggests that where a 'clean technology' approach is followed, developing new technologies from first principles, which reduce the inputs needed and the amount of pollution and waste produced, such initiatives can be very cost-effective (Irwin and Hooper 1992).

The green premium may also relate more to company pricing strategy than to economic necessity. Where green products are being placed alongside a company's conventional products, there is a danger that it will only accomplish a switching of some existing consumers to the greener version. If the greener version then replaces the conventional version, customers who are unconvinced by the arguments for greener products and who were loyal to the old version may well go elsewhere. Introducing a greener product at a premium price solves this problem. It differentiates itself from the existing version on the basis of eco-performance and price, allowing customers wanting to switch away from the conventional to do so, but in a way that also generates additional margin for the company.

Although assuming that greener products must cost more is misconceived, in many cases there are genuine cost burdens that greener products have to bear, because they are attempting to internalise some or all of the 'externalities'. The reality is not that these products are unusually expensive, but that conventional products are unrealistically cheap, because they are effectively subsidised by the environment. Unfortunately, marketing research is reinforcing the prejudice against green products by regularly asking consumers 'How much extra would you be willing to pay for greener products?' This is not a neutral question and it contains a very powerful message promoting the image of the environment as an additional cost burden on business and consumers. It would perhaps be more appropriate to ask consumers 'Do you want to continue buying products that are inexpensive because they damage the environment?'

Reconsidering Purchasing

Another assumption that is embedded within conventional marketing thinking is that consumer satisfaction is generated through purchase and ownership of a product.

A moment's reflection makes it clear that generally it is product use, not the purchase or ownership of the product itself (ostentation benefits notwithstanding) that creates the benefits that generate satisfaction. The greener economy of the future is likely to make this distinction clearer, with products that are used infrequently being increasingly rented or shared rather than individually owned. The growth in tool hire and car pools associated with residential developments are examples of the growing trend in product use without individual ownership.

The green challenge will eventually demand that the marketing focus is broadened to examine alternatives to purchase that can create satisfaction. Renting, borrowing, re-using or repairing products can all create satisfaction without outright purchase. In terms of consumer behaviour, the bulk of marketing academics' and practitioners' attention has been focused on the decision-making processes and behaviours that lead to purchase. In terms of the environmental impact of consumption, much of this is related to post-purchase issues such as product use, maintenance, re-use and disposal. At the moment, one only has to pick up any standard marketing textbook to see that 90% or more of the content relates to the build-up to, and completion of, the purchase. The greening of the marketing paradigm will require a much stronger focus on what happens after a purchase is made.

Rethinking the Marketing Environment

One of the benefits of a marketing orientation for a company is that it encourages the firm to be externally focused on the marketing environment. However, conventional marketing has followed the tradition of mechanistic economic models, which ignore the ecological contexts in which economic activity occurs (Capra 1983). So, although books on the marketing environment will dedicate chapters to the social, cultural, technological, economic and political environments, they can often overlook the physical environment that underpins them all, and on which they all depend. Figure 1 visualises the marketing environment as composed of layers of issues and interactions, with the physical environment as the foundation on which societies and economies are based. Understanding the sustainability challenge involves appreciating that instability and disruption to the physical environment will have effects across the other environmental layers. Although the physical global biosphere may seem distant to many companies' day-to-day activities, ultimately all business activity depends on it, and its continuing stability and viability.

Rethinking Market Structures

The conventional assumption about exchanges within markets is that they are linear. Goods and services flow in one direction, while money flows in the other. At one end of the chain, materials are extracted from the planet and, at the other end, the waste is buried in a hole in the ground. A sustainable economy will require many

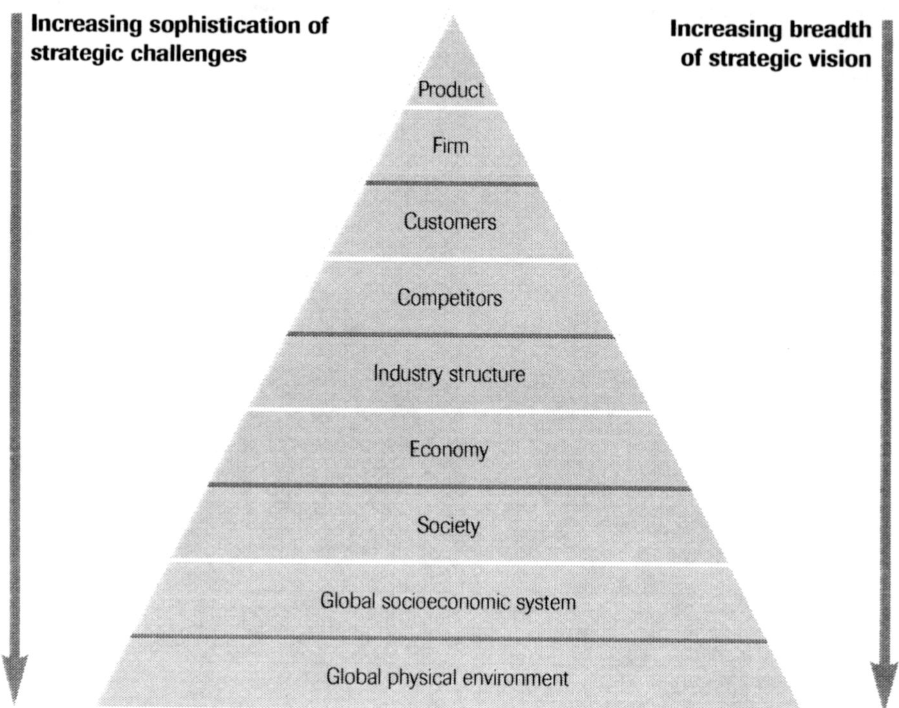

Figure 1: THE PHYSICAL ENVIRONMENT AS THE FOUNDATION OF THE MARKETING ENVIRONMENT

more 'closed-loop' supply chains in which products are returned for recycling, such as the one being developed by Xerox within their document processing business (Stenross and Sweet 1992). This approach casts a company's customers in a dual role also as suppliers, and calls for a much more partnership-based approach than the conventional concept of arm's-length economic exchange.

Rethinking the Environment as a Source of Competitive Advantage

Central to marketing is the development of strategies and actions to secure some form of competitive advantage. Much of the debate about the environment and marketing has therefore centred around whether good eco-performance can act as a source of differentiation to generate competitive advantage. Authors such as Porter and van der Linde (1995) and Elkington (1994) have argued that 'win–win' strategies exist which are environmentally superior and which, by stimulating innovation and tapping consumer concern, can create competitive advantage. Others argue that this is difficult to achieve in practice (Walley and Whitehead 1994). Products marketed on an environmental platform have often proved vulnerable to competitor tactics such as discounting, or attacks on the level of technical performance offered or on the credibility of the environmental claims. This is seen as a form of 'prisoner's dilemma' in which companies wishing to invest in improved eco-performance cannot for fear of being undercut by rivals. Capitalising on good eco-performance in search of competitive advantage is also made difficult by the attitude of the media. Companies such as The Body Shop and Loblaw have found that the media is more inclined to attack relatively good companies for their absolute shortcomings, than to highlight the poor environmental performance of conventional companies.

Although framing the greening of marketing in the positive light of achieving competitive advantage seems like a good way of winning friends and influencing people, there is a dangerous side to it. This ecopreneuring approach works on the assumption that greener products can be environmentally superior to conventional products while also being price-competitive and similar or better in terms of primary technical performance. It is a tall order for greener products to have to match or exceed their conventional rivals on all dimensions, while conventional products do not have to attempt to match the eco-performance of greener products and can take advantage of the subsidy that the environment provides them (by treating socio-environmental costs as externalities) to undercut their new greener rivals. The development of a new generation of win–win products will create some progress towards greener markets, but it is unlikely to achieve the magnitude of change necessary to move the economy towards sustainability. At some point, the debate will have to shift from an emphasis on the environmental benefits of new green products to grapple with the socio-environmental consequences of conventional products and consumption patterns. This is unlikely to be an appealing prospect for marketers who are generally happier stressing benefits rather than consequences, particularly when many greener brands are being marketed by companies alongside their conventional stablemates. While marketers are willing to engage in greener marketing only on a win–win basis, it will leave many green brands fighting for market share with their most effective weapon (highlighting the harm done by their conventional counterparts) ruled out of bounds.

Another reason why trying to frame the greener marketing debate in the context of competitiveness may be inappropriate is that, in practice, many companies in the environmental front line are part of global industries such as cars, chemicals and electronics, which are characterised by networks of international strategic alliances. These can assist in developing industry-wide solutions to common environmental problems. For example, Japanese and American electronics companies are co-operating in research to eliminate the use of CFCs as a solvent for the whole industry, rather than trying to gain individual competitive advantage.

Rethinking the Green Consumer

It seems logical for marketers, when faced with a population professing increased environmental concern, to respond by trying to identify 'green consumers' and finding out what motivates purchases of environmentally marketed products. If this can be done, and appropriate market offerings created, then the competitive advantage opportunities outlined by Porter and others can be achieved.

Academic researchers and market research agencies have put a great deal of effort into attempting to define and understand the relationship between peoples' environmental concern and their purchasing behaviour. Many factors have been proposed as influences on green consumer behaviour such as changing consumer values, demographic factors, knowledge of environmental problems and alternative products, perceived personal relevance and the ability of the individual to make an effective contribution. (For a model that integrates the majority of these, see Dembkowski and Hanmer-Lloyd 1994.) There has been a good deal of contradictory evidence in attempts to link factors such as gender, age or level of environmental knowledge to green consumption (Miller 1993; Peattie 1995). The difficulties in isolating green consumer behaviour reflect several factors:

- It overlooks the point made by Kardash (1974) that all consumers (barring a few who enjoy contrariness for its own sake) are 'green consumers' in that, faced with a choice between two products that are identical in all respects except that one is superior in terms of its eco-performance, they would choose the environmentally superior product.
- By attempting to relate a consumer's environmental concern to purchases, marketing researchers may be looking in the wrong place. Many of the most significant contributions that consumers can make towards environmental quality come in product use, maintenance and disposal, or in delaying or avoiding a purchase through a 'make-do-and-mend' mentality.
- Environmental improvements in products are often entangled with economic or technical benefits. Therefore, drivers may choose lead-free fuel for environmental or economic reasons, or they may choose organic food for reasons of environmental concern, personal health concern or simply for the taste benefits.
- Different answers are achieved depending on what is defined as constituting green consumer behaviour, and whether environmental concern is defined in general or specific terms. General environmental concern is often measured by researchers, but it is less easily related to products than specific environmental concerns (such as concern for dolphins translating into the purchase of rod-and-line-caught tuna fish).

Perhaps the solution to understanding green purchasing behaviour is to try to understand the purchase rather than the purchaser. If we accept Kardash's proposal that, all other things being equal, most customers would differentiate in favour of greener products, then understanding environmental purchasing behaviour (and often the lack of it) is assisted by looking at the extent to which other things are not 'equal'. Many green purchases involve some form of compromise over conventional purchases. The compromise can take a variety of forms including :

- Paying a green premium. This can be imposed by economic necessity where improving eco-performance increases production costs. Alternatively, it can be created by marketing strategies in which greener products aimed at green market niches are given a premium price irrespective of production costs.
- Accepting a lower level of technical performance in exchange for improved eco-performance (e.g. green detergents)
- Travelling to non-standard distribution outlets (e.g. specialist green retailers, such as the 'Out of This World' organic supermarkets)

Where there is a compromise involved in making a greener purchase, a key factor that will determine whether or not customers will pay more, or accept reduced technical performance, is the confidence they have in the environmental benefits involved. Customers will need to be confident that:

- The environmental issue(s) involved are real problems.
- The company's market offering has improved eco-performance compared to competitor or previous offerings.
- Purchasing the product will make some sort of material difference.

The framework presented in Figure 2 encapsulates this approach to diagnosing green consumer behaviour, and it can help to explain some of the inconsistencies in the research findings into green consumer behaviour. The majority of consumers profess concern for the environment, a desire to buy greener products and a willingness to pay for them or accept technical performance reductions. The numbers of consumers measurably changing their purchasing behaviour to buy green are much fewer, and this has generally been interpreted as a failure to back up intentions with purchase and a tendency to over-report social and environmental concerns (e.g. Wilson and Rathje 1990). This undoubtedly explains part of the discrepancy, but the missing element is the confidence that customers have in companies' green marketing offerings. A BRMB/Mintel survey found that 71% of UK consumers thought that companies were using green issues as an excuse to charge higher prices (Mintel 1991).

Attempts to relate environmental knowledge to green consumption have produced inconsistent results, but researchers have assumed that increasing environmental knowledge will lead to an increased desire to purchase green products. The reverse may be true, in that increasing environmental knowledge can actually reduce consumers' confidence in the effectiveness of market-based solutions for environmental challenges, and it may make them more aware of the shortcomings of products seeking to market themselves on a green platform.

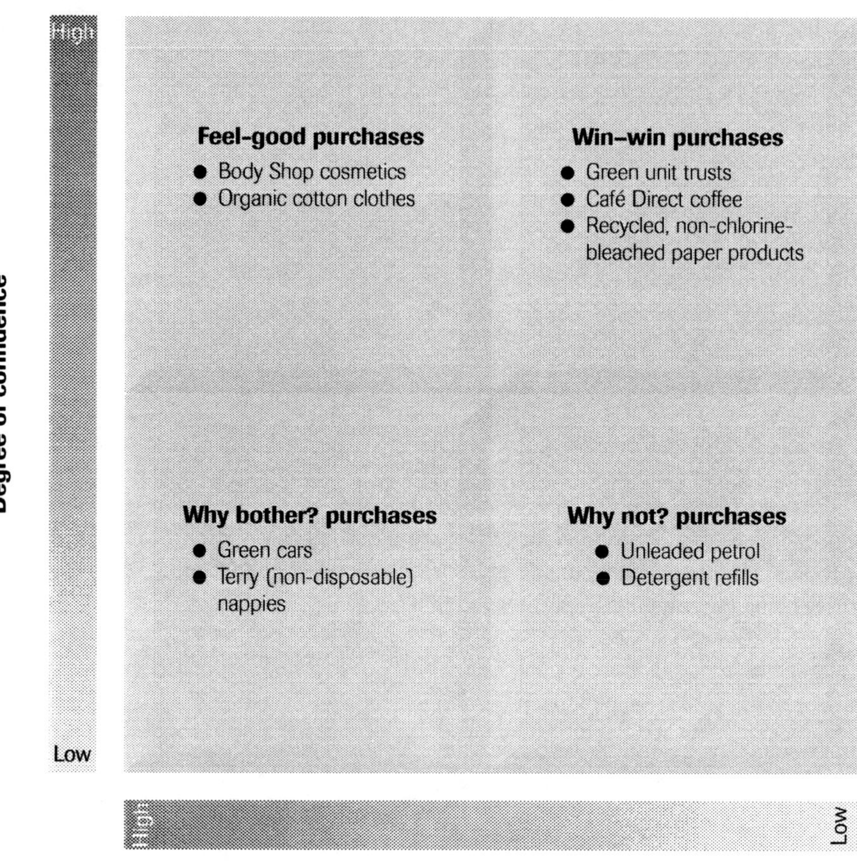

Figure 2: THE GREEN PURCHASE PERCEPTION MATRIX

Conclusions: Time for a Rethink

As we move into the new millennium, businesses are faced with many challenges, from the globalisation of markets to the development of electronic markets and virtual corporations. Such a landmark is perhaps an ideal moment to stop and reconsider where we are going. However, in the face of so many challenges, it is perhaps not surprising that the management community are downplaying the green challenge as 'just one among many key business issues'. Businesses, like people, are change-averse. One of the best ways of minimising the degree of change that greening will entail is simply to integrate it within existing ways of doing business. If we can turn the sustainability challenge into something that can be included within an existing total quality programme, or corporate image and community relations programme, or that can be focused on efforts to cut resource costs or seek out competitive advantage, it becomes something that is no longer a threat, and that managers can be comfortable dealing with.

This is an attractive proposition. Experience from the USA, where many high-profile corporate greening programmes have run into difficulties (known as 'hitting the green wall'), suggests that companies will have problems where the greening initiative does not integrate well with existing corporate strategies, cultures and even vocabularies (Shelton 1994). However, sustainability is something that we will not achieve by adjustment, and by changing the ideas and vocabulary of sustainability to bring them into line with those of business. Sustainability is an absolute concept like universal suffrage, although it is more complex to define and harder to measure progress towards. The history of British electoral reform shows that it took many steps to reach the position where virtually all adults have the vote. Each step that was taken was accepted by the establishment at the time as a means of reducing the pressure for more radical change to a manageable level, rather than to achieve a significantly new and fairer situation. Sadly, this appears to be the case with much of the commercial and political response to the pursuit of sustainability. Electoral reform would never have been achieved without the dedicated pursuit of the absolute concept of universal suffrage by those who believed in its equity and the need for reform. Similarly, sustainability will only be achieved if pressure to make concerted progress towards it is maintained, and is not diffused by environmentally related concessions made by governments and companies. As the American Ten-States Attorneys-General Task Force concluded in their *Green Report* into marketing and the environment,

> The tone, content and number of environmental claims might lead the public to believe that specific environmental problems have been adequately addressed and solved. This, in turn, could actually impede finding real solutions to identified problems by causing consumers to set aside their environmental concerns, making the assumption that these concerns had been addressed (National Association of Attorneys-General 1990).

Greener marketing activities have little point or meaning if they do not contribute to making the economy more sustainable. They also have little hope of succeeding if customers do not understand the issues involved or are not convinced by the greener market offerings that companies are developing. It is perhaps telling that, in a piece of market research conducted by Lancashire County Council in the UK, examining perceptions of 'quality of life' eight years after the Brundtland Report was published, the first conclusion drawn was that

> People generally are unfamiliar with the idea of 'sustainability' in its environmental sense. But once they understand it, they appear to identify positively with its values and priorities (MacNaghten *et al.* 1995).

It finally concludes by saying that

> Overall, whilst there is substantial latent public support for the aims and aspirations of sustainability, there is also substantial and pervasive scepti-

> cism about the good-will of government and other corporate interests towards its achievement.

One strength of the original *Greener Marketing* collection (Charter 1992a) was the practical focus of the majority of the contributions, which means that a rather philosophical piece such as this might appear out of place. However, its purpose is to highlight the fact that, although we may engage in a variety of changes to marketing practice that are all environmentally beneficial, ultimately businesses and marketers will not make substantive progress towards sustainability without the occasional pause for a little rethinking. Marketers stand on a huge raft of assumptions and beliefs about markets, the environment, customers, consumption, satisfaction, costs, prices and consequences. Rarely do they look down to check whether these are still valid in a world that, we can see with increasing clarity, is very different to the original neoclassical economist's vision of a space consisting of infinite resources and infinite waste disposal capacity. Without a greener philosophy and vision of marketing, the greening of marketing practice will be an uphill battle.

2 GREENING THE 'MARKETING MIX'

4

Achieving Sustainability

Five Strategies for Stimulating Out-of-the-Box Thinking Regarding Environmentally Preferable Products and Services

Jacquelyn A. Ottman

SINCE THE LATE 1980s when environmental issues resurfaced as a source of consumer concern, industry has responded primarily by cleaning up its operations ('pollution prevention') and refining its products; excess packaging, for example, was eliminated ('source reduction') or recycled content used in place of virgin. Business advantages were reaped from these initiatives, in that new laws and regulations could be pre-empted, and many costs of doing business could be avoided. From the standpoint of green marketing, initiatives could be communicated to stakeholders so as to enhance corporate and brand image; market share could be increased by appealing to the needs of green consumers; and new markets among environmentally conscious government agencies and businesses could be opened. Now, the challenges of sustainable development have been added to the business agenda.

To make the significant leaps in efficiency of resource use and overall environmental improvement required to address the challenges of sustainable development—and hence meet the needs of future green consumers—businesses will need to take more aggressive steps. In the same way that product redesign, using tools such as life-cycle analysis and design for environment, became recognised as ways of preventing pollution upstream in the process, achieving more significant environmental gains will require new types of products altogether, coupled with changes in consumer lifestyles. The pay-off to businesses that attempt to meet the sustainability challenge will include increased opportunities for innovation, competitive advantage in the form of significantly reduced costs, enhanced levels of customer satisfaction and the ability to redefine the rules by which one's industry competes.

While the need for radical innovation is clear and the rewards are potentially great, the direction for individual businesses is not well defined. Questions for forward-looking businesses include: What does sustainable development mean for our industry and our company? What are the implications for our existing products and services? How can we turn long-term threats into opportunities?

Some companies have already begun to tackle the challenges of sustainability. Their experiences suggest that answering these questions will require creativity—'out-of-the-box' thinking—and viewing one's business and the value it provides from a fresh perspective. The purpose of this chapter is to help other enlightened businesses navigate this next, most important phase of green marketing by providing five strategies supported by the pioneers' experiences to use in reinventing one's business to meet the challenges of sustainable development.

The Sustainability Challenge Defined

Sustainable development has been defined in the Brundtland Report as 'meeting the needs of the present generation without compromising the ability of future generations to meet their needs' (WCED 1987). Put another way, it means ensuring the long-term viability of the human species by living off the 'interest' of global resources rather than the 'capital', so there will be plenty of resources for future generations.

Consensus is growing that we are poised just before the breaking point when society's demand for resources will outstrip the availability of supply as well as the capacity of the planet to act as a sink for emissions. Such an occurrence seems inevitable if one considers that 80% of the world's resources are now consumed by 20% of the world's population. This 20% does not include the two largest countries in the world, China and India, which represent nearly half the world's population. These countries are fast adopting Western lifestyles with the attendant strains on natural resources and the emission-absorbing ability of the atmosphere.

Enabling developing countries such as China and India to improve their standard of living in a sustainable manner will require, by some estimates, that we improve the efficiency of resource use by at least a factor of four (von Weizsäcker *et al.* 1997). Achieving such efficiencies will require, among other steps, a redesign of the entire system by which consumers' needs are met. Many existing technologies and products will become obsolete, opening the way for opportunities for new products and new technologies that can efficiently meet the needs of all the world's people (see Fig, 1).

The Opportunity for Eco-Innovation

Most product-related innovation to date has focused on modifying single product attributes: for example, designing a product to be easily disassembled for recycling or made from recycled content. In addition, new products have been introduced using environmentally preferable materials or formulae. Examples include 3M's Safest Stripper, a water-based paint stripper, and Wellman's EcoSpun fibre made from recycled soda bottles. Whether examples of the former or of the latter, such activities

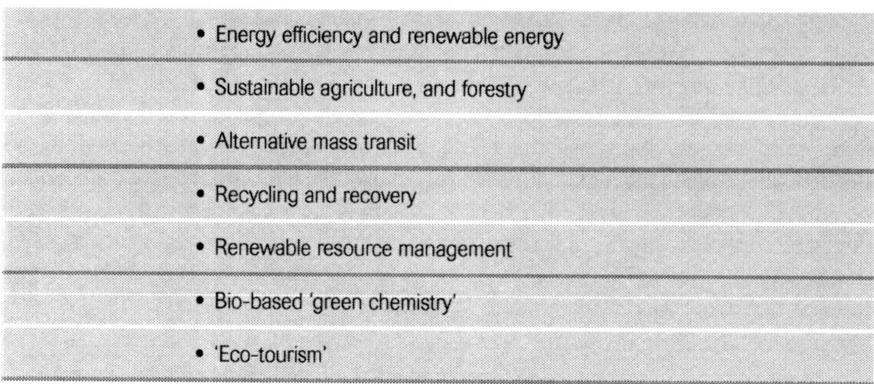

Figure 1: GREEN BUSINESS OPPORTUNITIES

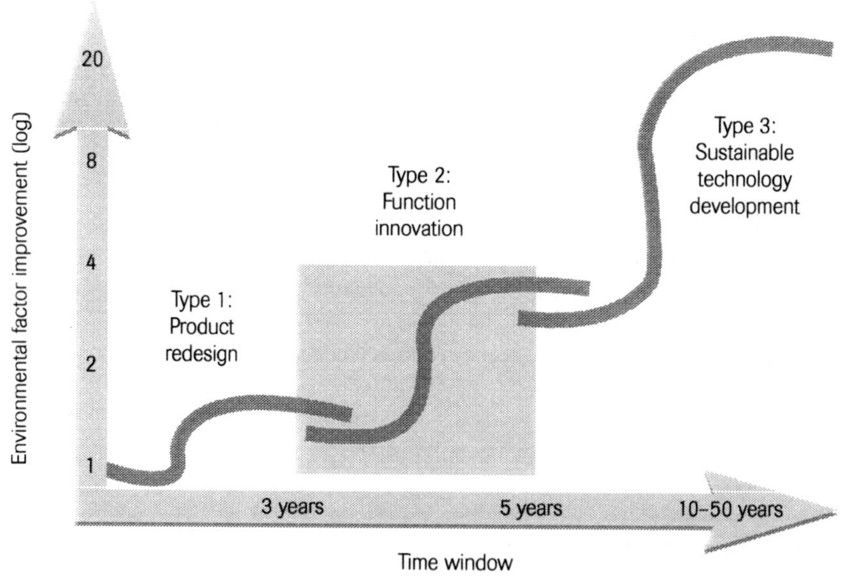

Figure 2: BARRIERS TO ENVIRONMENTAL IMPROVEMENT: THE NEED TO GO BEYOND EXISTING PRODUCT DESIGN

Source: Kalisvaart and van den Horst 1997

have been confined to existing product concepts. However, as illustrated in Figure 2, there is a limit to how much environmental performance can be enhanced by modifying existing products; at some point, the product concept itself becomes the barrier to further environmental improvement (Kalisvaart and van der Horst 1997). To achieve more significant gains, new product concepts representing functional substitutes to existing products must be developed. Examples of alternative product concepts—which can include services in whole or in part as substitutes for material products—include naturally coloured cotton versus dyed cotton, and e-mail in preference to voice mail. Many of these functional substitutes evolve over time as new technology becomes available: for example, electronic voice mail as a replacement for answering machines. As also depicted in Figure 2, even more significant improvements in environmental performance will only come about by changing technological systems such as energy and transportation.

While moving along this continuum, businesses can achieve progressively greater levels of efficiency which lead directly to enhanced profitability in the forms of savings on raw materials, energy and disposal costs. Reduced liability can occur from using less toxic ingredients and from less volume ending up in landfills. Using less toxic ingredients can translate into faster time to market, as some or all of the steps can be eliminated that are associated with what can be a cumbersome legal approval process.

These efficiencies can translate into an enhanced ability to meet consumers' needs. The resulting environmentally enhanced products and services can be viewed as approaching an ideal goal of 'zero waste'. This can result in higher-quality products delivering greater levels of customer satisfaction (Fig. 3).

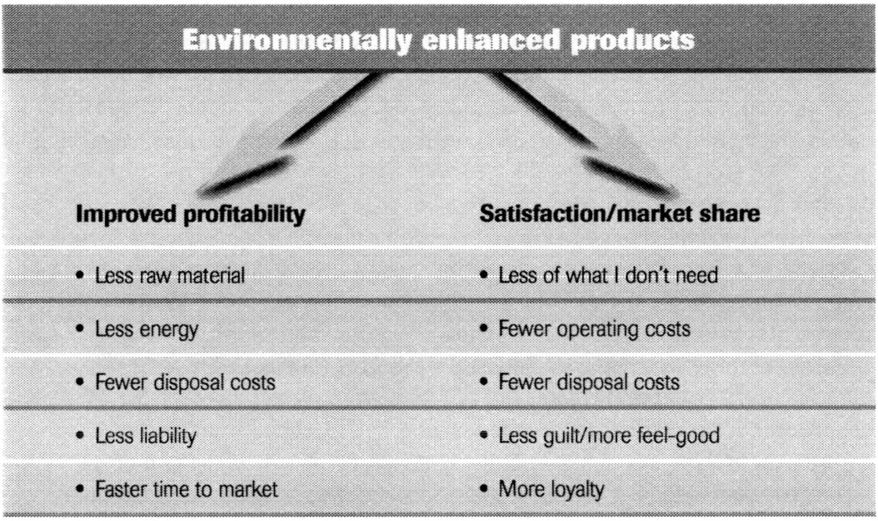

Environmentally enhanced products

Improved profitability	Satisfaction/market share
• Less raw material	• Less of what I don't need
• Less energy	• Fewer operating costs
• Fewer disposal costs	• Fewer disposal costs
• Less liability	• Less guilt/more feel-good
• Faster time to market	• More loyalty

Figure 3: ENVIRONMENTALLY ENHANCED 'ZERO-WASTE' PRODUCTS

Environmentally efficient products can lead to enhanced customer satisfaction via one or more ways. They represent an optimum fit between consumer's real needs and the product or service itself. Such products do not use any more technology than is required to do the job, nor do they come with features that are not desired by the consumer, such as all the 'bells and whistles' that automatically come with a new computer, for instance. As one in-market example of this, Dell builds each computer from the ground up to the exact specifications of its customers.

Environmentally enhanced products are cheaper to operate—such as computers that power down when not in use. They are also cheaper for the consumer to dispose of or recycle because they may not have unnecessary packaging, or are taken back by the manufacturer or a third party for recycling or re-use.

Given growing awareness for the link between product manufacture and use and their attendant environmental impacts, efficient products may also help reduce any guilt associated with material consumption; on the positive side, they can make consumers feel good about the products they buy—and the companies who make them.

The attributes and benefits described above can help generate customer loyalty. Church & Dwight, makers of Arm & Hammer products, estimate that the loyalty of their customers who appreciate their company's clean and green image and environmentally preferable baking-soda-based products translates into 5%–15% more revenues, or $75 million per year.

Creativity: An Opportunity for Everyone to Contribute

Einstein said, 'Imagination is more important than knowledge.' This is particularly true when it comes to addressing the challenges posed by sustainability. No one knows what a sustainable economy looks like: it must be created. Because creativity does not respect such factors as education level, job title or income, it represents an opportunity for literally every employee within a company, and all individual members of society at large to engage in the out-of-the-box thinking necessary for creating a sustainable society.

In the past, environment-related innovation has largely been delegated to technically oriented professionals working within the confines of existing product concepts and traditional business processes. However, the complexity of sustainability-related challenges requires a new, broader and more systematic framework for thinking about environmental-related business opportunities. This framework requires input from multidisciplinary corporate teams, as well as the various sectors of society. What will be critical to motivating and driving the efforts of such teams and co-operative entities are processes for envisioning a more sustainable society and developing a systematic plan for achieving this vision on local, regional, national and international levels. The first envisioning phase requires that the creativity of team members be stimulated.

What follows are examples of five strategies that can be used as starting points for companies attempting to reconceive of their business within a sustainable economy and promote out-of-the-box thinking regarding environmentally preferable products and service offerings within their firms. These strategies are part of a proprietary innovation process created by the author, called *Getting to Zero*[sm], to guide clients' environmentally sustainable product development efforts. When possible, they are illustrated by case examples of sustainable business solutions currently being implemented by pioneering firms.

Strategy #1: Set Outrageous Goals

Outrageous goals are the kinds of goal that sound virtually impossible to achieve. Such goals, when presented hypothetically for brainstorming purposes, represent an excellent way of bypassing incremental thinking quickly in favour of 'out-of-the-box' thinking (Everett 1997). For example, firemen can only race down a staircase so fast. Given the need to cut down significantly on response time to meet a fire, they have evolved the significantly faster method of sliding down poles.

DuPont and Xerox are two companies that understand the value of setting outrageous goals. DuPont has an environmental goal of 'zero waste'. Xerox's environmental goal is 'waste-free products from waste-free facilities'.

To test the potential of an outrageous goal to stimulate creativity, ask, 'What would we do differently if we had to reduce a certain impact on our business—e.g. water, energy, etc.—by 100% and still meet the needs of our customers? As an illustration, let us answer this question for washing machines.

A life-cycle analysis of a washing machine reveals that by far the greatest amount of environmental impacts occur during the in-use stage of the machine (as opposed to the manufacturing, distribution or disposal stages of the machine). Brainstorming ways of reducing the water consumed, the energy needed to heat the water and the waste-water effluent yields a number of alternative technologies and complementary products including: water- and energy-efficient machines such as those that rely on horizontal- (as opposed to vertical-) axis technology; washing machines that use only cold water accompanied by special cold water detergent formulations; 'dry' washing machines that might clean clothes via a waterless agitation technology or even the potential to 'wash' the clothes in the dryer via some type of heat-activated technology. Finally, if we were to be as creative as possible in applying 'out-of-the-box' thinking to this challenge, we might arrive at the solution of clothes that simply do not get dirty.

As the washing machine example illustrates, the impetus for creativity is embedded in the aggressiveness of the question. Furthermore, brainstorming with the somewhat outrageous goal of 'zero impact' can yield a spectrum of several new ideas, including some that may seem unlikely solutions at the present time, but may yet prove to be possible in the future, thus suggesting a direction for long-term technological development efforts.

Strategy #2: Think Like a System

Instead of making adjustments to specific features of an existing product in and of itself, the environmental performance of products can be more significantly improved through modifications to the system in which a product operates. The 'system' can be defined as the product value chain beginning with extraction of raw materials, processing and distribution through to the in-use phase and the product's eventual recovery or disposal. It can also be defined as a consumer use system—for example, other, complimentary, products with which a product is used or to which it is related. For example, in making sure that their customers get a good hot cup of coffee, producers can do a lot more than choose the best beans and package them well. They can also consider the hardness of the water, the kinds of cup that are used, as well as the sweeteners and whiteners available.

Products have a life-cycle that consists of a number of discrete stages, including: raw material extraction, manufacturing, distribution, in-use phase, and recovery or disposal. To date, refinements to many products have primarily focused on reducing impacts within specific stages—such as increasing the amount of recycled or recyclable materials, reducing energy or water consumption during the in-use phase, or by designing the product to be recyclable. However, possibilities exist to reduce impacts further by attempting to collapse two or more life-cycle stages, so to speak. For example, naturally coloured cotton is concerned with the combination of the raw materials procurement and the processing (dyeing) phases of the life-cycle. Xerox, who refurbishes its used copiers and sells them as remanufactured machines is actually combining the recovery/disposal and manufacturing stages of its own product's life-cycle (see Box 1).

A second tool to use in 'thinking like a system' is industrial ecology. Industrial ecology is a discipline that studies material and energy flows in the economy and ecosystems. It aims to redesign industrial processes according to nature's own processes, thus dramatically reducing pollution. When processes emulate nature, the 'waste' from one process becomes 'food' for another.

An example of industrial ecology is an industrial park in Kalundborg, Denmark. Begun in the 1960s during a water crisis, it has evolved over time to share additional resources so as to minimise waste and save money. At the park, an oil refinery, a pharmaceutical company, a gypsum-processing plant and an electrical power generation plant are situated together near a lake so that they can easily convert each other's steam heat, surplus gases and waste-water. There is even a fish farm that uses heated water from the power plant.

Although usually thought of as a tool for industrial processes, the concept of industrial ecology can help stimulate creativity in the design of products and product systems. A good example is a combination sink and toilet that was originally designed in the Netherlands to save space in a housing project. Along the way, the designer realised that he could save water as well by having the water for the toilet be sourced from the run-off from the sink. This run-off is saved in a cistern that doubles as the back of the toilet (*Journal of Sustainable Product Design* 1997). Considering that 30% of all the water used in households goes in flushing toilets, combining the sink and

TO ENHANCE their competitiveness against Japanese entrants, Xerox runs a corporate-wide Asset Recovery Management System. Copiers, which are historically leased rather than sold to commercial customers, are now designed for disassembly and recycling. This enables them to refurbish used copiers and sell them as remanufactured machines in Europe, as well as to use recovered parts in new machines in the US They also use recovered parts for servicing existing machines.

The supply of parts from returned machines gives Xerox a steady supply with which to service existing machines. Customers who waited up to six months for a part now wait only two weeks. Over a five-year period, this programme has helped Xerox save an estimated $200 million in parts, inventory and labour.

In Europe, adding the remanufactured machines to their product mix allows Xerox to compete in a lower price segment, which they could not previously afford to do. It also gives them an opportunity to sell their copiers to environmentally conscious companies under such names as 'EcoSeries' and 'Renaissance'.

A historical bias exists against remanufactured products, but Xerox has taken steps to overcome this by promoting their remanufactured machines as 'proven workhorses' and offering the same three-year total satisfaction guarantee that comes with their new machines. Xerox's research shows that consumer acceptance of the remanufactured machines has increased in the last several years.

Box 1: XEROX ASSET RECOVERY MANAGEMENT SYSTEM

Source: Davis 1996

the toilet into a unique new product system leads to significant water savings. The notion of recycling water within the household is called 'grey water'. We can expect to see more ideas inspired by industrial ecology thinking in the future. In Germany, all new homes must be designed so that rainwater can be collected from rooftops for landscaping purposes.

Many other opportunities may exist to conserve resources and energy by recycling the waste of one or more products within a system. For example, it may be possible for detergent makers to devise new laundry powders that would allow consumers to water their lawns with the run-off from the washing machine rather than send it down the drain, potentially polluting local streams and lakes. Doing so, of course, will require not only out-of-the-box thinking, but 'out-of-the-border' working, if you will. In order to be successful in developing such a product, detergent and washing machine makers will need to work with other professionals, including manufacturers of plumbing and gardening equipment and so forth.

It is also possible to conceive of one's product as a component in another product system. For example, Wellman, a maker of EcoSpun fibre, relies on the recycling stream of used soda bottles for its raw material.

Strategy #3: Dematerialise

The underlying notion here is to meet consumer needs with as few materials and as little energy as possible. Dematerialisation assumes that we can use services based on technology or know-how in place of material products so as to reduce global material flows and energy significantly. A measure of dematerialisation is 'material intensity (including energy inputs, either directly or indirectly) per unit of service obtainable from a product' (MIPS).

One question to ask in evaluating the potential to 'dematerialise' one's business is: Can our consumers' needs be met equally or better by providing them with a service as an adjunct to or replacement for a material product? For instance, in agriculture, grower needs can often be better met through integrated pest management services which make use of beneficial insects to overcome pests that are traditionally fought with pesticides.

Some excellent examples of dematerialisation are electronic voice mail as a replacement for answering machines, e-mail, virtual libraries including electronic alternatives to direct-mail catalogues, yellow pages directories and encyclopaedias, and telecommuting.

Dematerialisation, and especially the practice of selling services as an alternative to products, often requires that one redefine one's business. Volvo, for example, sees the writing on the wall for internal combustion engine vehicles and is redefining its business more broadly as 'transportation'. It also sees itself as a 'solution supplier' not a 'product supplier'. The company is now developing mass transit systems for China, and global positioning systems that can help vehicles and shipments get from place to place more directly (Friend 1998). IBM's service business is growing faster than its hardware business and it is more profitable, too (Hansell 1998). It is now capitalising on the superfast growing market for electronic commerce (Cook 1998).

Opportunities may exist for companies to add value to existing products by bundling them with a service. For example, fertiliser companies can redefine their business more broadly as 'lawn and garden care' and offer paid advice lines via telephone or Internet to answer questions of home gardeners or offer to test soil samples for pH levels. Similarly, detergent manufacturers can supplement their offerings with paid information services on how to remove tough spots and stains. How many other consumer products businesses could benefit by expanding their business definition from 'converting natural resources into products sold from a shelf' to 'using human resources to solve customers' problems'?

Strategy #4: Make it Fit

Einstein once said, 'Make things as simple as possible and no more.' From a sustainability standpoint, this can be translated into the notion of maximising the utility of resources by designing products to fit the real needs of consumers as closely as possible. It assumes that product design, materials and technology are appropriate to the scope and difficulty of the task as well as the locality (climate, resources, available solid waste infrastructure, for example). It makes a case for a technological counterpart to biodiversity: for example, a camel is better suited for desert transportation than a Mercedes-Benz!

One question to ask in uncovering ways of using existing products and technologies more efficiently is: How does our technology best fit within the entire range of alternatives that meet a particular need of consumers? The answer to this question might help develop the market for electric vehicles.

Some electric vehicles are currently positioned as substitutes for combustion engine vehicles. However, due to their relatively new technology and need for frequent recharging, they compromise economy and convenience. An opportunity may exist to develop the market for electric vehicles by positioning them as an adjunct to combustion engine vehicles instead—well suited, in a golf-cart-like design, for short hauls, commuting to the train station and the like.

This idea, in a modified form, is now in development. Lee Iacocca, former chairman of Chrysler, and Robert Stempel, his former counterpart at General Motors, have teamed up to develop an electric bicycle. Their target market is individuals in retirement communities and college campuses. And their goal is to sell one million electric bikes per year (Taylor 1998).

The strategy of 'make it fit' also suggests opportunities, Darwin-style, to meet needs based on specific local climate, topographical, resource or infrastructure conditions: for example, designing packaging that can be composted or recycled according to local solid waste infrastructure.

Strategy #5: Restore rather than Take

It is generally accepted that all products must use up resources and create waste. Thus, the current goal in eco-related design projects is to minimise impact on the environment. What if the paradigm shifted so as to design products with the goal of restoring the environment? Composting toilets help to restore nutrients to the soil the same way in which manure provides fertiliser for farmers' fields. Compostable golf tees made of cornstarch can be thought of as little vitamin pills for the golf course. Volvo has announced that, starting in spring 1999, it will incorporate special 'PremAir' catalyst systems to its S80 model car that are capable of destroying ozone created from other cars' pollution (*Associated Press* 1998).

Businesses who wish to utilise this strategy to generate out-of-the-box thinking regarding environmentally preferable product development can ask the following questions: What would it take for our product actually to benefit the environment or society? Can our marketing efforts help to educate consumers about key environmental and sustainability issues? Can our manufacturing be a source of jobs for the handicapped?

This strategy can lead to environmental marketing opportunities that can reinforce one's overall business strategy. For example, the US-based Hannah Anderson catalogue of children's clothing encourages customers to send back used children's clothing by offering a 20% discount on future orders. The company then sends these clothes to the poor in a programme that they call 'Hannadowns'.

In the UK, Trevor Baylis has developed the BayGen radio as a vehicle for bringing information about birth control and AIDS to people in third world countries, where

AT CANON, a corporate philosophy of *kyosei*—living and working together for the common good—guides the company toward cause-related marketing which reinforces the company's position as a market and environmental leader.

In the US, it began with the Clean Earth Campaign in 1990 which donated $1 to the National Wildlife Federation and The Nature Conservancy for each Canon toner cartridge returned to the company. The five-year effort resulted in the recycling of several million toner cartridges along with a corresponding donation. The success of the programme inspired Canon to deepen and enhance their cause-related marketing efforts.

Among other initiatives, the company now supports 'NatureServe', a comprehensive programme for sharing with the public The Nature Conservancy's scientific knowledge and expertise on natural resources conservation; and 'Expedition into the Parks', a programme with the National Parks Foundation to inventory and protect rare plant and animal species found in national parks; and 'Naturelink', a fun book on how families can enjoy the outdoors together, implemented with the National Wildlife Federation.

These initiatives help Canon USA show its environmental concern to its 9,800 employees worldwide, and serve as a model for other companies. The depth and scope of these efforts allow Canon to promote their participation credibly to all stakeholders via such vehicles as the PBS series 'NATURE' and adverts in *National Geographic*.

Box 2: CANON

Source: Personal communication with Russell Marchetta, Canon, Lake Success, NY

there are very few batteries. The radio relies on an old-fashioned clockwork mechanism, where the crank is wound for 25 seconds to obtain 25 minutes of playing time. To extend the societal benefits, the radio is made by disabled workers in South Africa (Chick 1997).

Innovative cause-related marketing programmes, such as those run by Canon USA, can restore the environment through education (see Box 2).

Conclusion

Sustainable product development strategies can make businesses more competitive and result in better products with enhanced levels of customer satisfaction over conventional products. Developing sustainable products requires that companies operate under a new paradigm that includes:

- Products do not have to be disposed of: they can be more useful to society and more profitable for business if they can be re-used or recycled into new products.
- Products do not have to be designed for obsolescence to be profitable: they can be more productive to society and the bottom line if they are durable.
- Consumers' needs can be profitably met with services rather than products, or at least an optimum combination of both.

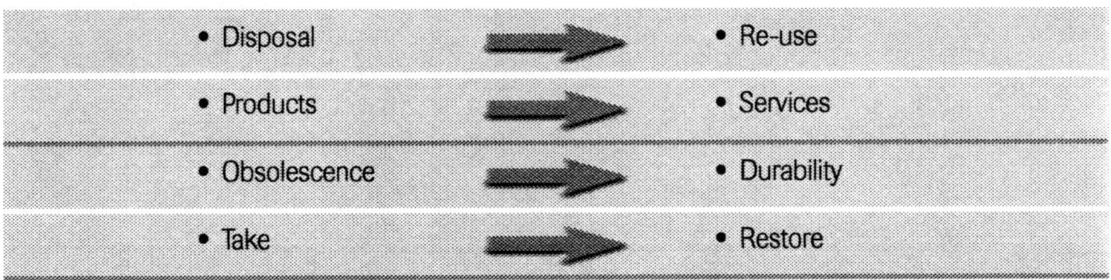

Figure 4: THE SUSTAINABILITY PARADIGM

Source: J. Ottman Consulting, Inc.

- Consumers will reward businesses that help restore the environment while creating the products consumers want (see Fig. 4).

To achieve more sustainable products and services, industry and other societal sectors will need to think and act in new ways. Businesses need to rethink the value their products provide and consider whether that value would not be better delivered wholly or in part by a service. Cross-functional, inter-disciplinary teams will be needed to consider system-wide effects of products. Companies may also need to transcend their corporate borders, teaming up with manufacturers of complementary products and designing more efficient systems capable of optimising the service delivery and minimising the environmental impacts of their original products. It may also involve acquiring expertise necessary to offer services as an adjunct to or substitute for material products.

The biggest strides in environmental improvements will come via the generation and implementation of alternative product concepts representing functional substitutes for existing products. Companies such as Volvo, which is willing to take a step back, redefine its business and acquire new technology and expertise necessary to serve its markets in more efficient ways, will reap big rewards. Companies that cling long term to established technologies and product concepts risk being displaced by competitors willing to take a risk on more efficient, even restorative alternatives.

Generating ideas and concepts for sustainable products and technologies requires creativity and 'out-of-the-box' thinking. Such thinking can be cultivated and stimulated by learning from natural processes and designs, by borrowing from established creativity approaches, as well as extrapolating from early in-market examples.

Finally, for such thinking to happen at all, it will need to be cultivated by appropriate organisational structures and reward mechanisms that encourage employees to behave differently—to think in new ways and to take risks with the promise of long-term rewards and the opportunity to contribute to addressing the significant challenges of sustainability.

Reprinted from *Corporate Environmental Strategy: The Journal of Environmental Leadership* Vol. 6 No. 1, Jacqueline Ottman, 'Achieving Sustainability: Five Strategies for Stimulating Out-of-the-Box Thinking Regarding Environmentally Preferable Products and Services', pp. 81-89, © 1998, with permission from Elsevier Science.

5

Eco-Marketing 2005

Performance Sales instead of Product Sales*

Frank Martin Belz

IMAGINE A DAY IN THE YEAR 2005 . . .

At 7.00 am on 9 January 2005, Priscilla gets up, takes a shower, prepares breakfast and reads the newspaper in the living room. Ever since she signed the 'performance contract' with the facility management company, her apartment has been warm and cosy. In fact, the performance contract actually guarantees a heated, comfortable apartment; she no longer pays for products such as electricity and petrol. In the past she paid around €10,000 annually on electricity and water bills. Due to ecological tax reform, energy and water prices have gone up slowly but surely. The service people from the facility management company put up solar cells on the roof, installed electricity meters, and adapted the toilet and the shower to save water. Within a short period of time, water and energy consumption had been reduced by no less than 50%. The performance contract lasts ten years. During that period, the facility management company recoups the savings to pay off its investments and to make profits; after that period, Priscilla pays less for her energy and water due to higher eco-efficiency.

After breakfast, Priscilla leaves the house and drives to the main station with the latest model of ZECAR, the zero-emissions city car run by fuel cells. At the main station she takes the next train from Frankfurt to Basel, where she has a business meeting. Her profession means she travels a lot. That is why she decided two years ago to sell her convertible and buy the 'mobility package' offered by the joint-venture

* This chapter is a contribution to the research project, 'From the Eco-Niche to the Ecological Mass Market' (1996–99), which is funded by the Swiss National Science Foundation within the Swiss Priority Programme 'Environment'.

mobility company. The 'mobility card' costs €100 monthly and allows her to use all kinds of public transport as well as the mobility company's car fleet.

In the afternoon, Priscilla arrives at the office. She is the owner of a company that organises sports events. She rents an office in the city centre with a beautiful view over the Frankfurt skyline and the River Main. For financial as well as environmental reasons, she leases the whole infrastructure of the office on a long-term basis (furniture, computers, copy and fax machines). The performance of products, not the ownership of products, is important for her. Many customers think alike. This new and growing market segment has led to a shift in business policy: furniture manufacturers, for example, started designing 'classical' furniture, the design of which is timeless and which is built to last. As part of the contract, the furniture is overhauled periodically, after which it looks as good as new. Computer companies have started designing modular computers, particular components of which can be replaced if necessary. As a consequence of the take-back law, which became effective in 1996 in Germany and 2003 in Europe, most electronic devices such as computers and photocopiers consist of recycled parts.

In the evening, Priscilla goes roller-blading with a friend. She borrows the latest equipment from one of the kiosks that are located along the Rhine. At home, she checks her home voice mail, which is a new service offered by the telecommunications company. Then she has an early night . . .

From 'Tweaks' to 'Leaps': Performance Sales instead of Product Sales

Many eco-products available on the market today represent small enhancements or 'tweaks' to existing ones: packaging is reduced and designed to be refilled or recycled; cars are equipped with catalytic converters and recyclable materials; washing machines save water and energy, etc. These achievements are quite essential and much-needed. However, replacing one bundle of goods with another is not enough to solve environmental problems (Ottman 1998: 87).

Current modes of living, especially in industrialised countries, are not sustainable due to the high level of material and energy consumption. Von Weizsäcker *et al.* (1997) argue that an increase of material and energy efficiency by a factor of four is necessary to double wealth worldwide and to halve resource use. The international 'Factor 10 Club' even proposes a much higher increase in resource productivity (Weaver with Schmidt-Bleek 1999). 'Leaping' instead of 'tweaking' is necessary to meet the ecological challenge. New modes of production and consumption have to be invented and rediscovered.

The imaginary day in the life of Priscilla in 2005 introduces some new modes of production and consumption that are vital for (eco-)marketing in the future and which may be crucial in seeking sustainable development: leasing, renting, sharing and pooling. The basic notion is to sell the performance of products instead of selling the products themselves: Priscilla does not necessarily need products such as solar cells, cars, computers, answering machines and roller-blades. What she really needs is to stay warm and comfortable, to be mobile, to communicate and to have fun. Consequently, she signs a performance contract; she has a mobility card; she leases the infrastructure of her office; and she rents roller-blading equipment instead of buying it.

Performance sales is a new way of looking at business (see Hansen and Schrader 1997 for the use of the term 'performance sales' in German). Other similar and associated terms are 'functional orientation', 'eco-efficient services' and 'consumption without ownership' (see Belz 1995 for an overview of the literature and different concepts). On the one hand, performance sales aims at tapping new business opportunities, especially in stagnating and declining markets. On the other hand, it aims at increasing eco-efficiency, i.e. resource and energy efficiency, by means of (1) extension of product use and (2) intensification of product use, which will be explained further in the following sections.

Extension of Product Use: From the Value Chain to the Value Circle

One means of tapping into new market opportunities and increasing the material and energy efficiency of products is to prolong the use of products. In many cases, products are used for a relatively short period of time, disposed of and replaced by new ones, which adds to the high level of material and energy consumption of industrialised countries. It follows the 'take–make–waste' rationale of the current industrial system which assumes indefinite supplies of resources and infinite sinks in which to dump waste.

There are different reasons for the quick replacement of old products with new ones. One is 'built-in obsolescence': manufacturers build products with an intentionally limited life. The result is that consumers buy more of the same products than they would if these products had longer lives. Another reason is 'psychological obsolescence': manufacturers promote new products as fashionable and in vogue; advertising is used to outdate products that are otherwise in good working order (e.g. clothes). Yet another reason is 'technological obsolescence', which occurs in the high-tech industry: the latest computers are advertised as faster and with more memory. Although these claims may be true, the average customer does not necessarily need the latest features and technology. These different forms of 'planned obsolescence' raise ethical as well as environmental questions (see Chonko 1995: 182-205 for the ethical dimension of product decisions).

If companies pursue a differentiation strategy offering high-quality products instead of low-quality, more disposable products, it may be beneficial for the customer, the company and the environment. High-quality products satisfy the needs of the customers for a longer period of time and build up the reputation of the company and its brands. As far as the environment is concerned, if a high-quality product lasts twice as long as a normal product, the resource productivity is increased by a factor of two. From this point of view, high-quality products are an important contribution in slowing down material throughput. In many cases, the environmental benefit of durability is hardly communicated to the customer at all. Take, for example, the Swiss shoe manufacturer Bally, which offers classic shoes for men. In

XEROX IS an international company with more than 90,000 employees and a revenue of $18 billion in 1997. It offers document-related business solutions, products and services. The three main product categories are light-lens copiers, digital products, paper and other products. Revenues from copying represented over 50% of total revenues in 1997; and digital products contributed 36% to total revenues in 1997 with high growth rates (Xerox 1998). Geographically, the most important markets are the United States (50% of total revenues) and Europe (30% of total revenues). Due to the high quota of rented and leased photocopiers, Xerox has always been responsible for the disposal of its products and the corresponding costs. That is one of the main reasons why it started thinking in terms of cycles during the mid-1980s. In 1987, Xerox introduced the 'Asset Recovery Management System' at its operating site in the Netherlands. The main economic goals were raw material savings and the reduction of disposal costs, whereas the main ecological goal was a high quota of remanufactured and re-used products and components.

One of the problems of closing the loop is the redistribution of products. The contract has a significant effect on the quota of redistribution. Here we have to differentiate between commercial and private customers: the majority of commercial customers rent or lease the photocopiers; Xerox remains the owner, maintains the products during use and redistributes them after use. Xerox is selling the performance of the product, not the product itself. Unlike commercial customers, almost all private customers buy photocopiers. In this case, it is difficult to control proper redistribution of the old machines. In Germany, there are three different methods of redistribution: (1) pick-up by Xerox customer service; (2) mail-in by the customer; or (3) take-back by the German electronic industry's recycling system. The new redistribution system is quite successful: between 1992 and 1995 the return quota was increased from 50% to more than 70%. Around 85% of the old photocopiers and components are re-used or recycled.

To market the high return and re-use quota, Xerox launched a new product line for remanufactured copiers in 1995: in Germany it is called the 'Green Line', in the Netherlands 'Eco-Serie', in England and Scandinavia 'Genus', and in Belgium 'Renaissance'. The remanufactured machines contain up to 80% used parts and, due to the raw material savings, they are cheaper than newly manufactured machines, which contain up to 25% used parts. The lower price of the remanufactured photocopiers attracts new price-sensitive market segments. The dual product strategy conforms to a unified quality standard: all remanufactured and recycled Xerox products such as copiers and toner cartridges meet the same quality and performance standards as its new products. To prove the level of quality to the customer and to counter the perception that remanufactured products are inferior, Xerox offers a customer satisfaction guarantee for the whole service life of all machines.

In Germany, revenue from remanufactured machines went up from 20% in 1992 to 25% in 1995; the introduction of the 'Green Line' led to an increase in total revenue of 5%. In 1995, Xerox Europe saved US$70 million in raw material and disposal costs. The overall savings between 1990 and 1995 were over US$200 million. Since then, the Asset Recovery Management System has turned from a cost centre into a profit centre.

With the new generation of products, which have been designed according to ecological principles, Xerox expects even more savings, gaining competitive advantage. The standardisation of components as a consequence of design for environment (DfE) leads to a high ratio (97%) of re-usable components of and cost savings in raw materials, disposal and customer service

Box 1: XEROX: LEASING PHOTOCOPIERS AND CLOSING THE LOOP

Source: Meffert and Kirchgeorg 1998: 693-743; and *http://www.xerox.com*

addition to its high-quality, durable shoes, Bally offers repair and maintenance services, which satisfy the needs of the customer and which prolong product life. These benefits could be communicated to the customer in terms of value per monetary unit instead of the total expenditure.

However, offering product life extension services is not sufficient. The important question remains: What will happen to the product after use? Landfill disposal bans for highly toxic products (e.g. batteries, electronics and refrigerators), rising disposal costs, rising costs for raw materials and energy, as well as extended product responsibility laws, make companies think in cycles. Product take-back programmes have become a serious alternative to landfilling (Ottman 1998: 79). On the one hand, these developments may be regarded as threats to existing markets and products; on the other, they offer new marketing opportunities. The product take-back programmes help build up long-term relationships with the customer. Thus, eco-marketing is closely related to relationship marketing (see Payne 1995 for relationship marketing as a broadened view of marketing).

A growing number of companies have started designing their products for repair, remanufacture and re-use: the German company Grammer AG guarantees to take back and re-use its 'Natura' office chairs with no cost to the customer. Returned chairs are disassembled, the separated parts are tested, and up to 90% of the old chairs is re-used in the manufacture of new ones. Sony, the world market leader in consumer electronics, announced that, from 2000 onwards, all Sony products will be designed in line with ecological principles, i.e. durable, repairable, re-usable, etc. By 2010 the company will completely cease disposing of its waste. And Interface Inc., the world leader in commercial carpet tiles, has set itself the long-term target of closing the production cycle. The 'Evergreen Lease' programme is a first major step into this direction: Interface not only produces the carpet, but also bears the responsibility of installing and maintaining it. Over the years, Interface selectively replaces worn and damaged areas and, most importantly, it is committed to remanufacture and re-use the old carpet tiles. Interface never passes the property rights to the user; it stays with the manufacturer, along with the liability for the used carpet tiles (Interface 1997). In 1995, when the Evergreen Lease programme was launched, the founder and CEO of Interface, Ray C. Anderson, openly admitted:

> . . . the economic viability of the Evergreen Lease for us depends on our closing the loop, i.e. being able to recycle used face fibre into new face fibre, and used carpet tile backing into new carpet tile backing; and we have yet to learn to do either economically. So, you might say, we're cantilevered a bit. But we will get there. It's a key to achieving sustainability (Anderson 1995).

One company that has already learned how to close the loop in an economic way is Xerox, which leases its copy machines (see Box 1). This case example shows that redistribution will become equally important to marketing as distribution; the competence to recycle old products and thus close the loop will become a critical success factor in the future. Renting and leasing are ways of prolonging product life and making sure that used products are returned to the manufacturer. Eventually, the linear 'throughput' economy is transformed into a 'circular' economy; the traditional value chain is turned into a value cycle (see Fig. 1).

Prolonged value chain 1

Prolonged value chain 2

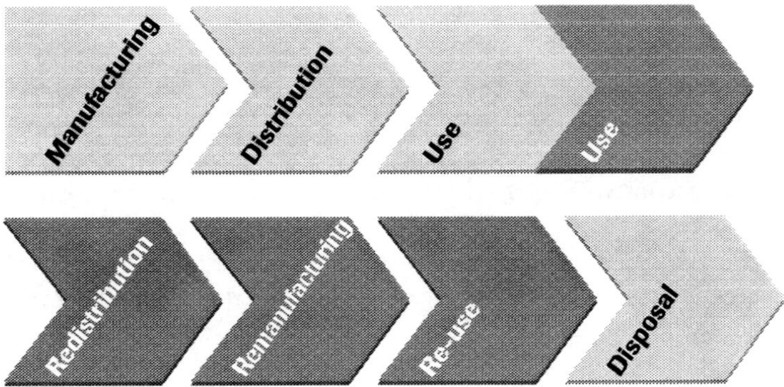

Value circle

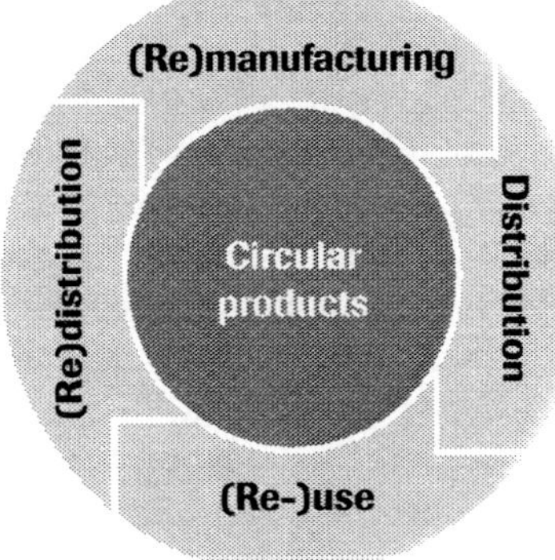

Figure 1: FROM PROLONGED VALUE CHAIN TO VALUE CIRCLE

The process towards a circular economy and value chain will not take place overnight. It takes time and money. Interface, for example, has invested over US$100 million in the creation of a value-added services network, which focuses on delivering flooring product services, not products themselves (Interface 1997). Interface has a clear vision of sustainability for its industry. It does not take the present flooring industry situation for granted, but seeks to transform it. To put it another way: Interface is 'competing for the future' (Hamel and Prahalad 1994). It is redesigning commerce by delivering services instead of products. Like other companies, Interface has already begun an ecological learning process. It may be argued that those companies that start progressing along the ecological learning curve today will gain competitive advantage tomorrow.

Intensity of Product Use: From Eco-Efficiency to Sufficiency

A second opportunity for tapping new market opportunities and increasing the eco-efficiency of products is to intensify their use. Products vary according to the intensity of their use. Some products are used quite often; others hardly at all. Take, for example, electric drills or gardening tools: these kinds of product are usually not used on a daily or even weekly basis; most of the time they are stored away. We may ask: Is it really necessarily for each household to possess an electric drill or for each gardener to own a set of gardening tools of her or his own? Isn't it less expensive and more ecological to rent or share such kinds of product? What is the ecological rationale behind it all?

The ecological reasoning is as follows: the more products that are shared, the fewer products have to be produced and the lower the environmental impact. If a circle of 10–20 gardeners shares a set of high-quality gardening tools, the potential increase of resource and energy efficiency is the equivalent of 'Factor 10', as proclaimed by Schmidt-Bleek and the 'Factor 10 Club' (Weaver with Schmidt-Bleek 1999). Regarding car sharing, around 10–12 users could potentially share a car; in this case, the resource productivity is said to increase by the a factor of 4–5 (due to the more frequent product use, the product's life is shorter!). Consequently, the shift from single users to multiple users reveals a high potential for environmental improvement (see Fig. 2).

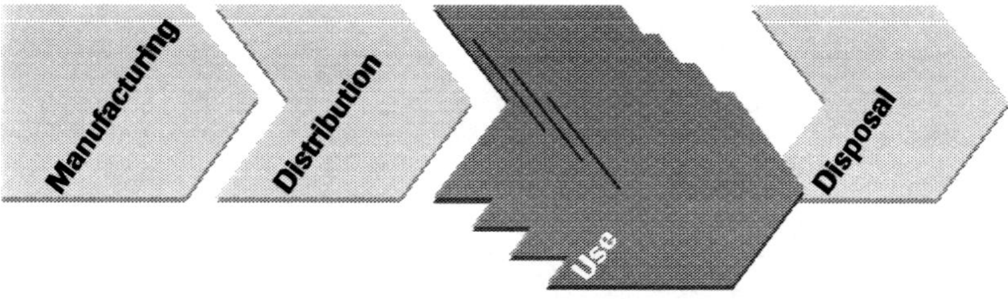

Figure 2: FROM SINGLE TO MULTIPLE PRODUCT USE

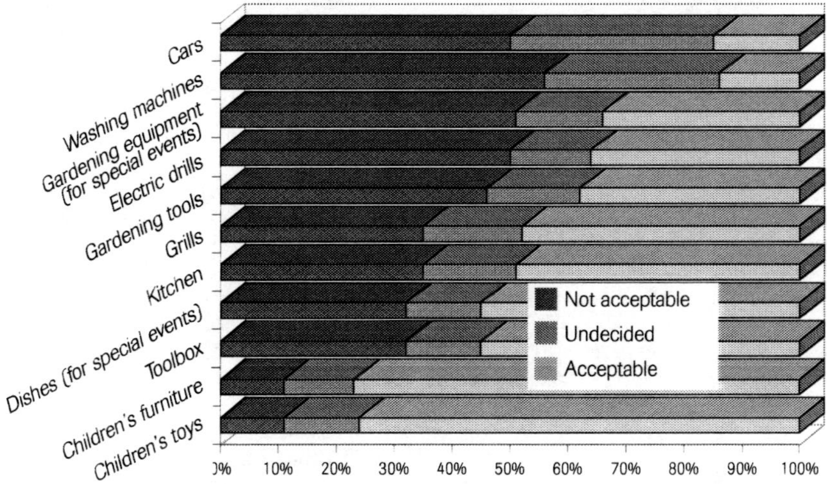

Figure 3: GENERAL ACCEPTANCE OF PERFORMANCE SALES WITH REGARD TO DIFFERENT PRODUCT CATEGORIES

Which products are suitable for sharing and which are not? The main characteristics are: a relatively high price, low product use, easy and instant access, as well as ease of use (Hansen and Schrader 1997: 102-103). In relation to the current business environment, products such as coffee machines and hairdryers, which are fairly cheap, do not qualify for performance sales. Other products such as cars or washing machines, which are more expensive, may be suitable for performance sales. A recent German study conducted by the University of Hannover indicates that general acceptance of the idea of sharing products is high among respondents (Schrader 1998: 11-13). However, level of acceptance differs between product category. Figure 3 shows that the willingness to share is fairly high for cars, washing machines, gardening equipment, electric drills and gardening tools; however, the willingness to share is rather low for children's furniture and toys.

Apart from environmental aspects, what are the advantages and disadvantages of sharing? What are the barriers that must be overcome to make performance sales successful? And what are the main target groups of eco-efficient services?

According to the German study, the advantages of sharing washing machines, for example, are: to save space, to save time and trouble with repairs and maintenance, to benefit the environment, to save money, and to reduce noise emissions. The main disadvantage of sharing washing machines is inconvenience: over 70% of the respondents suspected that they could not use the washing machine whenever they wanted to (Schrader 1998: 18, 24). For them, ownership is synonymous with convenience and freedom. However, this perception may be subject to change, as it is a

IN 1997 the two Swiss co-operatives ATG Auto Teilet Schweiz and ShareCom merged to 'Mobility CarSharing Switzerland', the largest car sharing company in Europe. In 1998 it served over 20,000 customers with 1,000 cars at 600 locations in 300 Swiss cities and towns. An increasing number of Swiss people are opting for car sharing: between 1993 and 1998, the number of car sharers has risen from 3,000 to over 20,000, with an annual growth rate of 50%. In 1998/99 the number of customers is expected to double.

Mobility CarSharing Switzerland has a wide range of different cars available from a number of strategically placed locations in Switzerland. Reservations can be made by phone or Internet, 24 hours a day, with minimal notice. The customers are charged per kilometre and per hour of use, according to vehicle type. The prices include petrol, depreciation, insurance, etc. The annual fee is CHF 100, which helps cover the company's capital costs.

What are the reasons behind this trend and willingness to forego individual ownership? Mainly two: first, car sharing saves money. Usually, car sharers combine different modes of transport (car, train, tram, bus, bicycle, etc.). Individual transport by car costs CHF 0.70 per kilometre, whereas public transport costs CHF 0.18. Consequently, by substituting car driving with other modes of transport, car sharers save money, which eventually adds up to several thousand Swiss Francs per annum. Besides, fixed costs such as depreciation, insurance, parking and breakdown service are shared by a greater number of people, which reduces the individual costs. Second, car sharing is good for the environment. If multiple users share a car instead of buying one, material efficiency is increased. Furthermore, the substitution of cars with trains and trams, buses and bicycles increases energy efficiency.

Are there any 'rebound effects'? Does car sharing make people drive more, since it is cheaper than car owning? Generally, car owners tend to consider only fuel costs when driving a car; often depreciation, insurance, maintenance, etc. are not taken into account. Car sharers, however, pay these costs per kilometre and per hour of use, which provides a strong incentive to limit use and/or to use other forms of transport. The high 'eco-efficiency' of car sharing goes along with 'sufficiency', i.e. users drive less (see also Meijkamp 1997: 24).

What are the critical success factors of car sharing? And why is Mobility CarSharing Switzerland the largest and most successful car sharing company in Europe? What is the vision for the future? As of today, Mobility CarSharing provides a wide range of different cars with easy access for a reasonable price. It offers an added value in the form of reduced costs and better environment. In the future, it plans to take the step of mobility sales one step further: currently, negotiations with rail and bus operators, taxi companies and other parties are under way to offer a 'mobility card' with a computerised touchless chip. Such a 'smart' card is convenient for the customer as it allows multi-module travelling that is billed on a monthly basis. Steps in this direction are the mobility packages 'Zuri Mobile' and 'Zuger Pass Plus' at a regional level, which have been quite successful.

Box 2: MOBILITY CARSHARING SWITZERLAND: SHARING CARS

Source: Wagner and Schmeck 1998; Mobility 1998; and *http://www.mobility.ch*

Pros	Cons
• Saving money (especially 'light users') • Saving time and trouble (e.g. repair and maintenance) • Saving space (e.g. parking lot) • Benefiting the environment	• Inconvenience • No social prestige related to the ownership of products • Less care and hygiene • High price (especially for 'heavy users')

Table 1: PROS AND CONS OF ECO-EFFICIENT SERVICES

result of habits and unreflected modes of washing behaviour. Unlike Germany, it is common to share washing machines in Switzerland. Each block of flats has a set of washing machines and dryers in the cellar which are shared by all parties. For reasons of liability risk and noise, most landlords do not even allow the installation of washing machines in individual apartments.

In the case of car sharing, the pros and cons are similar to washing machines. In addition, cars represent social status and prestige, especially for Germans. Table 1 lists some important pros and cons for eco-efficient services.

The marketing task is to overcome the disadvantage of eco-efficient services and to communicate the advantages. What are the main target groups for providers of eco-efficient services? The German study identified some characteristics of actual and potential customers: they are between 18 and 44 years old, have a high formal education, live in small apartments and are aware of the existence of eco-efficient services (Schrader 1998: 31-48). To move from a green niche to the green mass market, providers of eco-efficient services are well advised to identify the right target group, to communicate the main advantages of sharing more aggressively and to overcome the disadvantages of sharing by means of good and credible performance concepts. Mobility CarSharing Switzerland has been quite successful in doing this, as shown in the case example in Box 2.

Back to the Future . . .

This chapter has discussed some modes of production and consumption that will be vital for (eco-)marketing in the future: renting, leasing, sharing and pooling. The economic aims are to tap new market opportunities and gain competitive advantage; the ecological aims are to increase eco-efficiency by prolonging and intensifying product use. Due to the changing marketing environment (e.g. product responsibility laws), the forms of contract discussed, such as renting, leasing, sharing and pooling, may be expected to gain in importance in the beginning of the 21st century. Some companies such as Interface Inc. and Sony have already started competing for the future by building up core competences and strategic alliances around performance sales. Such core competences are, among others, local repair and service units, efficient redistribution of products, and the ability to remanufacture products in an economic manner. However, changing the present consumer society is not just about what kinds of product customers buy, but it is also about how they use them, i.e. general attitudes towards 'goods' and ownership. This might require a significant paradigm shift that is substantially beyond the ability of companies to effect on their own. Alliances with non-governmental organisations may help to bring about such a paradigm shift in the future. Besides, changes in the political system are badly needed (e.g. ecological tax reform) to give performance sales the chance to flourish on a broad basis and to move from niches to mass markets.

6

Innovation of Eco-Efficient Services

Increasing the Efficiency of Products and Services

Kai Hockerts

XEROX now promotes its products as 'document facility management *services*'. The German energy giant Siemens has invested heavily in energy-saving *services*. And Mercedes Benz is promoting vehicle management *services*. The underlying message of these activities is clearly that companies realise that it is more profitable to meet customer needs through services than by selling hardware. However, economic efficiency is not the only attraction, as service concepts contribute to environmental efficiency gains as well.

This fact has been realised by an increasing number of companies. However, for many, the belief still persists that 'this does not really apply to my business' and that eco-efficient services only have relevance for specific businesses and cannot be generalised. This is wrong. Whether a company trades in screws, washing powder or high-tech plant, eco-efficient services offer new opportunities for each business. This chapter will therefore not only present success stories of eco-efficient services, but will also provide a rigorous set of guidelines that can be applied to any existing product or service.

Indicators for Service Eco-efficiency

Before getting into the details of how eco-efficient service innovations can be achieved, the primary question is: Why bother about the environmental dimension at all? First of all, reducing ecological damage is a regulatory as well as an ethical obligation. But, more importantly, eco-efficiency is a strategic success factor. Four model efficiency categories will serve to clarify this.

The Longer Life Option

> I like the English method of caring about a car and thus maintaining it for a long time. I think the American system is wrong to reduce a car to scrap after only four years. The raw material situation on earth forces us to carefully maintain existing resources (Dr Ferdinand Porsche, in a letter to his nephew, Ghislaine Kaes, 1936).

Increasing a product's service life is not merely an environmental strategy; it may also offer financial gains. A product sold generates income only once—at the point of sale. A product offered that is combined with a service contract continues to provide a return for as long as it is in use. Therefore, much more profit can be drawn from the basic material product. Furthermore, high-quality long-lasting products bind clients closer to a particular supplier than is normally the case in today's 'throw-away' culture. This provides an opportunity for associated services such as maintenance or upgrading. One also has to consider that the longer service life of a material product is determined not only by physical properties, but also by technical development. Therefore products should be upgradable as well as easily repairable.

Energy and Material Consumption during Use

As well as extending a product's service life, energy and material consumption during use should also be minimised. A company *selling* products will only address this issue if its customers demand it. However, it seems that such features are often neglected by consumers, as demonstrated by the slow uptake of energy-saving light bulbs. Although the additional price of such bulbs is lower than the amount of money saved through lower electricity bills, they are not as widely used as one would expect. Generally, consumers do not consider costs of use, but only the retail price.

A service provider, however, will benefit from reduced maintenance costs, and, under the pressure of competition, will constantly strive to reduce them. From an environmental point of view, this helps conserve resources. However, the first conflict of interest crops up at this point: efficiency considerations may persuade people to stop using a product that is not yet at the end of its useful life.

A good example of the economic potential of energy conservation is the 'least-cost planning' strategy described in more detail below. This approach requires a fresh understanding of business strategy, as now the service (e.g. provision of light) and no longer the product (e.g. electricity) is the main focus (Stahel 1994: 178-90).

Revalorisation Potential

At the end of its useful life, a product is still relevant to the profit and loss account: most waste contains a value potential. If companies developed 'take-back' and service schemes, they may re-use parts of a product or sell it on into the second-hand market. Here again take-back and service concepts are useful. With the concept of 'extended producer responsibility' (EPR), the control of valuable raw materials (which until now has been left to consumers) rests with the provider. Additionally, many producers will be forced to take back their products anyway: legislation for the take-back of electronics and automobiles is in the pipeline in Europe and Japan and is likely to come into force by 2000 or shortly after. Well-developed service centres will be a key advantage once the take-back of old products becomes obligatory.

However, companies often look only at technical recycling of products (e.g. recycling of metals, plastics, etc.), and neglect to develop products that may easily be returned to biological cycles (Braungart and Engelfried 1992: 613-19). If products are designed to use biodegradable materials, for example, they do not have to be taken back at all: consumers can compost them on their own or introduce them into collection and composting schemes provided by their municipalities.

Effectiveness of Use

Finally, an important point regarding eco-efficient services is a product's effectiveness during use. This effectiveness may be increased by sharing and joint use (e.g. car sharing, copying centres, etc.). Canon, for example, has been very successful in franchising its copying centres, thus increasing the effectiveness of their photocopiers which are now used to a maximum—and which are also very well maintained and thus long-lasting.

A second opportunity lies in the fact that customers are not always the most competent people to operate a product. It can often be found that customers use much more of a certain product than is necessary or they accidentally misuse it and thus it breaks down sooner. Here, insurance and facility management services may help to increase efficiency. If, for example, under a facility management contract products are operated by trained personnel instead of less-informed customers, this may lead to higher efficiency of the product's use.

Eco-Efficient Service Concepts

> Services are anything sold in trade that could not be dropped on your foot.

This is how H.L. Freeman (1989: 329) loosely defines services. However, services and material goods are closely linked. This chapter will not examine those services, such as consulting or insurance, that are virtually independent of hardware, but rather those services that require a material product in order to be effective.

A main point this chapter addresses is the fact that, for any product sold, one can enter into a number of **institutional arrangements** with a customer. All too often, it is the same arrangement: in selling the product, all property rights are handed over to the user and thus the producer loses all interest in future returns derived from it. However, by reallocating property rights from the user to the provider, considerable efficiency gains may be achieved, as will be shown in more detail below.

But services are nothing without a customer. Therefore providers must actively seek **interaction** with their customers. Here both sides can benefit from better serving the need underlying a service or product. If these two dimensions are analysed in a matrix (Fig. 1), three typical service concepts can be identified. The typology may

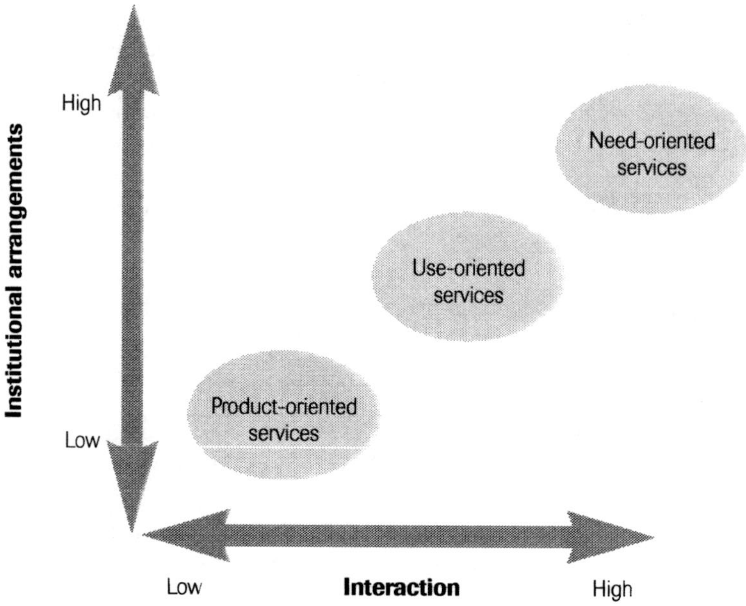

Figure 1: THREE SERVICE CONCEPTS

Source: Hockerts 1995: 31

differentiate between service concepts according to level of institutional arrangements as well as level of interaction. Product-oriented services make very little use of institutional arrangements and interaction and often address only one very specific point, whereas use-oriented services go further (the material product is not sold; the client pays only for its use). Ultimately, need orientation goes beyond product barriers, the most important aspect here being the delivery of a defined service: the service is no longer bound to a specific product.

In the following, some illustrative case examples of eco-efficient services are presented. For each concept, the function is described, two or three case examples are given and the eco-efficiency gains are analysed. Subsequently, a discussion is presented on how the innovation of eco-efficient services can be effectively managed drawing on these two dimensions.

Product-Oriented Services

Function of product orientation. The function of product-oriented services is to offer services that are *additional* to the product sold. Here a company capitalises on its internal expertise through **training and consulting** which helps customers optimise the application of a product. The provision of separate **maintenance and disposal services** should also be considered. In most cases, no one in the marketplace knows as much about a product's efficient disposal as the producer. So why not offer this knowledge to the customers?

These services normally need only a low level of institutional arrangements or interaction between provider and customer. They are an ideal first step in the business of eco-efficient services.

Case studies. The Wuppertal-based Herberts GmbH (Germany) is a specialised provider of car paint, traditionally selling only its paint to the automotive industry. However, realising the potential of rationalising the way in which customers use their product, they now offer an additional training service. This training shows the customer how to recycle paint overspray (i.e. the sprayed paint that is normally emitted in the air) and how to optimise paint application in general. In the wake of the outsourcing debate, Herberts even started to think about offering its customers the opportunity of completely taking over the paint application process within customers' facilities. However, such a radical 'facility management' step is still at a very early stage at Herberts.

Focusing on the risk management and disposal of its products, Dow Chemical introduced a new service through its subsidiary SafeChem (Voerde, Germany). This service includes safety consulting for the customer as well as a take-back and recycling guarantee for used chlorinated solvents, according to the German waste regulation (§3 HkW-AbfV). Dow Chemical offers its customers both the basic product without the service or additionally with the SafeChem take-back guarantee.

The Bavarian company Grammer AG (Amberg, Germany) offers a take-back and recycling guarantee for all its office chairs sold. The expenses for take-back and recycling are already included in the retail price.

Economic and environmental efficiency of product-oriented services. Product-oriented services take one specific issue of low environmental efficiency and address it with a tailor-made service. It may, for example, help the customer reduce the amount of consumption during use. Herberts's service, for example, recycles paint overspray, while Dow's assists its clients in reducing the loss of chlorinated solvents in the process, thus enabling a reduction of material throughput and the amount of waste and other emissions.

From an economic point of view, this concept is ambiguous: the more clients can re-use a product by overspray recycling or reduced losses, the less virgin product the producer will sell. However, both producing companies found that service income more than compensated them for the reduced product sales. Furthermore, their experience was that, with a steady service income, it was much easier to bind a client to their companies in the long term. As service concepts often go hand in hand with long-term contracts, the risk is reduced of a client changing supplier.

A guarantee to take back and dispose of the used product (as offered by both Dow and Grammer) allows maximisation of the **revalorisation potential**. It encourages the producer to consider the issue of disposal as early as the design and production phase. If the whole cycle is planned from the outset, used products can be recycled in sections and thus help cut material costs. Grammer also expects this service to assist in the development of expertise that they will need in the likely event of a legislated take-back requirement.

Use-Oriented Services

Function of use orientation. Use-oriented services go one step beyond product-oriented services. Here, the provider no longer sells a product but only its use. This strategy allows an increase in the financial return that can be derived from a product over its whole useful life. Possible options are the renting or leasing of products (so-called 'eco-leasing' or 'eco-renting'). A company might also offer **pooling or sharing** options, thus increasing the effective use efficiency of the product. The longer and the more intensively the product can be used, the higher the revenues generated from the service. This encourages a company to *increase* the service life of a product and to *reduce* maintenance and replacement costs.

Use-oriented services are a major step forward from the perspective of eco-efficiency. However, they require a higher level of interaction between provider and customer and a higher number of institutional arrangements. For example, the customer no longer has the right to sell the product or to adapt it. In the case of pooling or sharing, customers are even restricted in their right to exclude others from the use of the product (apart from with specific prior arrangements).

Case studies. A good illustration of the environmental and economic effects of use orientation can be found in the copying and printing business. Most producers of photocopiers have already begun leasing their products rather than selling them, the customer consequently paying only per copy made. This has encouraged nearly all producers to redesign their products, as they can increase profits if machines need less maintenance or are used longer.

A number of service providers offer car sharing or car pooling. Even Daimler-Benz has joined the contracting trend, offering a mix of leasing and long-term maintenance contracts for commercial vehicles. Its 'CharterWay' concept is described as focusing on 'renting transport capacity' rather than on selling trucks. This service appeals especially to smaller businesses who find it inefficient to own and maintain a permanent fleet of vehicles. Just recently, Mercedes announced its intention to offer a similar service with its new minicar, the SMART®mobile.

Economic and environmental efficiency of use-oriented services. The literature on outsourcing demonstrates the economic advantages of this strategy, so it is surprising that it is not more widely used either in the business sector or the private sector. However, the fast-growing market for car sharing in Switzerland (Mobility CarSharing) shows that barriers are crumbling.

But outsourcing offers more than just an economic edge. The environmental advantages of use-oriented services lie in the fact that companies will invest much more in increasing the service life of products. Also, decreased material and energy consumption during use may result, as such measures directly reduce the provider's maintenance costs. Finally, the service provider (as owner of the product) can realise the revalorisation potential of the disposed goods: at the end of the product's life the provider usually remains in possession and can thus re-use valuable materials or components.

Need-Oriented Services

Service function. Need orientation goes far beyond the two other concepts described above. Its function is the satisfaction of customer needs regardless of the material product, meaning that the service provider guarantees a certain result. However, the management of the product involved is no longer of any interest to the customer.

There are two types of need-oriented service: **least-cost planning** strategies are normally applied to consumable goods, while durable goods, on the other hand, are often offered through **facility management**.

Case studies. Least-cost planning was first introduced in the energy sector, where a number of providers understood that it may be profitable for them to promote energy-saving equipment. A decreasing demand through gains in efficiency allows the facility to increase its market share without having to build new power plants. The provider may even use its position in a monopolised energy market and slightly increase the rates, allowing it to benefit from efficiency gains that its customers realise due to more efficient equipment. If the supplier is not in a monopoly situation, it can agree in advance long-term contracts with customers to guarantee loyalty. As well as the energy sector, least-cost planning is now applied in water facilities and waste management facilities.

A very innovative service concept for pesticide management has been developed at Ciba Geigy (Basel, Switzerland), founded on the realisation that many farmers are highly risk-averse. Research has found that, rather than risking loss of yield, they use up to twice as much pesticide than is necessary. Aware of this optimisation potential, Ciba Geigy offered an 'integrated pest management' service. If they are correctly trained and assisted, farmers can save up to 50% of pesticide costs and at the same time even increase yields, as many crops react negatively to an excess of pesticide. In a more developed version, the pesticide provider could even offer a 'crop insurance' service, guaranteeing the customer that certain pests will not affect its yield. The whole pesticide application process would be carried out by the service provider.

Facility management is most commonly used in the area of building management. However, in principle, any durable good may be offered through a facility management concept. The German subsidiary of Xerox (Düsseldorf), for example, has introduced a 'document facility management'. The complete 'document duplication' service is, for example, being provided by Xerox at the headquarters of chemical producer Henkel KGaA, where all copying machines remain the property of Xerox and are maintained and even run by Xerox personnel. Henkel just pays for each copy.

Economic and environmental efficiency of need-oriented services. From an economic point of view, facility management is a logical consequence of outsourcing strategy. Least-cost planning means that the most economically efficient route is always chosen. Also, underlying contracts are normally of a long-term nature, thus closely binding the client to the service provider. Need-oriented services help to address the product's effectiveness in use and the risk aversion of many users. They also incorporate many of the advantages of product- or use-oriented services.

In sum, it can be said that any of the three types of service concept allows an increase in economic and environmental efficiency. However, implementation has to be inspired by a strong commitment towards corporate sustainability and customer satisfaction.

Innovation of Eco-Efficient Services

This leads to the question of *how* one may transform the products offered today into eco-efficient services. Let us recall the two basic dimensions:

- First of all, analyse—step by step—the **institutional arrangement** associated with the product. As can be seen, not all property rights have to be held by the user. By distributing the rights between user and provider, one can already achieve considerable efficiency gains.
- Then the **interaction potential** with the customers has to be considered, which allows both sides to benefit from interactive services.

To achieve this goal, a six-step approach is proposed (Table 1).

Step 1	Assign the eco-innovation team
Step 2	Brainstorming
Step 3	New institutional arrangements
Step 4	Eco-efficiency test
Step 5	Market viability test
Step 6	Validation: seek interaction

Table 1: A SIX-STEP MODEL FOR INNOVATION OF ECO-EFFICIENT SERVICES

Find the Innovators

First, an innovation team must be formed, which should include not only staff from the department that 'owns' the innovation project. 'Magic' ingredients derive only from a mix of different opinions and ideas, which could originate from marketing and sales, R&D and product design, environmental management, strategic development, as well as finance.

External participants may also be invited: customers and general interest groups could both offer insights. One possibility is to generate ideas in two separate meetings, the first session serving to gather ideas in co-operation with external stakeholders and clients, and the second assembling the 'hard' facts and excluding outsiders.

Brainstorming

For the second stage, it is essential that everyone is prepared for the exercise. A useful preparation is an initial meeting two weeks before the actual innovation workshop where it can be explained why it is believed that eco-efficient services will provide a business opportunity for the company. Here, everyone should be assigned a small task in order to prepare the topic. It is also important to identify clearly the product or service in question: it has to be decided what is going to be investigated and what is out of scope.

In beginning the actual innovation process, all are invited to write down the 'state of the art' of the product or service on a flipchart, gathering everything that has been undertaken so far in an attempt to innovate it. One should look at positive experiences as well as negative ones; the status quo is what should *not* be discussed. If something that is already in operation and which is effective is brought up, it is put on a separate board, 'Things that worked'. There is no need for further discussion on these: they cannot be classed as innovations. Write down ideas that have turned out to be ineffective on another board, 'Things that didn't work'. If proof exists for why something has not worked, time should not be wasted on further discussion.

The idea of the brainstorming session is to prepare everybody to think about *real* innovations: the new service ideas that nobody has yet thought of. Now the process really begins.

New Institutional Arrangements

> We keep [durable goods] by us even if we do not want them at the moment. But their utility will of course, be increased the more often we can arrange to use them, so that it is often better to hire, or to buy and sell, or to make various arrangements for common usership (Stanley Jevons, economist [Jevons 1871]).

As early as 1871, classical economic theory was aware of the value of institutional arrangements: an intelligent distribution of property rights increases economic efficiency. In the following, a systematic list of possible institutional arrangements is presented (Furubotn and Pejovich 1972: 1137-62). One blank flipchart should be prepared for each arrangement, onto which all ideas that the group comes up with are assembled. By creatively analysing the options gathered here, a wealth of possibilities for new and more efficient services can be found.

Profit retention arrangements. If a product is sold, all rights of profit retention are transferred to the customer. The basic idea of profit retention is to generate profit not only at the point of sale but over the whole product's service life, which can be achieved, for example, by eco-leasing or eco-renting. A quick look around the company will be of assistance here: have the photocopiers been bought or are they leased and on a pay-per-copy contract? In the latter case, imagine the effect an increased machine service life has on the provider's profitability: the longer its products last, the higher the profits.

On an investment level, the same result can be reached through facility management services that offer the whole management of a product or facility. For example, the management of a building or its heating can be sourced out to third parties that can offer this service more efficiently.

The focus is now on the product that is to be analysed. Usually, nobody knows more about how to operate or apply a product than the very people that are selling it. So why not take this aspect away from customers and bill them for each service unit? Companies may also start to think of themselves as contractors. In a contracting arrangement, a bonus could be factored in if the services aid the customer in realising higher efficiency gains (e.g. reduced energy costs in energy contracting).

Maintenance and training services. If a company chooses to sell a material product, there are still opportunities for offering additional maintenance services to ensure optimised product use. If the value of the service can be identified and documented, both sides will benefit. Many customers will also be interested in training and consulting services that offer expertise transfer. Companies should always remember: 'Nobody knows the product like we do.'

Revalorisation services. A product's life-cycle should not be thought of as one-way traffic: considerable potential lies in revalorisation, re-use and recycling. A take-back service, for example, allows a company to benefit from valuable parts and materials embodied in a product. To allow a better revalorisation of used products, disassembly considerations must be included as early as possible in the design phase. Take-back services provide a further advantage: they bind customers closer to the provider as they help them reduce their disposal costs. And, in addition, the contact at the point of disposal is important in selling new products.

Sharing and pooling services. An important consideration of property ownership is the right to exclude others from its use. If companies offer a service instead of selling a product they retain this right and may, accordingly, offer sharing facilities where different clients use the same product to increase its use-effectiveness. A second possibility is the product pool service. A pool contains a number of different product variations (e.g. from small runabout to limousine in the case of a car pool), so customers can choose the product that best reflects their needs at the time.

Risk arrangements. Finally, a company can offer a customer protection against the risks associated with using a material product. For example, it is customary for car rental companies to offer additional insurance against breakdowns or accidents. But how can a company provide its own customers with better protection against product failure? The crop insurance example described above is a case in point. Often, the risk-aversion of customers leads to considerable inefficiencies. Sharing the risk may not only increase environmental efficiency but also benefit the balance sheet—both the client's and the provider's.

The Eco-Efficiency Test

Thus far, the process should have generated a large number of ideas for service concepts. Each idea should now be reviewed for its eco-efficiency effects. The ideas should therefore now be posted on a separate sheet of paper. The four categories described in the first part of this chapter provide a good model for assessment of environmental benefit, helping to judge the effectiveness of each idea. What can the idea contribute to each of these indicators (see Fig. 2)?

Interdependencies should also be considered: does an increased service life diminish the re-use potential as technical progress reduces the re-usability of product parts? Who contributes in the first instance to the increased efficiency: the client or the provider? How can further efficiency gains be encouraged?

Market Viability Test

By this point, some genuinely good ideas for improving the environmental efficiency of the product/service packages should have developed. However, before the decision is made to run with a particular idea, the marketing department will certainly want

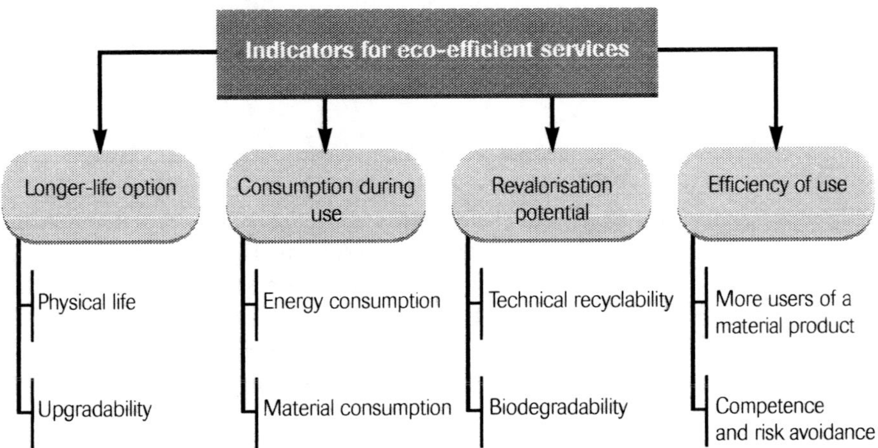

Figure 2: ECO-EFFICIENCY INDICATORS FOR SERVICES

to carry out numerous tests to identify its value potential. The following criteria allow a rough first assessment of the real market viability of new ideas. Using the marketing tasks approach (Tomczak and Reineke 1996), the business potential of innovations can be analysed. Those ideas that contribute most to these marketing tasks are probably tomorrow's winners.

- Does the eco-efficient service allow a company to **retain its existing clients**? How far does each idea contribute to binding customers closer and longer to the product/service. Any measure that reduces the mobility of the customer (e.g. long-term contracts, technically unique offers, etc.) reduces the risk of losing clients to the competition.
- Does the eco-efficient service facilitate the **acquisition of new clients**? Does the innovation allow clients to be won over from the competition? Which new client groups may be penetrated with this approach? Are there possibilities of addressing client sectors that have so far been beyond the company's scope?
- Does the eco-efficient service **increase the value potential of the existing product**? Here the product/service package should be examined more closely. The idea of a service concept is to increase efficiency, so what are the real cost advantages related to the idea? How will the bottom line be affected by the implementation of the new idea?
- What is the new service's **diversification potential**? Finally, time should be taken to consider the global influence the service innovation has on company strategy and general market positioning. Does the new service offer access to new business opportunities? What will the competition do?

If an idea contributes to all these four core marketing tasks, one can be confident of winning considerable support from the marketing department as well as from product managers. Nonetheless, the case should be well prepared and it will have to be proved that eco-efficient services will not only protect the environment but will also benefit the balance sheet.

Validation: Seek Interaction

Although institutional agreements help in achieving efficiency, a second important element has to be considered: the interaction between customer and service provider can heavily influence the nature of a service. In such cases, we speak of an 'external factor' needed in providing the service. In service provision, customers can interact more with the provider and thus influence the result of the service, which is not the case in product manufacturing. A consultancy, for example, will always work in a highly interactive manner. A car mechanic, on the other hand, would not normally appreciate his customers being involved in the service process (i.e. the car repair).

It is important to know how far service innovations are dependent merely on a company's own competence or whether input from the client is needed. Such

Some basic considerations include:	Strategy element
• What is the underlying problem to be addressed by the new service concept?	Identification of the eco-efficiency gain
• Which institutional arrangements will be needed between the company and client?	Institutional arrangements
• Is input required from the customer in order to provide the best service?	High/low interaction
• Which concept corresponds with the problem and the externalisation and interaction required?	Product, use or need orientation
• Which material product is applied in providing the service?	Technology
• How is the service communicated to customers and how is it made available?	Marketing communication
• Which customer does the service focus on?	Target group
• Which area will the service cover?	Geographical spread

Table 2: ECO-EFFICIENT SERVICE CONCEPTUALISATION

considerations will provide a feeling about what point clients need to be involved in the development of new services. Once all steps have been considered, a report has to be drawn up, starting with the five best ideas and describing them in detail. However, ideas that have been dropped (for the time being) should also be added in an annex: they may be revisited in later sessions. Now it is up to the company to decide how to conduct the process.

Outlook

Once it is decided to run with an eco-efficient service, it must be ensured that the idea succeeds. To translate a service concept into action, a company must adapt its corporate strategy. All resources needed to implement a new service have to be considered: if they are not yet available, they must either be developed or acquired. It should also be clear what is to be done and why. Answering the questions presented in Table 2 may provide first insights.

Eco-efficient services can be a major strategy in increasing both a company's environmental and economic efficiency. However, such concepts require a greater innovation effort than usual. Service concepts will lead to fundamental change in a company's marketing strategy.

Service concepts benefit, on the one hand, the environmental performance of a product as they increase its useful life while progressively reducing energy and material consumption. They also help improve the revalorisation potential and increase use efficiency. These environmental goals go hand in hand with an improved customer relationship that reduces the loss of clients to competition while new market segments can be addressed. However, the most important aspect from an economic viewpoint lies in increased efficiency. Reduced costs or increased profits will ensure that service concepts ultimately become a success story.

Companies that are reluctant to introduce service concepts on a large scale have the option of taking a step-by-step approach. They can offer 'product-oriented services' around their current products (e.g. maintenance or disposal services). Later, they can initiate renting or leasing arrangements for the same product to some of their clients. These 'use-oriented services' can be extended slowly and on a case-by-case basis. Once it is decided that services are indeed the future strategy for the company, 'need-oriented services' can even be considered. The company in this case no longer classes itself as a hardware producer but as a service provider, offering facility management, contracting and insurance services.

7

Management of Environmental New Product Development

Devashish Pujari and Gillian Wright

ACCELERATION OF THE SEARCH for environmental excellence is done by managing a 'holy trinity of challenges': Public credibility, money making and new product development (Philip Gross, President, NOVON Products, a division of Warner-Lambert; quoted in Piasecki 1995).

New products are an important part of environmental marketing, and the focus of environmental new product development (NPD) must be in improving the primary and environmental performance of a product rather than merely introducing cosmetic changes (Peattie 1995). However, the definition of an environmental new product has as yet failed to reach a consensus. Many authors have attempted to define a green product, using a variety of criteria (see Table 1). Coddington (1993) explains that, in developing environmental new product concepts, three sets of issues must be addressed. These are:

- Concept issues (the new product as a line extension, a reformulation of an existing product, or a completely new product offering)
- Pipeline issues (the environmental new product's compatibility with current or available production capabilities, as well as with principles of sustainable development)
- Strategic issues (analysing competitive offerings, positioning a greener product, leveraging brand recognition and developing marketing strategies)

Although these three sets can be analysed as conceptually distinct entities, all are closely connected. Management must adopt a broader perspective of a product's characteristics when addressing environmental issues, examining the complete life-

Schmidheiny (1992)	Simon (1992)	Elkington and Hailes (1988)
• Eliminate or replace the product • Eliminate or reduce harmful ingredients • Substitute environmentally preferred materials or processes • Decrease weight or reduce volume • Produce concentrated product • Produce in bulk • Combine the functions of more than one product • Produce fewer models or styles • Redesign for more efficient use • Increase product life-span • Reduce wasteful packaging • Improve reparability • Redesign for consumer re-use • Remanufacture the product	• Reduced raw material • High recycled content • Non-polluting manufacture/non-toxic materials • No unnecessary animal testing • No impact on protected species • Low energy consumption during production/use/disposal • Minimal or no packaging • Re-use/refillability where possible • Long useful life • Updating capacity • Post-consumer collection/disassembly system • Remanufacturing capability	• Not endangering the health of the consumer or of others • Causing no significant damage to the environment during manufacture, use or disposal • Not consuming a disproportionate amount of energy during manufacture, use or disposal • Not causing unnecessary waste, either because of over-packaging or because of an unduly short useful life • No use of materials derived from threatened species or from threatened environments • Not involving unnecessary use or cruelty to animals, whether this be because of toxicity, testing or other reasons • Not adversely affecting other countries, particularly the third world

Table 1: WHAT IS AN ENVIRONMENTAL NEW PRODUCT?

cycle impact. A product life-cycle stewardship orientation relates to considerations of a product's environmental impact from design, through manufacture, storage, packaging and use to disposal (Beaumont *et al.* 1993). Table 1 summarises different characteristics regarding what constitutes an environmental new product, as advocated by three key authors in the field.

As early as 1971, Cracco and Rostenne wrote: 'The concept of socio-ecological product has to extend our understanding that environmental consequences (the product's aggregate impact on everyone affected by its use) are more important determinants of its acceptability than either user satisfaction or corporate profitability.' Today, the concept is further expanded to that of the 'total product', which includes the activities of the producer organisation (Peattie 1995). This means that the company manufacturing the green product should also ensure that its other organisational activities are green, because this will affect the perception of the product in its totality (core dimension and augmented dimension).

Critical Success Factors in Environmental New Product Development

Management Issues

Environmental new product development policy. Existing products represent limited opportunities for environmental marketing because renaming or redescribing an existing product with a green angle has a relatively low market impact (Bernstein 1992, quoted in Peattie 1995). There is a need for an explicit and clear definition of greener product strategy and its role in the overall strategy of the firm. Having an explicit environmental new product policy can lead to greater success for environmental new product programmes (Dwyer 1990). This would involve defining product/market areas to be served, and thus formalising the necessary organisational structures for implementation.

Issues relating to reconditioning and re-use should also be part of environmental NPD policy. Simon (1992) describes environmentally driven product strategies as product displacement and creation; expanded product life-cycles; and functional integration strategies for new process development. Furthermore, innovations in product design that increase durability and reparability or reduce material inputs are also useful strategies for achieving source reduction of municipal solid waste (Keoleian and Menerey 1994).

Top management support and involvement. Top management support and involvement has a positive impact on new product development (Booz, Allen & Hamilton 1982). Research has shown that clear messages from top management to the whole organisation about the importance of new product development can be a critical factor in its success (Johne and Snelson 1988). A study by Pujari and Wright (1996) found that all companies they surveyed had at least one senior person co-ordinating all environmental activities within the company, and a basic management philosophy that environment is everybody's responsibility. Two companies within that study were also found to have a global environmental committee with the purpose of responding to environmental issues at a strategic level. Integration of environmental concerns into the business process basically involves fundamental behavioural change within a company, which will not occur without the clear leadership and active support of the company's top management. Senior management must recognise and communicate to the corporation that environmental concerns are the responsibility of all employees (Cahan *et al.* 1993).

Top management should also commit resources into NPD projects from the initial stages. As the business processes and the marketplace of the future will become more environmentally sensitive, investment now in developing appropriate capabilities is crucial (Hutchinson 1992).

The presence of an environmental manager/co-ordinator. The challenges environmental professionals are facing today are as varied as they are daunting (Cole

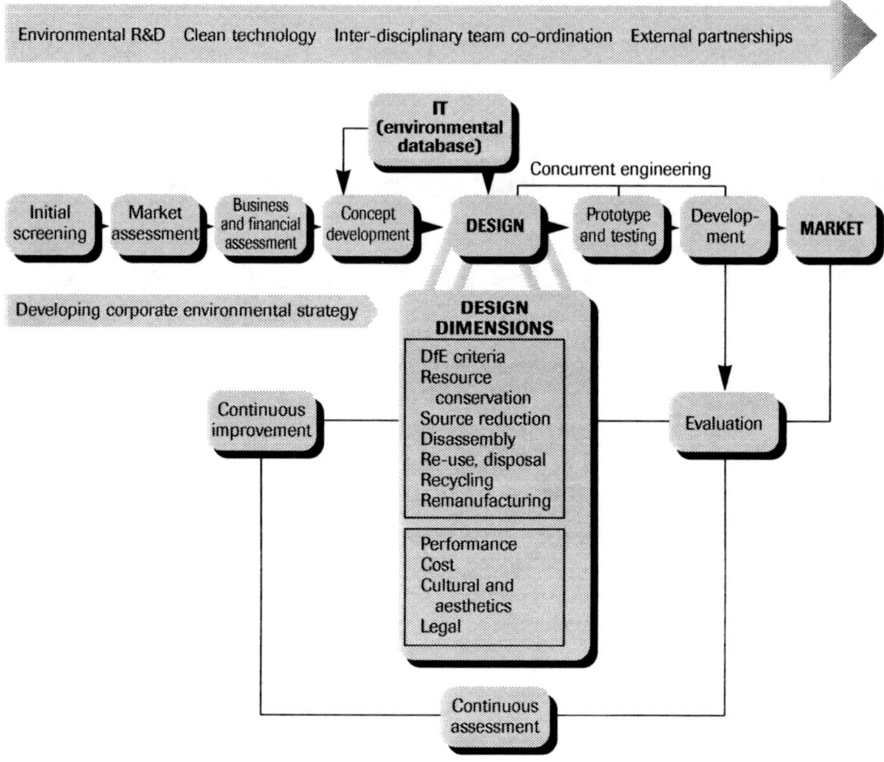

Figure 1: THE ENVIRONMENTAL NPD PROCESS

Source: Adapted from Cooper 1990; Keoleian and Menerey 1994

1993). The future of environmental managers will be intimately bound up with the evolution of environmental management culture in business (Barrett 1993). Many of the challenges with which environmental professionals grapple are the result of intransigent forces such as the economy, government action or inaction, regulatory mandates, and the inherent complexity of environmental issues. Other challenges are the result of the relatively recent development of the field and the subsequent lack of tried-and-tested off-the-shelf solutions.

It has been suggested that there are several scenarios in which environmental managers/co-ordinators work. These are: 'business-as-usual' (dealing with environmental pressures as they arise while creating minimal disturbance to line management); 'gamekeeper' (involving closer central control of the environmental performance of line management, and promoting the distinct role of auditing); and 'sharing the burden' (integrating responsibility for environmental improvement into line management) (Barrett 1993). In the 'sharing the burden' scenario, greater emphasis is placed on self-assessment against specific targets and objectives, with the environmental manager acting as a facilitator by providing guidance, data and verification. It has also been noted that most firms appear to be moving towards the central position, with a growing appreciation that the environment requires strategic management; and in larger companies it is more likely that there is a dedicated post for this role.

The role of environmental managers is shifting subtly from a purely technical focus to a more business focus. Research has found that most environmental managers have additional responsibilities besides co-ordinating environmental activities (Pujari 1996). At the new product development level, the environmental manager's role can be integrated in two fundamental ways. The environmental manager can either become an active member of an NPD project team, conducting an environmental review at every stage of the NPD process; or may participate indirectly by issuing guidelines for the NPD project team to comply with. Ideally, the environmental manager's role should be more integrative, providing inputs for technical solutions, facilitating environmental innovations and communicating effectively with NPD team members (manufacturing, R&D and marketing personnel). No doubt, this will be the biggest challenge for firms.

Process/Actions Issues

Companies should regard the introduction of environmental NPD as a **process-oriented change** in the mental attitude of a business and in the way things are done in that business, embedded in the process of product innovation (Crul 1994). Environmental NPD calls for a new entrepreneurial concept of the relationship between the product and the environment, along with the understanding of the innovation potential and the strategic value that such an approach can have for their business.

Though various models of NPD process have long been proposed, tested and applied, externalities regarding the environment have traditionally been totally neglected. On the other hand, research on environmental product strategy has tended to overlook the ever-increasing complexities and uncertainties of the marketplace. Environmental NPD requires management commitment to integrating environmental aspects into product innovation strategy by developing environmental databases and encouraging environmental R&D. In addition, it requires the creation of market opportunities for environmentally less harmful products by committing to environmental quality product certification and standards such as eco-labels.

Pre-development activities for environmental NPD. New product projects are born as ideas, move through screening, project definition and business and analysis stages, and eventually lead to product development. These early steps prior to product development are referred to as the pre-development or upfront activities (Cooper 1988). It has been suggested that those projects that have clear NPD project definition prior to development will henceforth become considerably more successful. Sufficient market analysis, research and sales forecasting and in-depth understanding of users' needs and wants are critical success factors for successful products (see Table 2).

Pre-development activities for an environmental product	Description
Initial screening	Generating new greener product ideas, and tentative evaluation or screening of these ideas. This can be done mainly through listening to 'green' stakeholders through periodic surveys; sales and service groups; creativity sessions; and through competitions and suggestion schemes. New environmental ideas may also be generated through environmental R&D innovation or can be acquired from external R&D and technological sources.
Preliminary market assessment	Resources are spent in gathering information regarding the feasibility of the environmental NPD project. This involves a non-scientific market appraisal about market size, growth, segments and competition, etc.
Market research	Understanding customers' needs and wants in relation to environmental value and the way they adapt to these by changing purchasing behaviour
Environmental and business analysis	Synergy with resources and skills in the areas of environmental R&D and clean technology, greener marketing, and manufacturing. Are there any outstanding needs regarding these environmental competences and skills?
Thorough financial analysis	Projection of return (including return on environmental quality), in terms of least environmental impact and financial return on investment on the environmental NPD project
Concept definition and development	Definition of target market is developed. The greener product's concept, the benefits the product will deliver, and the greener product's features and attributes. Consensus and experimentation on 'design for environment' (DfE) specifications and requirements is achieved. Competitive analysis is performed: identifying competitors' greener product offerings in their product portfolio, their strengths and weaknesses and their marketing, etc. Customers' perceptions of the competitive greener products can provide valuable clues as to what environmental aspects are to be designed in or to be designed out.

Table 2: PRE-DEVELOPMENT ACTIVITIES FOR AN ENVIRONMENTAL NPD PROJECT

Source: Adapted from Cooper 1988

Customer-led	Regulation-led	Market-led
• Minimising overall environmental impact during production, manufacture and use of the product • Re-usable or recyclable packaging • Prevention of adverse health effects from using the product • Providing re-use or recycling programmes (e.g. for batteries, cartridges and other consumables) • Using energy and materials efficiently • Providing upgrade paths that minimise the disposal of existing products	• Minimising overall environmental impact during production, manufacture and use of the product • Labelling products containing CFCs • Labelling environmentally friendly products (optional) • Collecting and re-using/recycling products and packaging materials • Establishing recycling programmes • Including recycled materials in new products • Eliminating specific harmful materials (e.g. asbestos, cadmium) • Information about the material content on the packaging	• Contributing to revenues, profits, growth • Providing competitive advantage • Increasing number of customers • Providing higher customer satisfaction • Decrease in customer complaints about environmental impact • Minimising delays in market introduction • No interruption in supply chain management • Eliminating barriers that prevent worldwide acceptance

Table 3: ENVIRONMENTAL NPD PERFORMANCE MEASUREMENT CRITERIA

Environmental benchmarking and product performance measurement systems. Due to the increasing importance of products' environmental performance, there has been much emphasis on developing 'in-company systems' for environmental performance measurement (Azzone and Manzini 1994; Hocking and Power 1993; Eckel *et al.* 1992). These measures of environmental product performance may be:

- Compliance- or regulation-related
- Risk-related
- Upstream impacts of virgin-resource use

It may be helpful for firms to develop metrics or a framework for environmental NPD performance measurement, for which there may be several sources. The criteria may be customer-, regulation- or market-led (Paton 1993), as shown in Table 3. The more precisely these criteria are quantified, the better the environmental NPD performance measurement.

According to Cooper and Kleinschmidt (1995): 'Benchmarking—both internally and versus other firms—provides the insights necessary to identify critical success factors . . . that set the most successful firms from their competitors.' James (1994) notes that, until now, environmental benchmarking has been impeded by the perceived sensitivity of environmental performance information and lack of comparative data. However, in the USA, Toxic Release Inventory (TRI) data has reduced the need for secrecy and created a standardised basis for benchmarking; for this, along with other reasons, several companies are now subscribing to the idea. One well-documented example is the joint benchmarking undertaken by AT&T and Intel (Klafter 1992; James 1994). In 1991 the two companies teamed together to benchmark best-in-class performance in pollution prevention, an exercise that involved preliminary data collection from five companies—3M, Dow, DuPont, Xerox and H.B. Fuller—identified as 'best in class' in their sector. AT&T and Intel both feel they have benefited greatly from the exercise, identifying areas of weakness and subsequently addressing them. This example demonstrates that a company can conduct environmental benchmarking outside its own industry sector.

Interface Issues

Cross-functional co-ordination. Environmental product innovation must cut across traditional functional boundaries and barriers and be driven by a cross-functional team approach. Investigations into new product success consistently cite interfaces between R&D and marketing, co-ordination among key internal groups, multidisciplinary inputs into the new product project, and the role of teams and team leaders (Cooper 1994). Poor co-operation or communication among different functional areas is noted as one of the main factors impeding product innovation; whereas one of the critical success factors for innovation and the NPD process is found to be effective functional integration, involving all the departments in the project, starting from its earliest stages (Rothwell 1992). Successful product projects feature a balanced process consisting of critical activities that fall into many different functional areas within the firm, including marketing and marketing research, engineering, R&D, production, purchasing and finance.

Life-cycle analysis in the environmental management literature also suggests the need for product improvement teams. Members of the improvement analysis team could be drawn from any of the participants in the earlier review stage, additional members of the company or organisation conducting the analysis, and selected outside participants. For internal product improvement analysis, this procedure is entirely consistent with current engineering concepts (Vigon and Curran 1993). Sullivan and Ehrenfeld's survey (1992) found that marketing is heavily connected to life-cycle analysis in active companies in at least two areas: as life-cycle team participants and as recipients of life-cycle study results. They suggest that, in order to foster environmental NPD innovation, life-cycle work should be integrated with marketing practices such as market research and other management practices encouraging teamwork, and can identify characteristics of a product that the firm can improve.

Customer and supplier relationships. A manufacturing organisation is extremely dependent on its suppliers, and the role of purchasing in the NPD process precedes even the formation of the engineering team (O'Neal 1993). Environmental NPD must be a co-operative venture between the primary developer and its key suppliers. Materials and components purchased from suppliers have a strong bearing on the quality, cost dependency, lead times, development cycles, development risks, environmental impact and market availability of a product. Bonaccorsi and Lipparini's study (1994) reports the benefits of supplier involvement in the NPD process, which are: reduced development costs, higher quality with fewer defects, reduced time to market, and supplier-led innovation. They point out that early involvement of suppliers in the innovative process is one of the main contributors to successful product performance. To source environmentally less harmful materials and components, it is essential to be aware of suppliers' environmental performance. Many writers have therefore commented on the need for monitoring/auditing/assessment of suppliers in value chain management. As the importance of the vendor grows, maintaining satisfactory relationships with them becomes a managerial and technical operation of the highest significance. Experience shows that long-term, collaborative arrangements mutually benefit both customer and supplier (Burt and Soukup 1994). Most recently, the Chartered Institute of Purchasing and Supply (CIPS) has published guidelines to assist procurement departments in working with their suppliers in identifying and implementing a partnership approach to the inclusion of environmental factors in their supply operations (CIPS 1995).

External partner relationships. In the 1990s, there has been an increasing emphasis on the role of co-operative strategies in developing new products and achieving organisational goals. There has also been a focus on how companies use competitive collaboration to enhance their internal skills and technologies. The term 'partnership' has become part of the terminology of leaders concerned with environmental quality, resource conservation and sustainable development (Long and Arnold 1995). In their seminal work, *The Power of Environmental Partnerships* (1995), Long and Arnold describe the environmental partnership as a constructive and voluntary collaboration among different stakeholders in environmental protection and natural resource management. They further define environmental partnerships as: 'Voluntary, jointly defined activities or agendas (focusing on a discrete, attainable, and potentially measurable goal) and decision-making processes among corporate, non-profit, and agency organisations that aim to improve environmental quality or natural resource utilisation' (1995: 6).

Many companies recognise that the creation of partnerships with their stakeholders is one of the most expedient and cost-effective ways of solving environmental problems and achieving goals (Dechant and Altman 1994). Moreover, as resource and environmental quality issues grow in importance, the rationale for partnering becomes more closely linked to organisational success. In a broader sense, individual organisations may develop some form of competitive advantage by seeking and forging successful environmental partnerships (Long and Arnold 1995).

The ever-increasing need for environment-related R&D will require alliances and partnerships with such different parties as research institutions, universities, solution-oriented environmental pressure groups and government bodies. In this section, we will discuss inter-organisational partnerships for environmental problem-solving at the product development level (mainly in terms of environmental R&D alliances and sharing and exchanging environmental information). Roome (1994) argues that environmental pressures impact particularly on R&D management because of the need to refocus products as a means of achieving greater environmental sensitivity. The pace of future demand for environmental R&D and technology will be driven increasingly by competitive advantage and competitive pressure, as investing in green R&D creates many opportunities for closing the recycling loop: from manufacturing processes for a single product to a collection and processing loop (Biddle 1993).

There is now a strong argument for an increasingly strategic outward-looking role for R&D managers, as their activities become more dependent on the dynamics of the environmental imperative. This calls for a triumvirate of collaboration involving industry, university and government, involving R&D managers in networks and collaborations beyond the normal axis of company R&D units (Rushton 1993; Roome 1994). Key ideas may include: the acceptance of the product stewardship concept; the development of more systemic views of products and processes; and the move towards action based on collaborative frameworks and networks. Close working relationships with government bodies can also be developed and maintained through their environment-related business initiatives and projects.

Several companies have taken the initiative in environmental partnerships, and these include WHYCO Chromium Company's partnership with IBM (for producing specialised chemical cleaning equipment without air and water discharge) and with General Motors (to develop new and more environmentally friendly multi-layer automobile finishes to eliminate cadmium in GM cars' exteriors) (Dechant and Altman 1994). Coca-Cola's partnership with Hoechst Celanese exemplifies another strategic R&D alliance, where the latter is a regular supplier. In addition to glass and aluminium recycling, Coca-Cola has invested in developing a two-litre soda bottle made with 25% recycled post-consumer plastic (Biddle 1993). There are also instances where companies have developed close relationships with reputable environmental pressure groups and succeeded in achieving organisational goals while solving environmental problems. One such documented case is the partnership of McDonald's with the Environmental Defense Fund, who worked in collaboration to analyse McDonald's solid waste problems.

However, despite recent progress, most environmental groups, which frequently differ in size, purpose and methods, often have antagonistic rather than co-operative relationships with business. Milliman *et al.* (1994) concluded that, in order to work together effectively, both businesses and environmental groups must have three critical organisational characteristics: compatible objectives, organisational adaptability, and perseverance.

There is also evidence of environmental problem-solving being achieved through a network of relationships among different firms in the value chain and outside. A study by Ostlund (1994) analysed three networks, each beginning with the objective of finding some solution to an environmental problem: substituting the use of CFCs in refrigerators and in the production processes of flexible foam and circuit boards. Examining the refrigerator industry network (from the perspective of one actor, Electrolux; see Fig. 2), an interlocked relational and technological system was

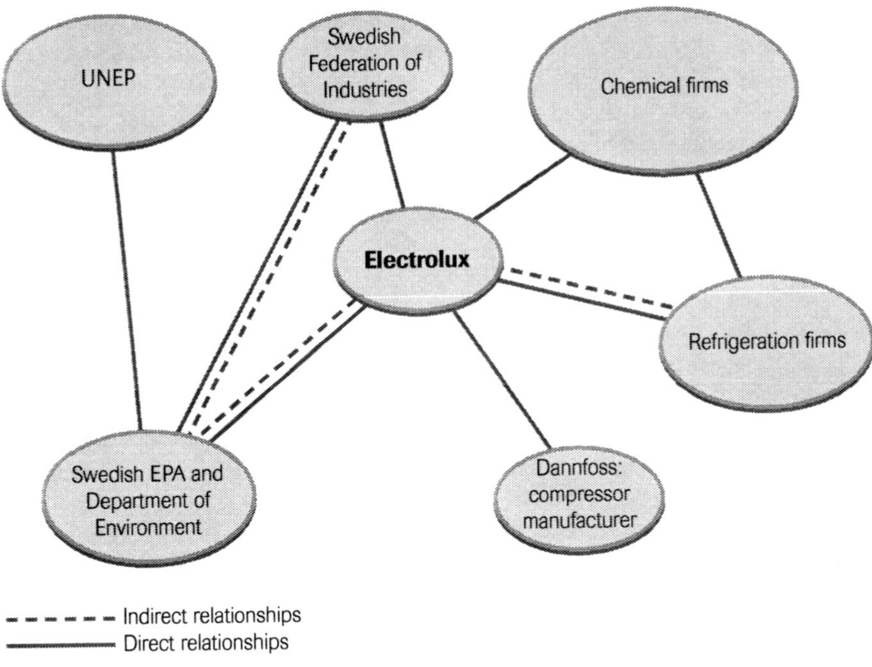

Figure 2: THE ELECTROLUX NETWORK IN REPLACING CFCS

Source: Ostlund 1994

found to be operating in the network dynamics. Electrolux, with the objective of replacing chlorofluorocarbons (CFCs), worked directly and indirectly with such various bodies as the United Nations Environment Programme, the Swedish EPA and Department of Environment, the Swedish Federation of Industries, chemical firms, refrigeration firms, and with Dannfoss, a compressor manufacturer. It was found that the common characteristic of the all cases studied was a highly integrated and standardised technological system. However, the cases differed in terms of the structure of their respective networks and the patterns of change. For example, the refrigerator industry network was mature and concentrated, with a few, important actors, and involved a high degree of interdependence throughout the production systems.

Environmental information. To determine the environmental impacts associated with a given product process, one needs to know the amounts and kinds of chemicals used, the human health and ecological effects associated with those chemicals and the amount of chemical released during the process. Rational, cost-effective provision of environmental management information remains elusive for almost all large enterprises (Ferrone and O'Brien 1993; Orlin *et al.* 1993). This lack of environmental data on specific product processes constitutes a significant impediment to companies' ability to design for the environment. An integrated database needs to be developed that provides the information for making design decisions that account for environmental risks and costs. The database needs to contain information on regulations, processes, and chemicals and materials (including human health and ecological effects). The database should be user-friendly and accessible electronically over existing networks. It could usefully be accessed by manufacturers, suppliers and retailers.

Wallace and Suh (1993) propose the use of information content, a subset of axiomatic design, as a basis for life-cycle design decision-making. The approach requires environmental issues to be explicitly recognised in familiar design terms (target specifications or tolerances). This information-based approach provides feedback about the likelihood that a design will perform satisfactorily and the amount of additional resources required in order to ensure that the goals are met. They further explain that efforts in this area are currently focused on the development of a general information-based decision-making framework and in accordance with a focus on environmentally responsible product design. The objective is to link the decision-making framework with environmental data and computer-aided design (CAD) applications to create an interactive life-cycle product design tool. It is important that environmental information is managed in a format that all members of the NPD process can understand. Thus all parties need to be able to speak the same language in terms of business and environmental issues.

Survey Results of the Bradford Study

In 1996, a large-scale study was completed at University of Bradford Management Centre which investigated environmental NPD process and performance in the British manufacturing industry (Pujari 1996). In all, 151 companies participated in this survey. The major objective of the study was to understand the underlying dimensions of the environmental NPD process and its performance and to identify the factors that lead to success.

A survey questionnaire was designed on the basis of a literature review and 14 in-depth interviews with environmental managers and product development executives in six international organisations. Only larger manufacturing companies were chosen. Data was collected through a self-administered postal questionnaire. A sampling frame consisting of 1,000 addresses of larger manufacturing companies operating in the UK was obtained from the database company, Market Monitor Ltd. It should be mentioned here that the sample was not classified according to industry, as the main concern was with the examination of the perceived driving forces for integrating environmental issues into product development and not in making inter-industry comparisons. The covering letter was co-signed by the Head of Environment, CBI (Confederation of British Industry). The survey was directed to environmental co-ordinators. The important role played by environmental managers/co-ordinators is increasingly being reported in professional journals (Barrett 1993; Cole 1993), being described as 'creative E-managers', 'pioneers', 'co-ordinators', 'champions' and 'thought leaders' with respect to the environmental orientation of their organisations.

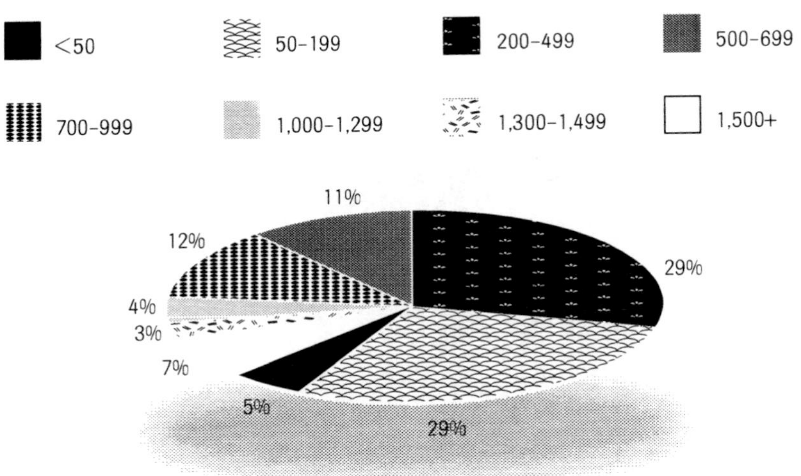

Figure 3: NUMBER OF EMPLOYEES IN THE SAMPLE

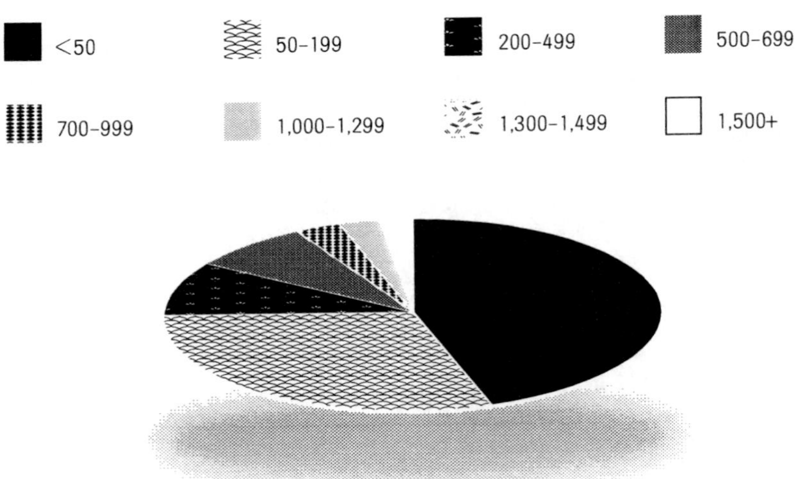

Figure 4: ANNUAL SALES TURNOVER ($ MILLION)

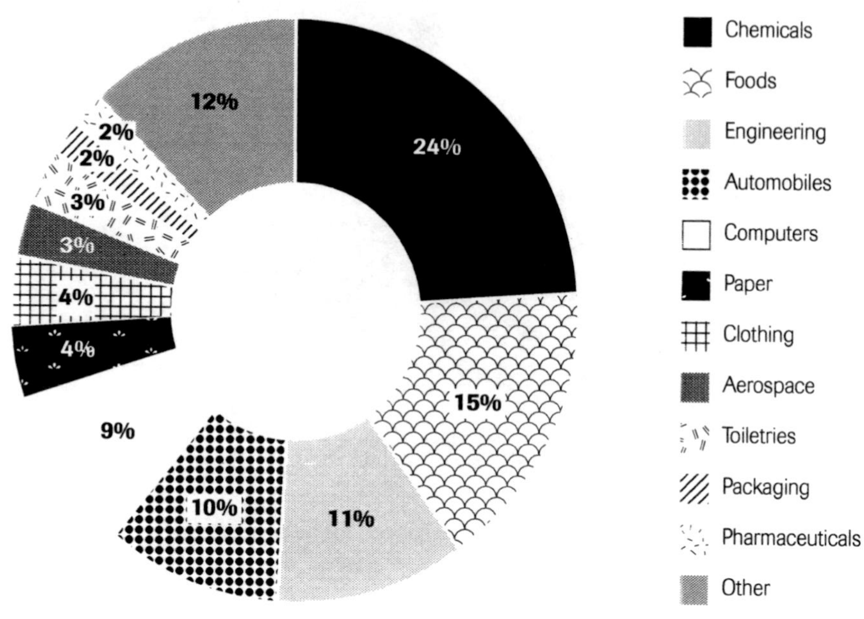

Figure 5: COMPOSITION OF THE SAMPLE

A database of 151 responding companies was created for analysis. Figures 3 and 4 show the numbers of employees and annual sales turnovers of the sample companies. The sample covered a wide range of industries, including automobile, chemical, textile, computer, electronics, food and drinks, furniture, paper, packaging, and cosmetics (see Fig. 5). Most of the environmental managers/co-ordinators have a dual role in their organisation, and therefore cannot be considered 'isolated' in any respect. Additional responsibilities of environmental co-ordinators/managers cover a varied range: technical, development, legal, quality, corporate affairs, R&D, operations, product safety, managing directorship, and training. The majority of respondents were graduates or postgraduates (over 80%). In terms of experience in environmental affairs in their present company, most respondents had been involved in the job for at least one year; 38% had been in environmental management for 1–3 years.

Major Findings

- One of the most significant findings of this study is that environmental NPD performance is not a single-factor solution. Two criteria of performance were identified in the research: 'market performance' (market share, new markets, competitive advantage and return on investment); and 'eco-efficiency performance' (reducing overall environmental impact, improving product quality and environmental image); and these are found to be related.

- Companies have begun to address environmental issues at product development level by developing environmental product policy. However, internalising the externalities, i.e. the environment, in the NPD process still has a long way to go.
- Proficiency in pre-development activities has an important bearing both on market and on eco-efficiency performance of environmental NPD.
- Only some of the respondent firms involve key suppliers in integrating environmental issues into the development process. Only a few actually share or exchange environmental information about materials and components with their key suppliers.
- Apart from sharing information at industry association level, there is no significant evidence of inter-organisational relationships for achieving environmental excellence.
- The environmental manager/co-ordinator's role is more like a 'gatekeeper', not as a real player, in the change process towards environmental orientation, particularly at product development level.

Implications and Conclusions

Several implications can be drawn from this study with potential significance to managers and researchers. Good housekeeping in pre-development activities has an important bearing both on market and on eco-efficiency performance of environmental NPD. These activities relate to detailed market research, preliminary market assessment, defining target market, thorough financial and business analysis and obtaining customers' views of product concepts. To achieve continuous improvement in the environmental NPD process, firms should employ an environmental benchmarking process and should measure the environmental performance of their products. Further, the role of the environmental co-ordinator in the process of environmental NPD is found to be crucial in achieving both commercial success and environmental excellence. Firms who are interested in proactive environmental responsiveness in NPD activities should seek to integrate further the role of environmental professionals in the environmental NPD process. Furthermore, this role should have a business focus rather than a narrow environmental view.

Other environmental NPD activities and organisational issues that are found to be critical for better market performance include a high degree of functional co-ordination, effective environmental information management and product experimentation in environmental NPD. Top management support and involvement is found to be important for the environmental excellence of environmental NPD. Furthermore, close customer and supplier relationships have been found to be crucial in achieving environmental excellence. These relationships may take the form of a sharing and exchange of the environmental impacts of components and materials, involvement in the design for environment process, joint R&D efforts, and supplier evaluations/assessments. Other significant factors are: the existence of an environmental policy explicitly addressing the environmental issues of product development, a focus on packaging issues, co-operation at industry association level, and environmental costing.

The Bradford study has also led to the recognition that previous research in conventional NPD has an important role to play in the success of environmental NPD. Since the environment touches every sphere of the business activity, and is a technical and sociopolitical issue, there is a strong case for forming an inter-disciplinary environmental task force for every new product design and development project. The main function of this task force should be to perform an internal environmental audit for the project and co-ordinate the fulfilment of the company's environmental design and development guidelines. Environmental auditing on a project-by-project basis would seem to be more effective than doing so on a large scale for the activities of the whole business. The biggest challenge for academics and practitioners is, however, 'integration'—integration of the environmental paradigm with the conventional paradigm of NPD. This is particularly true for firms wishing to respond to environmental challenges in a proactive way. Though the general framework of NPD and many traditional NPD activities are found to be relevant to environmental NPD, managers and academics must meet the challenge of 'balancing' market success and environmental excellence. This is because of the unique requirements of environmental NPD: reducing the overall environmental impact at all stages of the product's life-cycle.

Products that deliver real environmental improvements must do two things. First, they must have a lower overall environmental impact than the alternatives (i.e. there must be measurable improvements over the product's life-cycle of conserving raw materials, improved manufacturing process, distribution, use and disposal). Second, they must be bought by consumers in preference to products that have a greater environmental impact. If they do not sell, they do not deliver their environmental benefit (Hindle *et al.* 1993). According to Bernstein: 'The most obvious response to green consumers is the launch of brands specifically geared to meet their needs, but where they are also competitive in terms of their cost or technical performance, they may also be marketed to non-green consumers' (1992, cited in Peattie 1995: 188). Moreover, there may be some other strategic marketing variables that may prove to be key success factors for a green product strategy, such as exceptional company warranties. Customer loyalty can also be created through take-back and recycling strategies, or even product leasing.

Of special interest to practitioners may be relationships and partnerships formed for the common good of their businesses and the environment. These partnerships may be with relevant stakeholders such as suppliers, customers, environmental pressure groups or even competitors. Closer relationships with different stakeholders are advisable, not only for environmental reasons but also in order to achieve competitive advantage. A recent study (Polonsky and Ottman 1998) concluded that there is a limited amount of formal interaction between a firm and its stakeholders. It is argued that adapting a co-operative approach should ensure that the objectives of both the stakeholders and the firm are met.

Finally, firms need to think in strategic terms about the nature and scope of their core business. For example, a car manufacturing company needs to decide whether it is in the business of making cars or a provider of means of transport integrating the goals of sustainable development. The challenge will be about getting involved in symbiotic business relationships. The formation of environmental networks among producers, suppliers and waste management providers could allow industry to address environmental problems more effectively. For example, the ICER (Industry Council for Electronic Equipment Recycling) in the UK is actively involved in developing its National Recycling Plan. A German example deserves a special mention in which car manufacturers in Germany (particularly in Munich) are in the process of developing an 'Integrated Transport System', the concept of which goes beyond environmental management within the company and is seeking to attain the ultimate goal of sustainable development.

Appendix: Some of the Environmental NPD Projects Encountered in the Bradford Study

- Personal computer
- Laser printer toner
- Water-based paints
- Greetings card
- Weatherproof clothing
- Minibus
- Biocides
- Emulsion coatings
- PET bottles
- Preservatives
- Agrochemical
- Air braking equipment
- Non-mineral oil defoamers
- Cartons
- Water-based resins
- Bleach
- Liquid toilet cleaner
- Steering wheels
- Leadless glazes
- Cables
- Textured wall coating
- Textile
- Ceramics
- Zinc die-casting
- Naturally derived flavourings
- Powder-coated window fittings
- Recycled cable film
- Recycled cases/ducting
- Cellular rubber products
- Microcircuits
- Wire-wound resistors
- Paper bags
- Luminaires
- Staples
- Drinks cans
- Detergent
- Wall coverings

8

Designing and Marketing Greener Products

The Hoover Case*

Robin Roy

UNTIL QUITE RECENTLY, the usual technical response of business to environmental problems involved measures to reduce pollution and wastes *after* they have been produced; for example, by installing factory waste-water treatment plant or equipping cars with catalytic converters. However, from the late 1980s onwards, some companies began to shift their attention from these 'end-of-pipe' approaches towards developing 'cleaner' manufacturing processes, which generate less pollution and waste in the first place or make more efficient use of energy and materials. Then, with the growing understanding that many environmental impacts arise from the choice of materials in a product and its use and disposal, attention began to turn to the design of **greener products**.

There are several pressures that might stimulate a firm to incorporate environmental criteria into the design of its products, including:

---

* This chapter is based on research by the Open University Design Innovation Group conducted by Stephen Potter, Mark Smith and the author with additional research by the author and a BBC producer, Cameron Balbirnie, carried out for the production of a video entitled *Green Product Development: The Hoover 'New Wave'* for an OU undergraduate course, 'Innovation: Design, Environment and Strategy'.

Thanks are due to all the firms that participated, and especially to the following for their assistance during the production of the video and for information provided subsequently: Angus Dixcee, former Director of Engineering Development, Hoover Ltd; Barry Mayes, former Director of Engineering, Major Appliance Product Division, Hoover European Appliance Group; Tim Morgan, Engineering Manager, and Caroline Knight, Communications Manager, Hoover European Appliance Group.

- Commercial and competitive pressures and market opportunities
- Environmental regulation
- Cost savings from use of different materials and/or manufacturing processes
- New or improved technologies, components and materials
- Internal and external pressures from corporate environmental policy, staff concern, environmental campaigns, ethical investors or insurance companies

Several studies (e.g. Vaughan and Mickle 1993; Green *et al.* 1994) have shown that the first two pressures above are usually the most important in stimulating firms to develop greener products. Commercial/market factors include pressures from retailers or purchasers, prospects of expanding market share, and the actual or expected introduction of green products by rival firms. Regulation includes both existing and anticipated legislation and voluntary agreements.

This chapter is based on information obtained in a study of 16 British, American and Australian firms that had developed and marketed 'greener' products which have a reduced impact on the environment arising from their materials, production, use or disposal (see Smith *et al.* 1996, summarised in ENDS 1996a). The firms included a small enterprise making rechargeable batteries and portable lighting, a medium-sized company making cleaning products, and two large multinationals making compact fluorescent lamps.

The study aimed to identify:

- What made the companies take environmental factors into account?
- Which environmental factors were considered important, and why?
- Whether the 'greener' products were commercially successful
- Whether any product development changes were needed to address environmental factors
- Whether the projects led to changes in the companies' strategy and environmental practices

The study produced a number of interesting findings. For example, most of the firms did not set out to produce a 'greener' product. Instead, they aimed to develop a product that would perform better, create a new market, enable them to increase or maintain their market share, or respond to market demands or regulatory pressures. Environmental factors were thus incidentally, or deliberately, taken into account in pursuit of commercial and market aims, or in response to existing or anticipated environmental regulation.

Given the commercial focus of the projects, it was not surprising that, with one exception (a manufacturer of volatile organic compound [VOC]-free paints), the firms in this sample did not attempt to sell their 'green' products primarily on their 'environmental friendliness'. This was because some firms did not initially recognise the environmental advantages of their product. However, the main reason was that the firms understood that the environmental performance of a product is generally less important for market success than attributes such as technical performance, reliability and cost. This sample confirmed the findings of other studies of green products (e.g. Wong *et al.* 1995) that, to be commercially successful, the products had to be competitive in terms of performance, quality and value for money. Environmental performance only entered the list of customer requirements once such levels of product performance, quality and cost were attained. As a manufacturer of energy-efficient condensing central heating boilers observed, 'Environment is a "comfort factor", a reinforcement to the basic motivation of buying because running costs are low' (Roy *et al.* 1998).

Nevertheless, this study showed that the development of green products can be a very worthwhile activity. The majority of green product development projects were commercially profitable, with the small firms creating new niche markets with innovative 'green' products, such as lead-free air gun pellets or furniture made from forestry thinnings, and the larger firms developing their consumer or commercial markets with environmentally improved products such as energy-efficient refrigerators, lamps and washing machines.

In the light of the commercial aims of these projects, another unsurprising result was that most of the companies adopted an incremental 'green design' approach to the incorporation of environmental factors in product development. This focused on one or two environmental issues of particular concern to the firm—most often materials choice and the environmental impacts of production, followed by concern for reducing the energy and pollution impacts of the product in use and with recycling materials at the end of product life. In this study, none of the firms routinely adopted a systematic 'eco-design' approach to product development (Ryan *et al.* 1992) in an attempt to reduce environmental impacts over the whole life-cycle of the product from 'cradle to grave'. However, some of the firms had broadened from exclusively considering environmental impacts of the product during either the production or the use phase to considering the environmental impacts of its materials, distribution and disposal, indicating a learning process involving a shift from 'green design' towards 'eco-design'.

The rest of this chapter focuses on a particular product development project which involved such a shift from green design to eco-design: the development by the UK-based Hoover European Appliance Group of its 'New Wave' range of washing machines (Roy 1997). The New Wave range was in production for five years from its initial market launch in 1993 and were the first—and for over three years the only—products to be awarded an EU Eco-label. In addition, Hoover won a 1996 Queen's Award for Environmental Achievement for the design and manufacture of the New Wave range. The case study also discusses the replacement of New Wave by a completely new range in 1998, following the sale of the Hoover European Appliance Group to the Italian white goods manufacturer, Candy, in 1995.

Case Study: Development of the Hoover 'New Wave' Washing Machine Range

The New Wave project originated in the late 1980s when Hoover found itself with a range of washing machines that were losing market share. Although the range had been updated a number of times, 'new' models relied on core designs that had not changed significantly since 1967. In 1989, Hoover felt that a completely new design was needed.

The company had already decided to buy in pre-coated steel for manufacture of washing machine cabinets, thus replacing the slow, labour-intensive and dirty process of welding and spray-painting sheet steel. Developments in materials meant that the outer tub could now be moulded from reinforced plastics, rather than fabricated from steel. There was therefore an opportunity to develop a new product range and simultaneously introduce new manufacturing processes.

Hoover also recognised that a growing number of consumers were demanding 'greener' products and that overseas competitors, such as AEG and Zanussi, had already taken a lead in this market by introducing water- and energy-saving machines in the mid-1980s. The company had traditionally held a significant share of the UK washing machine market, but only a small share of the continental European market where in some countries, such as Germany, environmental factors were crucial to sales. Hoover was aware too of EU plans to label products, such as its new washing machines, for their environmental performance.

Environmental performance was, of course, only one aspect of the planned new machine's appeal. Hoover wished to move upmarket from its traditional position in the washing machine market with its new range, and detailed market research was commissioned to define the features wanted by purchasers in different European markets and countries. According to Hoover, New Wave was 'a statement of modernity' and 'aimed at a very different audience' (Roy 1996).

The Product Development Process

Although development of the New Wave involved serious consideration of environmental factors for the first time, apart from a move from the former linear product development process to a more 'concurrent' team-based project, no changes to the product development process were required. No environmental specialists were involved and the team 'learned as it went along'. Nevertheless, the environmental objectives added some difficult design problems, the solution of which was aided by the establishment of a closely knit product development team.

At the start of the New Wave project, senior managers from the engineering, marketing, manufacturing and finance departments met to agree the business and market specification for the new range. This formed the basis of the technical specification.

Environmental impacts were a key factor in the specification. Hoover recognised—before this had been shown by formal life-cycle analysis—that any reduction in the

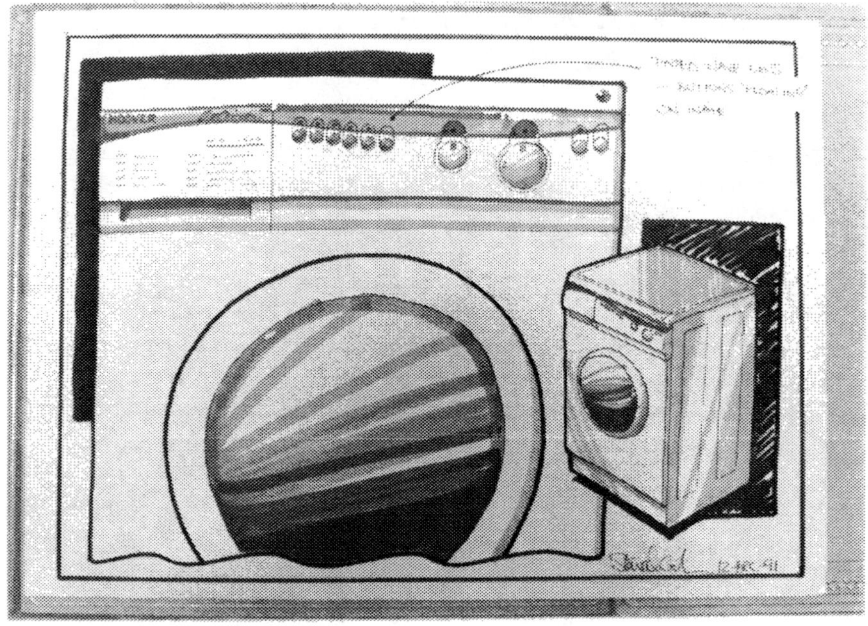

Figure 1: EARLY INDUSTRIAL DESIGN SKETCH SHOWING 'WAVE-FORM' CONTROL PANEL CONCEPT

Source: Hoover European Appliance Group

environmental impacts of the new machine would depend mainly on minimising water and energy consumption in use.

The next step was to convert the technical specification into feasible design concepts. This involved three parallel tasks:

- Researching how to reduce water, energy and detergent consumption
- Deciding how the machine was to be engineered and manufactured. For example, the best method of joining the pre-coated steel panels for the cabinet had to be decided. Replacement of the enamelled steel outer tub, previously fabricated from some 50 components, by a single-piece plastic tub contributed to a substantial reduction in the number of parts in the machine.
- Design of the visual aspects of the machine, including consideration of ergonomic principles. In particular, the Hoover industrial design team were concerned from the outset that the new machine should have a stunning visual appearance that would enable it to stand out from competing products in the market (see Fig. 1).

Research, design and development. In order to reduce the amount of energy, water and detergent consumed by the machine Hoover asked its R&D group to

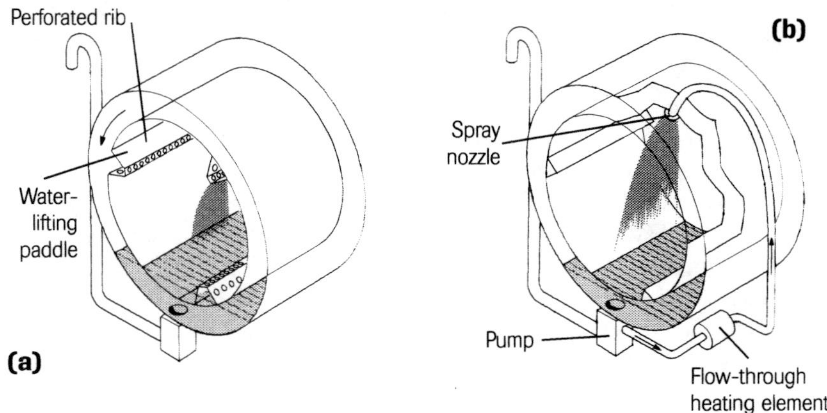

Figure 2: DIAGRAMMATIC COMPARISON OF:
(a) PASSIVE SPRAY PADDLE SYSTEM (SIMILAR TO THAT USED ON HOOVER MACHINES);
(b) PUMPED SPRAY SYSTEM (SIMILAR TO THAT USED ON ZANUSSI 'JETSYSTEM' AND OTHER WASHERS)

Source: Open University

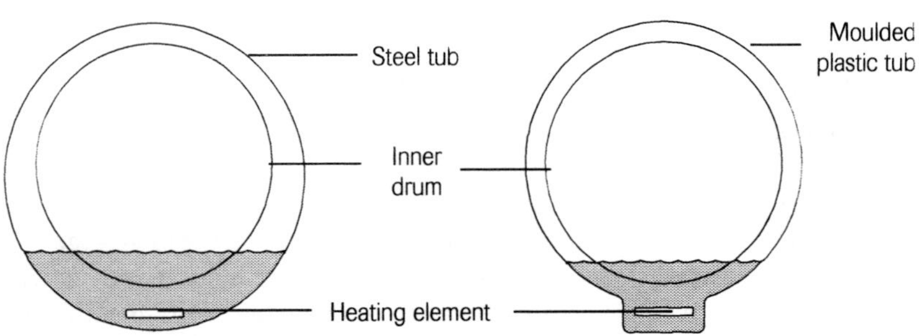

Figure 3: (a) TRADITIONAL OFFSET STEEL OUTER TUB AND INNER DRUM ARRANGEMENT; (b) PLASTIC TUB WITH MOULDED-IN SUMP, SHOWS THE REDUCED AMOUNT OF WATER REQUIRED TO COVER THE HEATING ELEMENT

Source: Open University

conduct a feasibility study for a radically new wash process. This research produced several new concepts. These included: 'spin-wash' (slow-speed spins during the wash to wet the load with a reduction in water level in the tub); 'front-fill' (the idea of filling the machine from the front and using the wash load as a filter, to reduce the loss of detergent flushed into the sump); 'spray paddles' (the idea of perforated agitator paddles in the drum to scoop up water from the base of the drum and shower it over the clothes; see Fig. 2a). The spray paddle concept would avoid infringing the patents on pumped systems used by Zanussi and other manufacturers (see Fig. 2b).

The new wash process required parallel work to develop an electronic control system to allow more precise control of the wash cycle than was possible with the electro-mechanical timer traditionally used on Hoover machines. In addition, the decision to use a plastic outer tub allowed a sump to be moulded in, thus allowing the drum and tub to fit more closely than previously and the machine to operate with less water (see Fig. 3b).

The research concepts then had to be developed into a practical design. This involved Hoover product engineers developing and testing key elements of the conceptual machine. They found that getting a consistently good wash performance required increasing tub water levels and reducing mechanical action, plus development of an effective spray paddle system. Subsequently, a simple float valve was added to the base of the tub to fully eliminate detergent being flushed into the sump during filling.

In order to reduce costs, speed assembly and improve quality, further parts reduction was achieved by substituting snap fits for screws and other fixings on components such as the door. As a result of applying this approach throughout the design, the new machine had one-third fewer parts than the previous range.

Environmental Policy and Eco-labelling

The original impetus for the environmental aspects of the New Wave project was the growing European market for energy- and water-saving washing machines. However, soon after the project had begun, Hoover senior managers began to consider what the company should do to respond to environmental issues more generally. An environmental consultancy was commissioned to establish an environmental policy for Hoover. Their report resulted in the board issuing an Environmental Mission Statement in April 1990, with an Environmental Affairs Committee responsible for implementation. The Hoover Environmental Mission stated that the company aimed 'To adopt the best practical environmental methods in the design, production, packaging, use and disposal of its products, whilst continuing to improve their benefits to the consumer' (Hoover 1990).

Also in 1990, the European Union announced its Eco-labelling scheme, with washing machines among the first categories of products to be labelled. Like the Hoover Environmental Mission, the Eco-labelling scheme is based on a cradle-to-grave approach to product environmental impacts and, in 1991, a life-cycle analysis (LCA) was commissioned to establish criteria for the washing machine Eco-label. The study showed that over 95% of the environmental impacts of washing machines occur

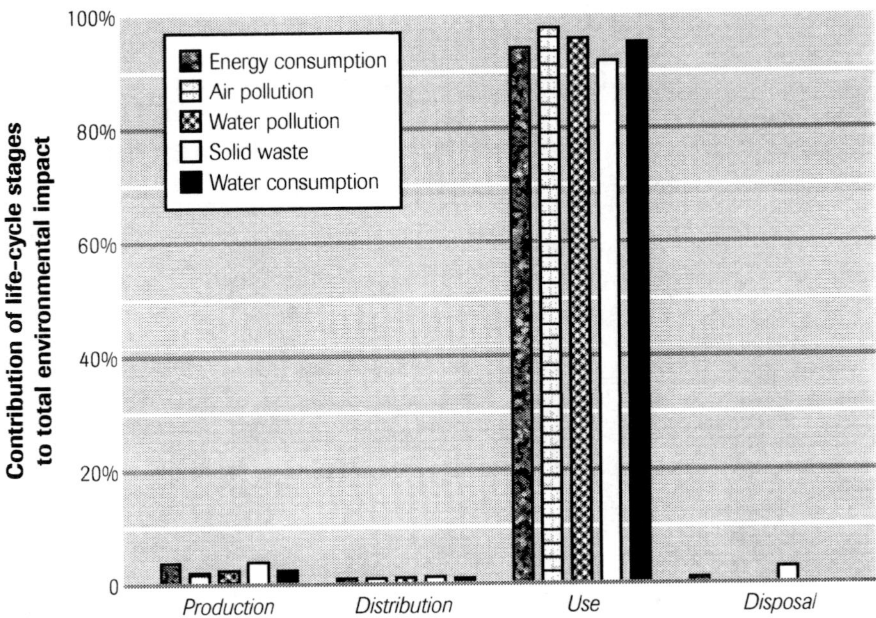

Figure 4: LIFE-CYCLE ANALYSIS OF WASHING MACHINES, SHOWING THE PERCENTAGE CONTRIBUTION TO TOTAL ENVIRONMENTAL IMPACTS AT THE PRODUCT LIFE STAGES FROM CRADLE TO GRAVE

Source: After PA Consulting 1991

during their use (see Fig. 4). Another LCA of washing machines, which included the impacts of detergent manufacture, concluded that some 80% of environmental impacts occur during the use phase (Deni Greene Consulting 1992).

The LCA for the Eco-label confirmed the focus at Hoover on reducing the water, energy and detergent consumption of New Wave. The LCA study also indicated other areas of environmental impact that may have had to be considered before the launch of the range, if it was to meet the Eco-label criteria. These criteria, according to Hoover's former Director of Engineering Development and a member of the Eco-labelling committee, 'were developing at the same time as we were developing our machine'.

◿ Eco-design

Hoover therefore began to consider the environmental impacts arising from production, distribution and disposal of the New Wave.

Production. Use of pre-coated steel for the cabinet afforded Hoover major savings in energy consumption (previously required for welding and drying) and eliminated toxic emissions of VOCs from the factory. But the company was careful not to claim environmental improvements until the 'export' of emissions to the supplier of the pre-coated steel had been evaluated. This subsequently showed that a real overall environmental benefit had been achieved—for example, because pre-coated steel is roll-coated involving less energy and VOC emissions than spray-painting.

Environmental benefits also arose from the elimination of welding and enamelling from tub manufacture, and from the reduction in the number of parts in terms of the amount of materials and energy needed to make a machine.

Distribution. Hoover examined the advantages and disadvantages of a cardboard pack versus a polystyrene pack shrink-wrapped with polythene. Both cost similar amounts and could be recycled, but the company came to the conclusion that—despite controversies about its use—polystyrene performed better and had the edge on environmental grounds, being lighter to transport and using much less water and energy to manufacture.

To reduce transport costs and fuel consumption, Hoover commissioned a new design of trailer for transporting its washing machines. This enabled more machines to be carried in each load, and significantly reduced the number of vehicle movements required in bulk distribution.

Disposal. A reduction in the number of fixings in the machine—adopted mainly for production reasons—made the New Wave easier to take apart for recycling. Recycling would also be facilitated by reducing the variety of plastics used and identifying them by type.

Marketing the New Wave

The New Wave range was first launched in February 1993 but, due to a delay in introducing the Eco-labelling scheme, the award of an EU Eco-label to the range did not occur until November 1993.

By 1997, the Hoover New Wave and its successor range were still the only appliances to have been awarded an Eco-label. Use of the Eco-label is voluntary and involves payment of registration and licensing fees. Although other washing machine manufacturers have models that qualify, they have not applied for an Eco-label because they do not consider the marketing advantage is worth the cost (ENDS 1996b). Another factor was the introduction of mandatory EU energy labelling in April 1996. The Energy Label for washing machines gives a ranking for energy efficiency, wash and spin performance, plus information on water consumption (see Fig. 5), thus rewarding manufacturers of the most efficient machines, unlike the simple pass/fail criterion of the Eco-label.

Nevertheless, Hoover believes that its environmental approach, supported by the Eco-label, was an important factor in selling the New Wave, especially in the environmentally aware German market, in which the company doubled its share in 1994, and in enabling the company to enter other environmentally conscious markets such as Denmark and Austria.

Environmental factors are generally less important in the UK market. In the absence of any other independent endorsement, the New Wave was initially promoted mainly

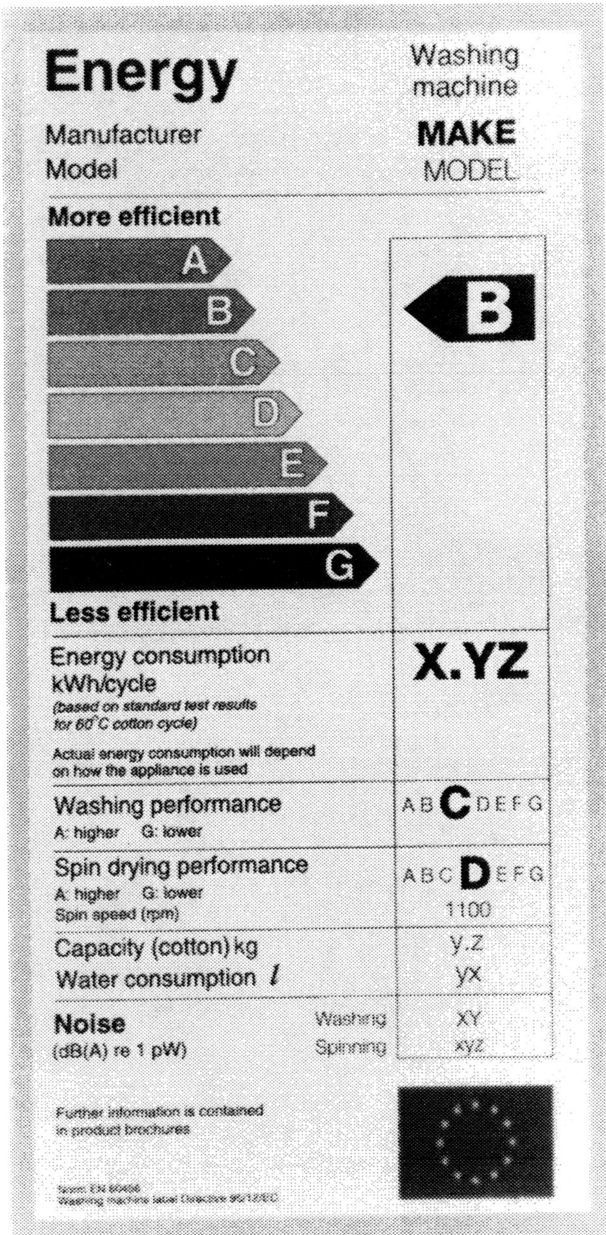

Figure 5: THE EU ENERGY LABEL FOR WASHING MACHINES, WHICH RATES ENERGY EFFICIENCY, WASHING AND SPINNING PERFORMANCE ON A SCALE FROM A TO G, BECAME MANDATORY IN APRIL 1996.

HOOVER New Wave:
The new generation of washing machines

HOOVER:
Environmentally friendly and backed up with the EU seal of approval

What do you expect from your washing machine? The latest technology, of course, a wide choice of programme settings as well as an easy to use machine. That goes without saying. But you're also concerned about the environment. With a machine from the HOOVER New Wave Generation you know you're making the right choice.

The EEC has awarded the New Wave Generation HOOVER machines the first "EU Environmental Symbol" – the ECO LABEL – for environmentally-friendly technology. All of which makes HOOVER the first manufacturer in Europe to be honoured with this much sought-after distinction.

For you what this means is that your New Wave machine has achieved technical excellence. Tests have shown that durability and advanced technology are the best guarantees of reducing the damage to the environment caused by the excessive use of electricity, water and detergent. Compared with previous HOOVER washing machines, the New Wave models use around 40% less electricity, 36% less detergent and 31% less water.

The EU's environmental symbol certifies that along with environmentally-sound manufacturing, the minimum of packaging and recyclable plastic parts are used for New Wave machines.

So there you have it.

The reassurance that your washing and the environment are in safe hands.

Our top-of-the-range model from the latest generation of HOOVER Auto Washers:

New Wave Plus 1500 Electronic Model AC 184

- Metallic pewter control panel
- 500/800/1500 rpm spin speeds
- Six-stage Eco Wash System Plus
- 50 programme combinations available
- Quick Wash programme
- Variable temperature selector
- Digital programme display
- Rinse Hold
- 9 hour delay start option
- Electronic load sensor
- Water consumption 68 litres
- Electricity consumption 0.9 kWh (Hot fill 60° C wash)

3

Figure 6: PART OF A 1995 HOOVER BROCHURE FOR THE NEW WAVE RANGE, SHOWING THE TOP-OF-THE-RANGE 1500-RPM SPIN SPEED NEW WAVE PLUS MODEL PLUS THE EU ECO-LABEL SYMBOL

Source: Hoover European Appliance Group

on its money-saving aspects. However, in later sales material, the 'environmental friendliness' of the range and the independent backing offered by the award of the Eco-label were strongly featured, targeting environmentally conscious consumers in the upper-middle segment of the UK washing machine market. Hoover also introduced a model called 'New Wave Plus', again promoted with the aid of the Eco-label, to compete with top-range German machines in the European market (see Fig. 6). In 1996, the New Wave range was succeeded by an incrementally improved range called 'New Wave Plus 5', with increased capacity and lower water consumption. This range also gained an Eco-label, plus a 'B' rating for energy efficiency on the compulsory Energy Label (Fig. 5).

Extending the product family. The £15 million investment in the development and manufacture of the New Wave was also employed to produce other washing machines aimed at the price-sensitive volume UK and Southern European markets. By modifying the New Wave design, mainly by substituting electro-mechanical for electronic controls, Hoover developed the lower-priced 'Soft Wave' and later 'Eco-Wave 5' ranges. These also saved energy, water and detergent compared to previous models, but did not quite satisfy the Eco-label criteria because they lacked features, such as spin wash, that require electronic controls. These ranges had the same mechanical components as New Wave and were assembled on the same lines.

Lessons

What lessons about designing and marketing 'greener' products can be learned from the New Wave project?

Balancing design attributes. Any successful greener product must balance environmental performance against the many other design attributes—performance, reliability, appearance, etc.—wanted by the market at which the product is aimed, and do so at a competitive price. This point was highlighted in the study of 16 greener products (including the New Wave) mentioned earlier. This study showed that, to be commercially successful, the products had to be competitive in terms of performance, quality and value for money before environmental factors entered the list of customer requirements.

Integrated product development. Incorporating environmental objectives into the product development process does not require a fundamental change to that process. What seems vital is that the green product development process is carried out in a concurrent, integrated manner. Adopting a concurrent, team-based approach, such as that introduced by Hoover for the New Wave project, means that the marketing, engineering/industrial design, production, financial and environmental aspects of product development can be considered by team members from the planning and specification stage onwards.

From 'green design' to eco-design. Designing for the environment is relatively new activity and, for most companies, will involve a learning process. Hoover began the New Wave project with the aim of producing a design that would meet the growing market demand for washing machines that require less water, energy and detergent in use. The choice of materials and production processes was mainly determined by performance, cost and manufacturing efficiency considerations. Following the introduction of the Hoover Environmental Mission, and the company's involvement in formulating the washing machine criteria for the EU Eco-labelling scheme, environmental impacts from the production, distribution and disposal phases of the life-cycle were considered. Fortunately, analysis showed that the new materials and production processes used less energy and generated fewer emissions than the previous processes. However, it is possible that, had the system been designed from the start with the environment more directly in mind, greater improvements might have been achieved. In other words, during the New Wave project, Hoover learned the benefits of moving from a 'green design' approach focused on selected environmental issues to an 'eco-design' approach aimed at balancing environmental impacts throughout the life-cycle.

Continuous improvement and innovation. Designing for the environment is a dynamic process involving continuous technical change. Data published in *Which?* magazine (Consumers Association 1995, 1996) and elsewhere (e.g. GEA 1995) indicate that the New Wave was one of the best European machines available in the mid-1990s in terms of energy consumption, and among the better machines in terms of water consumption. Nevertheless, Hoover is having to continue to improve its designs to keep up with the evolving standards of environmental performance set by competitors and the Energy Label. As is described in the next section, Hoover managed to further reduce energy and water consumption in the new washing machines developed to replace New Wave.

Beyond New Wave

In May 1995 the Hoover European Appliance Group was sold by its US owners, Maytag Corporation, to the Italian Candy Group. After a transition period, significant changes both in the design of the Hoover washing machine range, and in the processes used to manufacture it, were introduced in 1997/98 by the company's new owners. These changes were the result of joint development work by Candy and Hoover engineers and designers which brought the laundry products and production processes at Hoover into line with those in other parts of the Candy Group.

Production. The main change was a reintroduction of production plant to make washing machine cabinets from welded and painted steel, replacing the pre-coated steel plus mechanical fastening process installed for the New Wave and Soft Wave ranges. New plant, costing £17 million and involving a production system similar to that used prior to the New Wave project (but of course using more modern equipment) was reintroduced for several reasons: first, to be compatible with other parts of the Candy Group, thus affording flexibility in switching production between factories according to demand; second, because it was more economic, given advances in production technology such as the use of powder coating and automated welding; and third, to provide a cabinet that was strong enough to allow a new tub suspension system to be employed in the new range of washing machines being designed to replace New Wave.

Product design. Four new washing machine designs were launched in 1998 to replace the New Wave and associated ranges. New Wave was in any case due for replacement to take account both of the new production system and of changes in the market. The market changes included growing price competition and demands for further reductions in energy and water consumption, especially since the introduction of mandatory energy labelling.

As well as a need to reduce production costs and improve reliability, the new designs were driven by the desire to obtain higher ratings on the Energy Label than achieved by the New Wave/New Wave Plus 5 models. In the top-of-the-range 'Quattro Easy Logic AA' (see Fig. 7), this was achieved by several changes to the wash process developed from concepts used on other Candy upmarket machines. These include a Candy patented pumped water recirculation system, in which water from the sump is sprayed to the centre of the drum to minimise energy, water and detergent consumption, and a new electronic control system with higher-speed spins during the wash programme plus control of hot/cold water fill to minimise energy use. The plastic moulded tub, spray paddles and the spin wash concept from New Wave were retained, but front fill was abandoned.

As indicated by its name, the Quattro Easy Logic AA achieved an 'A' rating on the Energy Label for both energy efficiency and wash performance, compared to 'B' and 'C' ratings respectively for New Wave. The new machine's water consumption is 49 litres per 60°C cotton wash compared to 69 litres for New Wave.

Other models in the range, aimed at more price-sensitive market segments—the mid-price 'Quattro AB' with electro-mechanical controls and the lower-priced 'Performa Eco BB' without the pumped recirculation system—perform better than or equal to New Wave in terms of energy and water consumption and wash performance. Only the lowest-priced base model, 'Performa', has a lower 'C' rating for energy efficiency, but nevertheless improves on the performance of the equivalent Soft Wave and Eco-Wave 5 designs.

Market. Although the New Wave range had helped Hoover break into environmentally sensitive export markets, sales were not as great as had been expected, at least in the UK. The new laundry range has been designed to meet the demands of both the upper and the more price-sensitive sectors of the market, as well as providing the Quattro AB model, priced around £399 and rated 'A' for wash performance and 'B' for energy efficiency, for a growing, environmentally aware middle market.

Environment. As indicated above, the production system for the new Hoover laundry range was introduced primarily for strategic and economic, rather than environmental, reasons. Nevertheless, the new electrostatic powder coating plant uses no solvents and produces less waste and emissions than the old wet paint plant

Figure 7: HOOVER TOP-OF-THE-RANGE QUATTRO EASY LOGIC AA WASHING MACHINE, LAUNCHED IN 1998, HAS AN 'A' RATING ON THE ENERGY LABEL FOR ENERGY EFFICIENCY AND WASH PERFORMANCE PLUS A LOW WATER CONSUMPTION OF 49 LITRES/CYCLE

Source: Hoover European Appliance Group

used before introduction of the New Wave range, and complies with all environmental regulations. Packaging and transport have not changed from the environmentally improved systems introduced for New Wave. The new machines have not been specifically designed for ease of disassembly and recycling, although all plastic components are marked as before. The main environmental benefits arise from the greater energy efficiency and lower water consumption of the new Quattro and Performa machines compared to the ranges they replace. As some 90% of the environmental impacts of a washing machine arise from its use, these improvements almost certainly outweigh any possible increases in environmental impacts arising from other parts of the life-cycle.

Although for many years Hoover was the only company to have been granted an EU Eco-label for its washing machines, the company decided not to apply for an Eco-label for its new laundry range. Hoover considered that the Energy Label is more relevant, while the few marketing advantages of the Eco-label did not justify the costs involved. Whether this will mean the end of the Eco-label, at least for white goods, remains to be seen. One indication was the closure of the UK Eco-labelling Board in 1998, plus a review of the whole scheme.

Some Conclusions

The Hoover New Wave and other examples show that there are several pressures that may stimulate a firm to incorporate environmental criteria into the design of its products, but that commercial and market considerations and environmental regulation, such as energy labelling, are generally the most important. However, the existence of an 'environmental champion' at a senior level within the firm, such as existed at Hoover at the time of New Wave, can be crucial in encouraging a firm to respond to these pressures and in enabling it to move beyond simple 'green design' approaches to consider the environmental impacts of the product over the whole life-cycle.

Environmental factors tend to become a consideration in customer purchasing decisions only after other factors such as technical performance, quality and cost have been met. But, for well-designed products, energy efficiency, avoidance of toxic materials, etc. can provide an additional attraction to customers, especially in certain niche and export markets, as well as helping to differentiate a product from the competition. This provides a commercial incentive for companies to develop greener products.

There are greater barriers to the adoption of comprehensive 'eco-design' measures such as increasing product life and re-using components from repaired or discarded products. Although it may be argued that there are environmental advantages in designing products to last longer, the issues are complex. Hoover, for example, kept to its standard design life for the New Wave, arguing that, due to improvements in technology, designing for a longer life was unlikely to be environmentally beneficial. In common with the rest of the domestic appliance industry, Hoover did not plan to take back and re-use components, believing that such components would be outdated and unacceptable to consumers. One way in which this problem could be approached is by designing a long-life product platform, which could be 'upgraded' with the latest components at the end of its initial life (Goggin 1994). However, the adoption of such radical eco-design approaches depends on market acceptability, legislative change, and perhaps on the introduction of new patterns of production and ownership, such as leasing.

9

Eco-Innovation

Rethinking Future Business Products and Services

Colin Beard and Rainer Hartmann

IN ORDER TO FIND A SOLUTION, first we have to 'see' the problem, and a problem is clearly a market waiting for a solution. But what we 'see' and what we cannot 'see' determines the nature of the solution; this is important, as the environment is the biggest problem market looking for new innovative solutions (Beard 1996).

Charter (1998: 5) refers to sustainable product development design (SPDD) as

> being concerned with balancing the environmental, ethical and social aspects in the creation of products and services . . . To create sustainable products and services that increase stakeholders' 'quality of life', whilst at the same time achieving major reductions in resource and energy use, will require a significant emphasis on stimulating new ideas through higher levels of creativity and innovation.

It is this stimulation of new ideas that forms the basis of this chapter. We highlight the need to cultivate an innovative learning culture: one that moves on from the prevailing tendency to focus on environmental impact reduction, which by its very nature focuses on the continuation of differing degrees of negative environmental impact. Impact reduction is currently a process that, as we discuss later, is largely driven by engineering and scientific analysis. At present, many businesses are looking to reduce the environmental impact of their products, or services or activities, but, as we shall argue in this chapter, logically the results of mere **impact reduction** will be disappointing.

We argue that **learning to learn** is a crucial ingredient in innovation, and so it is important to generate more questions than answers—both **seeing** and **questioning** are crucial to innovation. The chapter is not written in such a way as to offer a focused view or to provide simple answers, but our examples will generate uncertainty about many things that we currently accept as everyday facts. We hope to provoke new thinking patterns; such thinking is the precondition of creativity.

If creativity meets with action in the market, it may result in real innovation. Innovation is the implementation of a creative idea, and the answers may be simpler than we think—if we think differently, accept risk and confront the traditional barriers that prevent innovative thought. We offer innovative examples of both products and services, from toothbrushes, pens and razors to telecommunications companies, and from roofs and walls to leisure products. But these examples will be confusing rather than neat, messy rather than clean, illogical rather than logical, ideas not answers. Many of the great innovations around the world have come about as a result of initial rejection or failure: they have succeeded by accident rather than design.

The Concept of e+ Products and Services

Going 'green' has so often been perceived as a trade-off between the profit motive and corporate citizenship and, as a result, some organisations remain distanced from true 'sustainable development . . . which meets the needs of the present generation without compromising the needs of future generations to meet their own needs' (WCED 1987). The earth 'bank' balance is clearly in debt, as previous and present generations have taken considerable amounts of earth resources to provide for their needs and wants—dumping the same resources back into the earth to be absorbed as 'waste'. Thus the usual business cycle produces two environmental negatives (***e–***). Who then is going to reinvest in the natural capital and produce the elusive ***e+*** products and services that contribute to the environment? The answer is that business may well do so.

As business moves beyond confrontation and compliance, we see that the marketplace is already beginning to offer a new generation of products that indicate that business could in future become a strong positive environmental force. Patagonia, for example, the outdoor clothing company that developed the manufacture of fleeces from shredded plastic bottles, recently commented that they wish to develop a business that will 'contribute to the formation of an economy which restores the ecological health of the planet' (Brown and Wilmanns 1997: 28). Clearly, businesses need to find ways of reinvesting in natural capital so as to restore the planet for future generations. Indeed, business may well be the only force powerful enough to reverse planetary destruction. We need to know more about our natural capital, our global bank accounts (see e.g. Beard and Egan 1998).

Current estimates by the Centre for Ecological Analysis at the University of California label nature as worth £20 trillion p.a. That is twice the worldwide total Gross National Product (GNP) of £11 trillion p.a. (Schoon 1997). Extrapolations by Weizsäcker *et al.* suggest that global waste is currently in the order of £6 trillion p.a. (Weizsäcker *et al.* 1997: xxi).

Thus we are likely to be 'wasting' more than half of the world's GNP at present: much of this will be the result of inefficiencies in product design. Some suggest that

'Business should sack the unproductive Kilowatt-hrs, tonnes and litres rather than their workforce' (Weizsäcker *et al.* 1997: xxiv). Our view is that the markets are now ripe for eco-innovation.

Business entrepreneurial creativity can and will focus on finding solutions; they love nothing more. Business will eventually produce more products and services that actually generate an environmental benefit, or an ***e+*** effect. In order to do this, we need to go beyond design for recycling, repair, re-usability and recovery (Fiksel 1996; Burall 1991; Fussler with James 1996); here lie exciting opportunities to significantly rethink product design principles, not only to produce more with lower levels of natural resource consumption, but to design more products that can create an environmental contribution or payback to the current natural stock of resources. But first we must manage and develop eco-innovators.

So what is creativity and innovation?

> An innovator is a dreamer who does (Gifford Pinchot at Institute of Personnel and Development (IPD) Conference, Harrogate, UK [Pinchot 1996]).

As environmental problems create a demand for solutions, so creativity and innovation are likely to be key employer skills that will be much sought after. In the UK, the Economic and Social Research Council (ESRC) Innovation Programme defines innovation as 'the creation, successful exploitation and impact of new ideas at all levels—the economy, sector, enterprise, workplace and individual' (IPD 1997: 1). The environment, the source of both wealth and innovation, is noticeably absent as an integral part of this definition.

Creativity is about the quality of originality that leads to new ways of seeing; however, while inspiration accounts for 1% of our thinking and energy, perspiration, in the form of implementation, is likely to require 99% of effort as virtually all good ideas are destined to die. One person alone is unlikely to produce a successful innovation; it is more often a series of innovative acts that bring exciting new products to market. More importantly, as Pinchot says, 'we cannot compel innovation—we have to evoke it' (Pinchot 1996). We have moved from a skills era to a knowledge era and we are now moving into an innovation era. Sinetar, in her excellent chapter on the survival of creative people in large organisations, and author of *Developing a Twenty-First Century Mind,* notes that business is now 'in love with creativity. With the same fervour with which it courted MBAs in the nineties, American industry is now trying to lure entrepreneurs into managerial positions' (Sinetar 1992: 109).

Weizsäcker *et al.*, in *Factor Four*, comment that

> changing the direction of progress is not something a book can do. It has to be done by people . . . motivation needs to be experienced as compelling and urgent by a critical mass of people otherwise there won't be enough momentum to change the course of our civilisation (1997: xix).

Eco-innovation may require the involvement of a greater diversity and critical mass of people with the confidence to overcome the notion that

> the development of environmentally responsible new products was regarded as a very difficult and complicated task which went beyond the expertise and experience of the majority of personnel . . . [except] . . . the scientists and environmental experts (Dermody and Hammer-Lloyd 1996: 377).

When brainstorming first emerged as a popular management tool, people were unsure of its value. Brainstorming is nothing more than an attempt to bring a critical mass of creativity together, where participants are encouraged to suggest all of their ideas no matter how outrageous, without initial discussion and criticism. Generating pools of creativity can prevent the early termination of new ideas. With technology making communication so easy, idea exchanges can so easily be set up where anyone can enter the eco-innovation arena.

Creating an Innovative Culture for the Development of Products and Services

In developing and sustaining innovation, the way people are managed has a crucial part to play (IPD 1997). Managing innovation is concerned primarily with managing the intellectual capacity of people; it is a key managerial and human resources development (HRD) role to provide a culture and climate for innovation and creativity. Gifford Pinchot, at the 1996 IPD conference, described the struggle that many of the world's creative people have had to face in order to get their products fully developed. Creativity and innovation, he comments, are often suppressed, as typified by Trevor Baylis who could not initially acquire support and funding for his 'Freeplay' clockwork radio idea, yet this product has now 'been endorsed and received many accolades from more than 20 international humanitarian organisations' (Chick 1997: 56).

'The way we do things around here' was the basis of the definition of culture that Gareth Morgan used. It is something that is learned—an emotional or mental response to experiences (Schein 1985). Yet many everyday behaviours that are 'observed' in the workplace are relatively easy to change. But people cannot really see the deeper behaviour, and so helping people to *see* is a first step in the development of eco-innovation. These behaviours are like ships on the surface of the ocean: it is sometimes easy to *see* all the activity on the surface. But they are underpinned by the values and beliefs of the workforce that lie under the surface of the sea. Finally, the culture could be regarded as the sediment deposited on the sea bed. This is less easy to address and few companies have achieved the innovation stage where environmental 'goals are truly institutionalised in all parts of the company' (Post and Altman 1998: 91). This metaphor is illustrated in Figure 1.

Managing and promoting eco-innovation should be a process that systematically moves down these three layers at the right pace; rushing into layer 2 or 3 too quickly

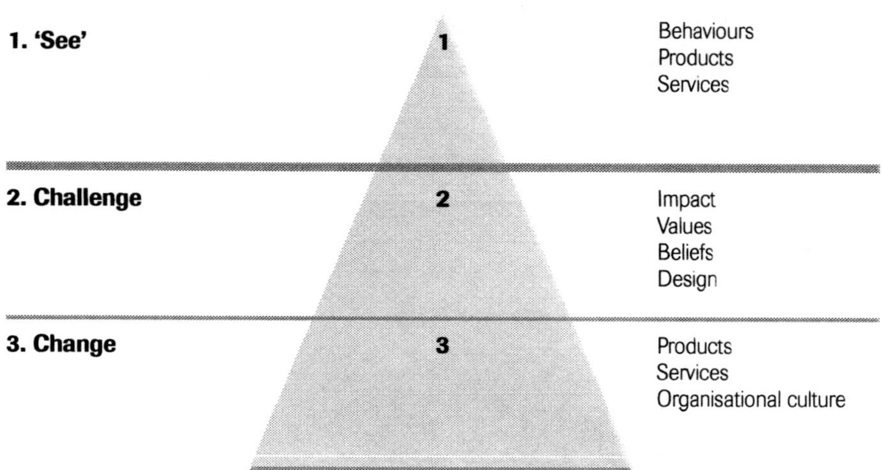

Figure 1: THREE STEPS TO A CREATIVE CULTURE

Source: Adapted from Beard and Hartmann 1999

is likely to be counter-productive. Our 'throw-away' consumer society does not encourage us to think about waste. British Telecom staff did not 'see' the 160,000 nickel cadmium batteries being disposed of to landfill sites, but their audit team eventually created this picture *for* the company. Having got the picture, the company worked with their suppliers to develop new rechargeable batteries. Kent County Council Education Department did not 'see' the hundreds of thousands of plastic biros dumped into the earth, or their total paper waste accumulating in numerous offices until the auditing process started. Developing stakeholder partnerships can also accelerate the innovative development of new products and services (see also Polonsky *et al.* 1998b). In this process of **seeing, challenging** and **changing**, ideas can be generated by everyone, and the use of 'green teams' is becoming an important phenomena in eco-innovation (Keogh and Polonsky 1998b; Strachan and Moxen 1998). But managers will increasingly need to recognise that the talents of good innovators require tolerance and different management techniques in order to be nurtured. Belbin (1981), in his classic study of teams, includes a role of **innovator** and describes behaviours associated with such people as unorthodox, independent, imaginative, original, clever, loner, dominant, socially bold, uninhibited and forthright—just what some managers don't want to hear! But innovators can also be over-sensitive, prickly and poor finishers—hence the powerful role combinations with 'shapers' and 'finishers' (Platt 1988)—easily bored, uninterested in social matters, and comfortable with ambiguity (Sinetar 1992). Sinetar also comments that managers must be able

Adapter	Innovator
Prefers disciplined and methodological approaches	Approaches tasks from unusual angles
Prefers to solve rather than find problems	Discovers problems and avenues for solutions
Refines current practices	Questions basis of current practices
Means-oriented	Ends-oriented
Capable of extended detailed work	Low tolerance of routine
Sensitive to group cohesion and co-operation	Little interest in or need for consensus
	May be insensitive to others

Table 1: THE ADAPTER–INNOVATOR INVENTORY

to identify creative talent. The adapter–innovator inventory (see Table 1) can, alongside such analysis techniques as Belbin questionnaires and other instruments, also help identify potential innovators (Kirton 1976).

Future innovation cultures might offer more freedom, be less controlling, open to change, have low fear of mistakes, will offer promotion for solutions, new products and innovations, and will use idea-generating techniques on an everyday basis.

The Eco-Innovator: A Chimera of a Scientist, Artist and Economist?

What kind of people are required to drive eco-innovation? The creative thinker is more like an artist or inventor, yet science and legislative matters dominate the business environmental agenda (see e.g. McDonagh and Prothero 1996). Managerial systems thinking is currently focused on compliance and scientific/technical requirements rather than eco-innovation and opportunity. Some commentators have indeed remarked that sociologists have not entered the environmental debate and that the 'gap between "lay" perceptions of environmental crisis and official scientific, technical and policy discourse, is of course, one area for sociological explanation' (Redclift and Benton 1994: 1). Scientists, particularly engineers and chemists, head many of the environmental units of top European companies and their prevalent concern is with environmental damage limitation of products and services. In the post-Brent Spar analysis of the battle between Greenpeace and Shell, senior managers at Shell commented that, as scientists, they were trained to look at 'facts' not emotions: losing touch with the 'soft' agenda cost the company dearly. British Telecom (BT) has tried to overcome such problems by including representatives from many sectors of society, including schoolchildren, who represent the concerns of the next generation, on their liaison panel within their environmental management systems (EMSs) (Tuppen 1993). Can innovative products and services be designed *with* the next generation rather than *for* them?

Scientists and scientific tools may thus be significantly influencing product design—creating new products with efficiency and environmental impact reduction in mind (***e–***) rather than innovation and effectiveness for sustainable design. Edward de Bono comments that the brain evolved to find out what is, not to develop what can be. 'Mere analysis of data will never give you a new idea, because all you're doing is drawing on the repertoire you already have' (Pickard 1996: 34). Product design is influenced by the environmental impact of a product which in turn is often measured by life-cycle analysis (LCA) tools, but there is limited concern for lifestyle change (LSC). Product LCA often involves detailed and expensive scientific analysis, which is not always possible for many companies because of the sheer volume of their products. Fiksel, for example, highlights his work with Johnson Wax and the concerns of their designers in adopting tools that might interfere with their work or 'slow down' the design process. Johnson Wax thus chose to 'de-emphasise rigorous life-cycle assessment (LCA) tools and to work with a more streamlined approach' (Fiksel 1998: 50). B&Q, the British DIY retailer, has over 44,000 products from 500 suppliers (B&Q 1995), and it is virtually impossible to identify accurately the specific environmental impact of every product. But B&Q is perceived as making great strides in environmental management: it has, for example, recently completed a detailed evaluation of the global warming effect, per item, of individual products such as hammers, shipped from Taiwan into the UK.[1] B&Q, however, acknowledges that, while it is now becoming more concerned about the social impact of product production and design, in some overseas countries social factors are less easy to quantify, or influence.

Measuring the environmental impact of a product for design purposes can be a simple process, however. Such a system is the ABC–XYZ approach, adopted from financial controlling (Vollmuth 1994). While the environmental impact [**I**] is valued by the figures A, B and C, the quantitative importance [**Q**] is valued by the figures *x*, *y* and *z* (see Fig. 2). The importance of action increases from the lower right to the upper left in the diagram. The diagram shows that, if a product has a high impact but exists in very low quantities, it might be much more effective to focus attention on lower-impact items that have their impact multiplied by large amounts.

Some of the energy currently directed to the meticulous and detailed analysis of impact reduction could be redirected towards the creation of more design innovation for sustainable products. This may require a multidisciplinary approach using a wider range of human talent and using the best bits of right-brain creativity and left-brain logic. Over-reliance on science and analysis could be creating design paralysis. This can be illustrated by the comparative LCA for phosphates in washing powders conducted by the Öko-Institut Freiburg, Germany. The study took four

---

1 B&Q 1995, and personal discussions with Dr Alan Knight, B&Q.

Impact ➤ Quantity ▼	A	B	C
x	Ax High impact + high quantity	Bx Medium impact + high quantity	Cx Low impact + high quantity
y	Ay High impact + medium quantity	By Medium impact + medium quantity	Cy Low impact + medium quantity
z	Az High impact + low quantity	Bz Medium impact + low quantity	Cz Low impact + low quantity

Figure 2: THE SIMPLE IQ TEST

years at an estimated DM 1 million. By completion of the study, phosphate-containing powders had been generally phased out! (Hartmann 1997). Detailed analysis data is often data for defence, denial or justification—but it often fails to convince people.

Changing Mind-sets: Eco-Innovation in the 'Services'

Eco-innovation is also happening within the service sectors. The following two industrial sectors, telecommunications and leisure, provide interesting examples of new ideas that try to incorporate and focus on bringing environmental benefits into their services.

Telecommunications Companies

A service sector that is widely thought to have little or no impact on the environment is that of the telecommunications companies. They are perceived as providing a telephone 'service', and as they do not manufacture 'products', they have been traditionally thought of as environmentally benign, unlike the extractive industries such as oil companies. Deutsche Telekom AG, for example, is the world's third-biggest telecommunications company and it is working to reduce the impact of its activities; however, like many other national telecommunications companies, it buys and uses over 200,000 products, has a procurement budget of £8 billion and is listed 24th in the league of world debtors—above nations, such as Turkey and Norway. It is a big company and therefore has a big environmental impact. But European telecommunications companies—as service industries—were excluded from initial participation in the European Eco-Management and Auditing Scheme (EMAS). However, new research into international telecommunications companies (Beard and Hartmann 1999) has revealed extensive and largely unknown environmental negative impacts through consumption of considerable quantities of global resources. The so-called **1% phenomenon** was shown to apply to telecommunications companies in a number of different countries. Some telecommunications companies use 1% of their nation's electricity, consume 1% of national paper or produce 1% of national GDP. Japanese Telekom consumes 1% of all Japan's paper. While British Telecom plc (BT) has for a long time been at the forefront of environmental management and impact reduction in Europe, this company accounts for a phenomenal 1% of the UK GDP and consumes 1.2% of the UK industrial electricity supply. BT consumes 2.9 million m^3 of water and 70,000 tonnes of paper, of which 57,000 tonnes are used in telephone directories!

However, the rate of change in this sector is greater than any other business sector, and telecommunications companies are now busy reducing their environmental impact as a result of technical developments, and the global forces of trade liberalisation, privatisation and competition—i.e. market forces. Telecommunications companies are using this as a marketing opportunity; they are set to impact considerably on travel and lifestyles and might have a significant positive effect on the environment—through changes in working practices as well as impacting on both indoor and outdoor leisure activities. The recently held futures exercise by the European Institute for Research and Strategic Studies in Telecommunications, chaired by BT, concluded that 'Telecommunications have the potential to make a major contribution to sustainable development' (BT 1997). Some telecommunications companies are selling the message of dematerialisation quite hard, receiving support, for example, from Sara Parkin of Forum for the Future, and guest speaker at the Conference on Telecommunication and the Environment in Frankfurt:

> For me a sustainable society means ecological security; trust in justice and governance; appropriate technologies which add value to what people do rather than replace them; satisfying work; a safe supportive community and a shared sense of purpose and values. Telecommunications can help achieve all of these—for example, by allowing us to move information rather than mass, by enabling more people to work and participate in their communities and by making information available and useful to all (Sara Parkin, quoted in BT 1997).

BT produces a homepage document on the Internet called 'Telecommunications, Technologies and Sustainable Development', in which Ian Ash, Director of Corporate Relations, comments on the potential environmental benefits of the shift from products to services:

> Consider for a minute: the social and economic implications of consumers carrying out much of their domestic business such as banking, shopping and booking holidays over the telephone network; how lifestyles might change if we could receive education, medical consultation, films, music, newspapers and other information 'on-line'; and how our towns and

villages might be affected if the majority of office based people worked at neighbourhood tele-centres or even from home (BT 1997).

Telematics and the environment may appear contradictory fields—but the European Commission has now recognised that they are indeed complementary. Telematics can contribute to better environmental management and protection (Karamitsos 1998). What we do know is that the impact of future innovation in telematics will have a significant but unpredictable impact on our lifestyle: an impact that will depend on the creativity and social competence of all of us.

Outdoor Management Programmes

Increasingly, a number of companies are using the outdoors in their training and development programmes and the rise in customer inquiries has led the Institute of Personnel and Development to produce a *Guide to Outdoor Training* (IPD 1997). Outdoor management development (OMD) programmes are currently being offered by over 200 providers in Great Britain. The vast majority of such training programmes take place in the hills and mountains of our National Parks, and the outdoor management development industry has now developed new environmentally friendly products. Providers of management development programmes that use the outdoors are now realising that the environment is not only a location but a new product: by undertaking conservation projects such as rebuilding dry-stone walls or replanting forests rather than traditional climbing and abseiling as the basis of their process skills development, they actually contribute to producing a positive payback to the natural environment.

New programme designs are now introducing innovative recipes as a result of the deconstruction of the traditional original product, using a selection of **environmental activities** or **community activities** as substitutes for traditional recreational pursuits such as canoeing or climbing (Beard 1996). The new product has a sense of end-product 'reality', rather than simulation, and as such seems to have a significant effect on the motivation and learning of the participants. There are many potential benefits in introducing environmental problems and projects into development or training programmes. One large supermarket chain experimented with such projects involving community environmental works and found that managers were highly motivated to carry out the work and were keen to learn. Their training managers dubbed this impact as the 'Anneka Rice Effect'—after the television show hosted by Anneka where people rang up in response to a good-cause challenge, and offered their help in kind or with products for special charitable projects. They were interested to know why people gave so much personal energy to projects that are 'perceived' as doing good work, resulting in a high emotional response to their experience. This phenomenon still requires more research, but the reconnection with the community and the environment appears to be of significant value. This benefit of using nature to create natural thinking has a lot more development potential for future working practices.

This, then, leads us on to another of our strange thoughts. For entrepreneurs, it is likely that innovation and creativity is a form of play, and that 'they experience their work as a calling or dedicated vocation' (Sinetar 1992: 113). The idea that work can become play or a leisure activity is a crazy idea—or is it? The main product of one of Britain's largest environmental/conservation charities, the British Trust for Conservation Volunteers (BTCV), involves turning environmental work into an outdoor leisure activity. Its 'Natural Break' working holiday programme is the largest of its kind, giving over 6,000 people every year first-hand experience of practical conservation activity. This is the organisation's most profitable and popular service product development: people pay to work and landowners pay to get the work done. Many thousands of people each year pay to work—work paid for twice over and enjoyed as leisure!

Innovative Patterns of Thinking: Barriers and Opportunities

Eco-design often starts with the notion that everything has a negative impact on the environment (***e*–**) and this may be a mental barrier, as we have shown with LCA. Inventing and designing products and services that 'contribute' to the environment

'Parties are a time to drink and create fertiliser', as was the case with the organic gardener who asked the men to urinate in a swing bin to the rear of the house. Strange behaviour, maybe, but the opportunity to collect free fertiliser was one the gardener couldn't afford to miss. Now translate the idea to motorway stations and urine could be seen as an opportunity for profit rather than a cost of disposal. A Swedish company (Servator AB) entered this market recently and now offers composting toilets on the Internet.*

A new breed of scrap merchant is also emerging: opportunist recycling in the wake of the growth in telecommunications. Deutsche Telekom, for example, reported that they produce about 1.5 million tons of electronic scrap per year. What is interesting is that just one ton of electronic scrap has the potential to yield:

- 200 kg copper
- 80 kg iron
- 1 kg silver
- 0.5 kg gold!

One major recycling company in Germany is said to turn over DM 2 million of gold per day (DTAG 1995).

'Saving paper doesn't save trees.' That will be the case in the future. If paper is not made solely from trees, then the old well-known saying does not hold true. The Body Shop, for example, supported a community initiative in India to make paper out of the water hyacinth, an abundant river weed.

Box 1: MYTHS AND LEGENDS: INNOVATION OR IDEAS?

Source: Beard and Hartmann 1997b

will require high levels of innovative thinking. So how can our patterns of thought be changed so as to release more innovation in design? The short examples in Box 1 are simply offered to inspire creative thought.

But very different forms of design thought patterns may emerge and flourish, for example, in times of war or when resources are very scarce. Frugality produces different innovative patterns of thought; constraint drives a 'design for necessity'. 'Elegant frugality' is the key design principle of Lee Eng Lock, the Singaporean designer of the world's most efficient air-conditioning system (Weizsäcker *et al.* 1997: 53). Design with constraints often forces creativity.

Figure 3 offers the notion that different patterns of environmental design may emerge under conditions of wealth as opposed to poverty or frugality—shown on the horizontal axis. If resource scarcity will drive future innovative design, then there is much to learn from people who have experienced such conditions. Innovation lethargy or laziness may also be a major barrier towards the product greening process: gimmicks are relatively easy to design, as is reinvention. In the North, for example, people have often attempted to export inappropriate design and technology to the South, failing to adjust or produce different kinds of eco-innovative thinking. Mies, in her chapter on 'The Myth of Catching-up Development', comments that 'virtually all development strategies are based on the explicit or implicit assumption that the model of "the good life" is that prevailing in the affluent societies of the North' (1993: 55). We might learn from the emerging creativity patterns of the merger of East and West in Germany.

Industry often sees environmental legislation as a threat to competitiveness, rather than the view that it could drive creativity and enterprise. Dominant thinking patterns are thus embedded in the top left-hand quadrant of Figure 3 (the square); here, it is thought, lies opportunity and the most competitive position. This erroneous mind-set often pushes eco-innovation too far in one direction. The bottom right quadrant (the circle) is thus perceived as the position of low design opportunity, with the needs of frugality subsumed by a 'catching-up' mind-set. Yet this is not necessarily a less competitive position, and here lies opportunity for new forms of socially and environmentally positive innovation. Carl Rogers, the pre-eminent American psychologist, and known as the originator of 'client-centred' therapy, delivered an interesting paper back in the 1950s called 'Towards a Theory of Creativity'. In it he offers a definition of creativity:

> the creative process is that it is the emergence in action of a novel relational product, growing out of the uniqueness of the individual on the one hand, and the materials, events, people, or circumstances of his life on the other (Rogers 1967: 350).

He noted a distinction between 'good' and 'bad' creativity, with very different resulting social values (Rogers 1967).

The car industry, for example, should be awash with creativity (Weizsäcker *et al.* 1997). However, many of their design ideas have been a survival response to increasing regulatory pressures to reduce the impact of transport on the environment. 'Whereas German and Japanese car makers have gained some early advantages, particularly

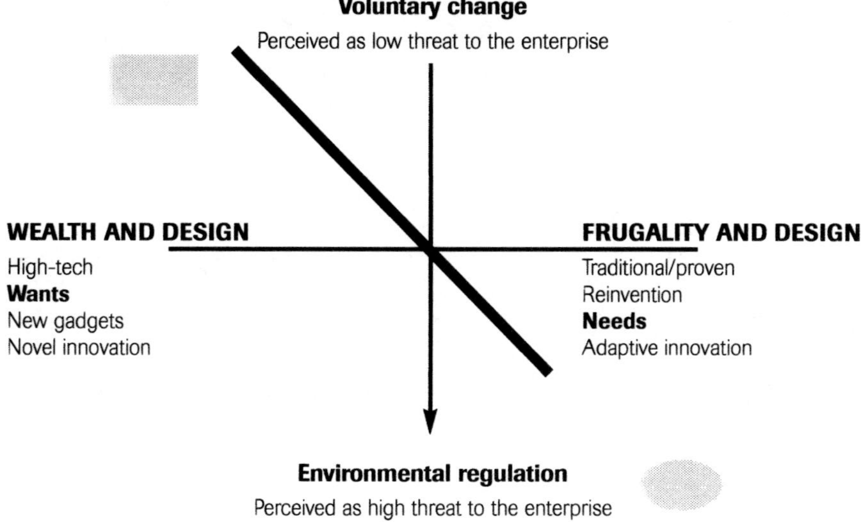

Figure 3: BARRIERS TO SUSTAINABLE DESIGN: THE SQUARE AND THE CIRCLE

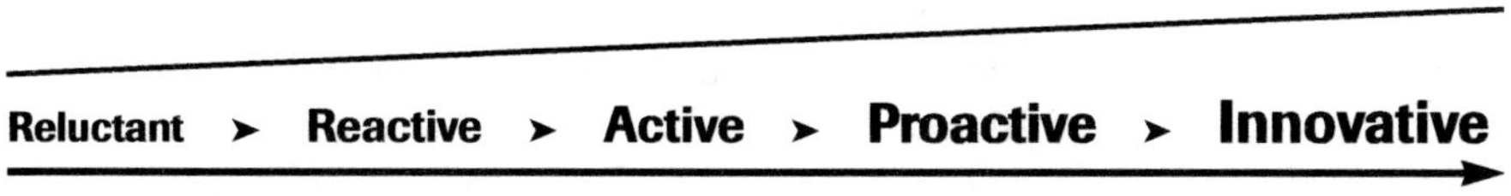

Reluctant ➤ Reactive ➤ Active ➤ Proactive ➤ Innovative

Figure 4: SHIFTS IN BUSINESS ATTITUDE TO THE ENVIRONMENT OVER TIME

CAR PRODUCTION in the German Democratic Republic provides an interesting example of creative thought. With waiting periods of about 20 years for an ordered car, and with car life-cycles running at 8–10 years, a whole new business emerged rebuilding the 'Trabant' (the 'mini' of the East) at Wartburg, buying and selling all the parts for this car and rebuilding them after accidents or after 20 years on the road. Special refashioned body panels were being sold to make a 20- or 30-year-old car look new and fashionable. The Trabant's body, because of a shortage of sheet metal, was made from a steel frame with screwed-on panels that were made of pressed cotton soaked with synthetic resin (Beard and Hartmann 1997b). This 'cotton' was often, in fact, second-hand clothes felted and pressed into the shape of car body panels.

Box 2: FRUGALITY AND INVENTIVENESS: THE TRABANT

IN THE 1950s, a small group of Shell engineers working on a mileage marathon produced a 150 miles-per-gallon (mpg) car. Forty years on, with 500 million cars on the road and 50 million models rolling off the production lines in 1996,* the average fuel efficiency for cars is still only 38 mpg.

GREENPEACE, in under two years, showed industry that any car produced today can be built with a 50% fuel reduction for about the same price with no compromise on safety. There were lots of declarations of interest by industry and other institutions but little action. Greenpeace eventually gave a loan of £1.1 million to the engineering company in Switzerland who developed the Twingo SMILE car (**SM**all **I**ntelligent **L**ight and **E**fficient).†

In spite of all these prototypes and the constant demands by environmental pressure groups for a so-called 'three-litre car' that does almost 100 mpg, it was not until the introduction of a special road tax incentive in Germany in 1997 that SEAT, the Spanish affiliation of the German Volkswagen Group, actually started producing such a car.

CARS ARE DENOUNCED as a major source of pollution, yet with some basic changes they could become street air cleaners, hoovering and filtering the polluted air and possibly cleaning the streets. The shift from unwelcome to welcome is not too far away: another ***e+*** example in the making.

* *The Economist*, 22 June 1996

† Brent Spar Conference, 1996

Box 3: CARS AND SOLUTIONS CAMPAIGNING

under pressure from consumer and pressure groups and regulation, . . . US car makers chose to fight regulations' (Porter and van der Linde 1995). Regulation, rather than voluntary change, can be perceived either as a threat to business or in practice as a positive driving force creating opportunities for innovation and sustainable design providing future success in the markets. The development of an eco-defensive culture is best avoided. A *Harvard Business Review* article commented that 'Businesses spend too many of their environmental dollars on fighting regulation and not enough on finding real solutions' (Porter and van der Linde 1995). Spending energy on fighting regulation often results in a misplaced notion that regulation reduces competitiveness. Learning to find 'solutions' to the environmental problems is not going to be easy; however, losing energy by fighting regulation is wasteful. Figure 4 illustrates some of the phases that can occur as business moves from energy-consuming resistance to environmental issues. Moving on to the very productive eco-innovation stage can take time.

Finding New Solutions

The UK New Economics Foundation is advocating a National Development Fund for Social Entrepreneurs as a growing number of businesses are now emerging as a potentially powerful positive social force, doing more with their purchasing power and innovative energy. At the same time, campaigning organisations are now refocusing some of their energy on what has become known as the '**solutions**

agenda' (Millais 1996), shifting from their detective approach with its associated politics of blame to partnerships with business that support innovative problem-solving. Greenpeace, for example, is investing large sums of money through its **'solutions campaigning'** to lead the way in some areas of eco-design. It argues that business has been dragging its feet on sustainable product design, and it is determined to close the implementation gap by persuading industry to use known but unused or suppressed technologies. More environmental organisations could refocus and help business see environmental problems in need of solutions—and ask for help. Indeed the **solutions strategy** was summed up by Lord Peter Melchett, head of Greenpeace, in 1997, when he said that it was not enough simply to say there is something happening to the environment. Melchett stressed that people have to say what needs to be done to solve the environmental problems (Bolton 1997).

Simple Natural Examples: Seeing it Differently

If B&Q have 44,000 products, there may be great opportunities for redesigning the vast majority of them. Many simple ideas have yet to be included in product design. Let us look at one or two everyday products. We have for years used razors with disposable 'heads', but many people continue to use toothbrushes that are disposable in their entirety.

The photographs in Figures 5 and 6 show the potential for redesigning products. The toothbrush design moves from complete brush disposal (i.e. the whole brush becomes waste material) through to partial product (head) disposability; yet disposable toothbrush heads are still to reach the UK shelves despite their popularity in Germany. These simple ideas remain to be discovered and implemented in many other products. Pens and highlighters are also poised to face a design change with more use of biodegradable card and wood if plastic products continue to remain more harmful to the environment. In turn, the consumption of greater numbers of cardboard pens will generate a stronger market for paper recycling, which experienced a decline in the UK in 1998 due to the slump in the value of waste paper (Nuttall 1998).

We often fail to learn effectively from natural evolution; nature rarely influences our patterns of thought. The earth systems have had some 4.6 billion years on the design board. The stock markets actually confirm that much of the earth's clever design and current resource wealth still exists in the central equatorial regions where, unlike Europe, the drawing board was not wiped clean by an ice age 15,000 years ago. Many of the new **emerging markets** are related either to genetic and plant/animal stock, species patenting, biotechnology or to general development in areas where tropical natural resources are traded for infrastructure development. Eli Lilly is one of the world's largest drug manufacturers, and has made large profits out of two potent anti-cancer drugs developed from plants such as the Madagascar rosy periwinkle ($130 million per year worldwide in the mid-1980s). Then there is the case of the outdoor clothing industry, which has for many years struggled to produce the ideal material that keeps us warm and dry—but doesn't 'sweat'. Plants came to the

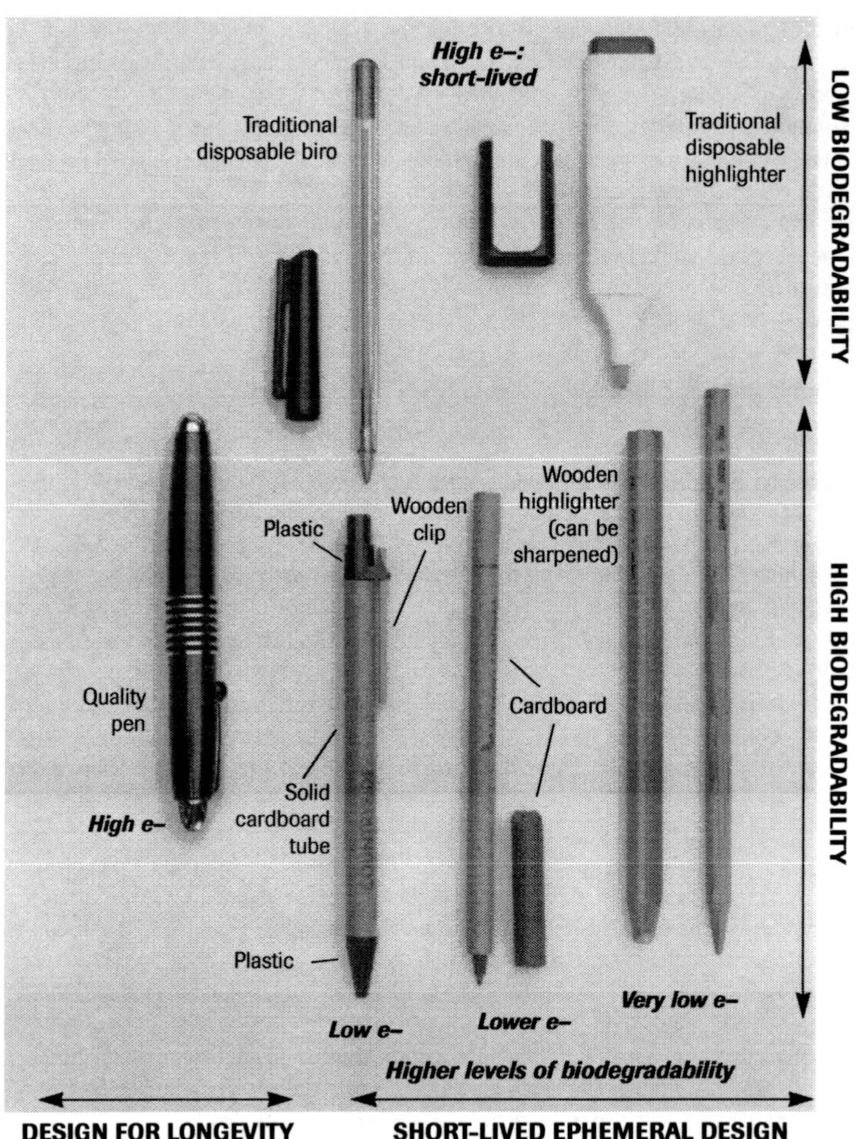

Figure 5: ENVIRONMENTAL DESIGN ASPECTS OF PENS AND HIGHLIGHTERS

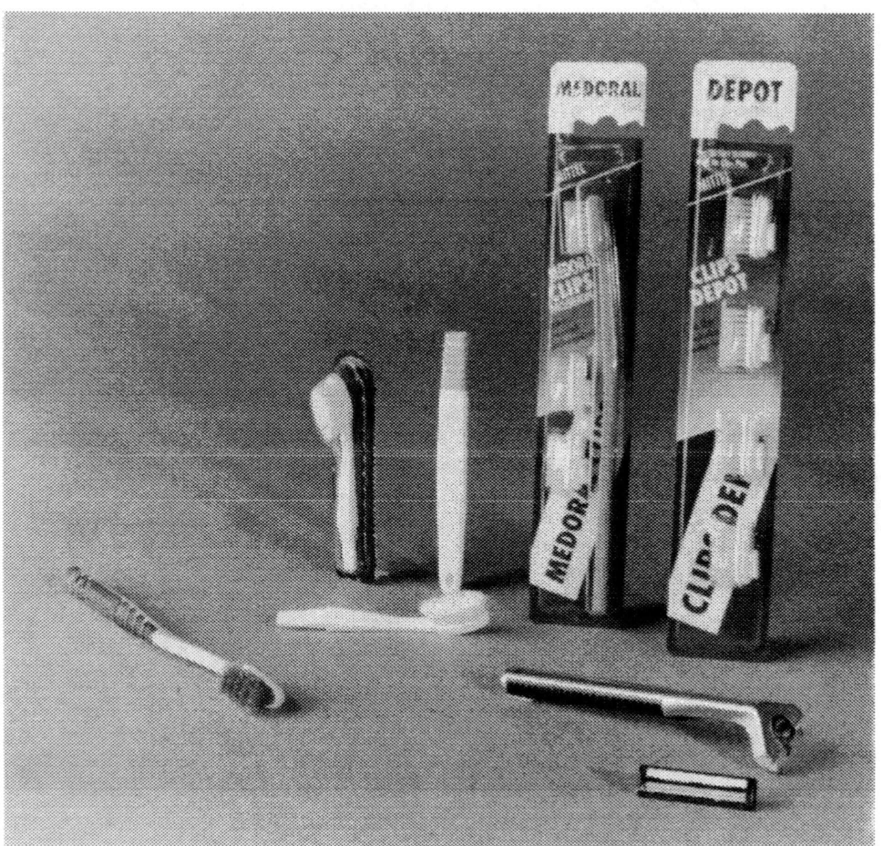

Figure 6: TOOTHBRUSHES WITH DIPOSABLE PARTS AND A RAZOR WITH A DISPOSABLE 'HEAD'

rescue. 'Gore-tex', 'Sympatex', 'Supplex' and 'Pertex' have all attempted to solve the problem. But, in 1996, a new product was designed (partly funded by the UK government) called 'Stomatex' which warms and breathes by following the natural design of plant stomata. Sadly, the copyright fee to nature will not be paid. On the other hand, the outdoor fleece jackets that Patagonia makes from recycled plastic bottles (as mentioned above) contain fibres that are small enough to feel soft to the touch.

Thus we see existing economic rules clashing head on with natural ecological principles. As business communities increasingly grapple with their understanding of natural systems, waste markets will develop alongside stock markets, and linear economic models will be replaced by cyclical ecologically focused models as we move from 'cradle-to-grave' to 'cradle-to-cradle' concepts. Waste stock exchanges have recently emerged on the Internet—thus encouraging the waste material flow to re-

Impact reduction: air-conditioning and CO_2

For all the talk of road traffic pollution, it is buildings that account for about half the total man-made CO_2 emissions. If, as seems likely, governments find themselves having to make a serious attempt to meet post-Kyoto emissions reduction targets, then the need to encourage natural forms of heating and cooling for living and working environments will be high on the list of priorities . . . it is quite possible to achieve a comfortable internal environment through careful design and the use of natural materials (Cottam 1998).

. . . and new patterns of thought?

Many very exciting and innovative rooftop ideas have already been generated such as solar power, photovoltaic roofing tiles, rainwater collection, living/growing roofs and so on. But the roof is not where the new innovation will happen; the roof is the least interesting place to think about in the high-rise concrete forests currently growing at an alarming pace in tropical zones such as the Asian Pacific. Perhaps new construction ideas could contribute yet more if we focus on the most significant surface areas: not on the roof, but on the sides or walls. The 'Twin Towers' in Malaysia, completed in 1996, reach up to the sky with 88 floors and end 451.9 m from street level. This means that the walls present a huge surface area; and here lies the creative opportunity. The primary energy on earth is produced by green tissues in plants. If such photosynthesis systems could be incorporated within construction materials in order to produce energy and/or biomass, could it be possible to produce energy and biomass just from sunlight, water and the CO_2 in the air?

Scientists have recently developed a chlorophyll-like substance, which, if closed between two sheets of glass, produces roughly the same amount of electric power as silicon-based photovoltaic cells when exposed to sunlight. This new system is at an early developmental stage, but can possibly produce much cheaper electricity than silicon photovoltaic cells.* Blue-green algae are also bred in huge glass tubes or plastic bags producing high amounts of biomass. A constant flow system, incorporated within the panels of a skyscraper, would provide a large amount of biomass, which could be used either for nutrition, bio-fuels or sugars, for the raw material of future plastics, and fix CO_2 in our cities at the same time. Airborne pollutants could be filtered out by air-conditioning systems supplemented with suitable filters. **Living coatings** could also be introduced onto concrete to produce an ***e+*** effect.

Shifting design principles

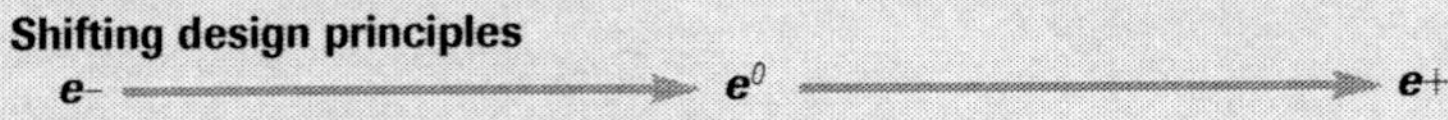

* Open University, 'Renewable Energy', UK BBC 2, 27 November 1996.

Box 4: ECO-INNOVATION: ROOFS AND BUILDINGS

WHY ARE WASHING MACHINES unable to filter out detergents rather than releasing such chemicals directly into water systems? Such filter systems to collect household water from baths and sinks—and potentially from washing machines and dishwashers—were recently reported (Stansell 1997). Products that produce cleaner output water than the input source are possible; such a washing machine would be a good example of an ***e+*** product. Hoechst AG, for example, claims that the cooling water from the River Main which it has to clean prior to use is returned cleaner than it formerly was. Water is certainly precious; for Europe as a whole, 53% of water abstracted is for industry, 26% for agriculture and 19% for household use—and one-third of European countries have relatively low availability of water, i.e. less than 500 m^3 per person per year (Industrial Environmental Management 1997). So, as water prices rise, we will re-examine this environmental resource 'product' and the purpose of the household or factory roof will change. People are not really sure what water costs, but at 60–70p per cubic metre compared to £2,600 for the same amount of beer, it seems good value!

Box 5: WASHING MACHINES AND WATER

enter our business systems rather than end in the ground or in the ocean. Following strict German legislation for refuse handling, recycling and incineration, recycling facilities have mushroomed within the last few years. This has created both change and opportunities for the refuse industry. Only a few years ago, refuse was transported around Germany, ultimately being exported or illegally dumped in order to dispose of it. However, operators are now actively seeking refuse waste for reprocessing, and a refuse war has emerged and disposal fees could drop significantly (*Der Spiegel* 1996). Resource exchange agencies will in the future be offering a profit to four distinct actors: those who get rid of their waste; those who buy in their recycled material; the recycling stock agent who charges a commission; and the environment. This kind of systems synergy will also influence future product design.

Conclusion

Many future products leaving the factory gates may actually generate environmental 'benefits'. A more detailed approach to the three steps to improve creative thinking is now given in Table 2. In order to generate change, it is important to see the product first by taking a closer look and breaking it down into its constituent parts. For example, we use washing machines that dispose of water and detergents into the drainage system and into the water supply, yet we could filter this out and recycle or dispose of it separately. We use toothbrushes that could be redesigned quite easily to reduce the disposable elements. We use plastic pens and highlighters when card or wooden versions could be made to do the same job, as well as creating a market for the current excess of waste paper. We could design cars to clean the polluted city air; we can design buildings that can contribute to pollution control and clean air production. Design is getting interesting and we have many examples of design for disassembly, design for recycling, for repairing, for reducing pollution and so on. But design for ***e+*** is new.

For every product and service, a market niche exists: opportunities for redesign with the environment in mind. In the future, there will be more focus on new eco-markets: shifting from simple environmental impact reduction to zero impact (zero emissions) and ***e+*** products and services. The whole basis of product or service design will be challenged as we look at new ways to contribute, in any way, towards removing and repairing environmental damage. 'Business-as-usual' will leave little environmental capital for the future, resulting in a bank account that is unlikely to sustain future generations. More eco-innovators will be needed by business, and managers will generate a greater understanding of innovator behaviours. A climate of creativity and innovation can be built into a new culture, particularly if we recognise that many barriers exist in the form of traditional mind-sets. New product and service drivers will come into play in eco-innovation. We argue that:

- Eco-innovation will generate new markets for waste.
- Nature is a powerful designing force.

Metaphor	PRODUCT	PEOPLE	BARRIERS
1. SEE THE PRODUCT JUST **LOOK!**	What does it look like now? What is it made up of? Is each component ***e+*** or ***e–***? How environmentally unfriendly is it? Use the ABC–XYZ analysis	Audit Product analysis Deconstruct Observe Collect information Watch people	No time to look Activity trap Environment as a threat to competition Can't see
2. REVIEW **CHALLENGE** ITS DESIGN	Deconstruct it! Does it have to be like this? Can it be different? What do we think about this? What are our ***e*** values at home with the family and our children? Transport these to work products	Challenging public and private beliefs and values Preferences Seeing it differently Visioning, fantasising, brainstorming Challenging existing thinking Creating new thought patterns. Take a risk: use imagination and humour	Resistance to change Environment as 'soft' Environment is not my worry It's not in my interest Yes but . . . Pessimism
3. **CHANGE** ITS DESIGN . . . OR THE PRODUCT	New products Natural designs Natural ingredients Natural payback Reducing ***e–*** Moving to ***e+***	New lifestyles Downshifters Teleworkers Job sharers Local social responsibilities *Your* business as a social force	Innovation lethargy

Table 2: THREE STEPS OF INNOVATION: SEE IT—CHALLENGE IT—CHANGE IT

Source: Beard and Hartmann 1997b

- **Solutions campaigning** is focusing entrepreneurial effort on eco-solutions.
- Creativity and innovation require art and 'play' as well as science and logic (left and right brain).
- Scientific and technical solutions can create dependency in people in the push for new solutions.
- The social sciences have much to offer the eco-innovation debate.

The redirection of entrepreneurial spirit within business can create a new social force: looking after the earth's natural capital for future generations.

10

Green Alliances

Environmental Groups as Strategic Bridges to Other Stakeholders

Cathy L. Hartman, Edwin R. Stafford and Michael Jay Polonsky

THE NATURE of business–environmentalist relations is changing (Lober 1997). Traditionally, environmentalists believed that the most effective means of enforcing corporate environmental responsibility was to adopt a singular, adversarial posture toward business (e.g. Murphy and Bendell 1997). In turn, firms viewed environmentalists as important stakeholders that needed to be considered, but kept at arm's length. While many firms still see environmental groups as a potential strategic threat, environmental groups can aid green marketing initiatives through various types of 'green alliance': collaborative business–environmental group partnerships that pursue mutually beneficial ecological goals (Stafford and Hartman 1996). In green alliances, environmentalists assist marketers *directly*, by providing expertise and technology, and *indirectly*, by influencing and brokering corporate relationships with other stakeholders to support the firm's overall green marketing programmes (Polonsky 1996).

This chapter proposes that understanding, encouraging and managing these indirect relationships created by environmental groups can build 'strategic bridges' between the marketer and other stakeholders for corporate benefit. Strategic bridging refers to situations in which a third party links diverse constituencies together to address some broad-based problems that require input from multiple stakeholders, such as corporate environmentalism (Westley and Vredenburg 1991). Consider the strategic bridges created by Greenpeace between its corporate partner and other societal stakeholders in its campaign against the use of chlorofluorocarbons (CFCs) in Germany.

In 1992, Greenpeace championed 'Greenfreeze', an environmentally friendly hydrocarbon refrigeration technology, as a substitute for Freon, a leading CFC damaging to the ozone (Beste 1994; Kalke 1994). The 1987 Montreal Protocol on Substances that Deplete the Ozone Layer mandated elimination of most forms of

CFCs by the end of the 1990s, and scientists from Greenpeace and the Hygiene Institute, Dortmund, collaborated to develop the hydrocarbon technology as a viable alternative (Beste 1994). Although this hydrocarbon technology had been around since the 1930s, appliance makers had not considered it due to its potential flammability. While modern refrigeration advances had eliminated this risk, major German appliance manufacturers were not interested in an old-fashioned, widely available technology that could not be patented to create a competitive advantage (Vidal 1992). Only the former East German manufacturer DKK Scharfenstein, later renamed Foron Household Appliances, was willing to consider experimenting with the hydrocarbon technology. However, unification had left Foron on the verge of bankruptcy. Foron, like many former East German firms, had been left with older, outmoded and less efficient technology which left them under-prepared to compete in a Western free-market system. As a result, the plant came under the control of the German privatisation agency, Treuhand. However, Foron's financial condition was such that, if investors could not be secured, the firm would be dissolved. With its engineers eager to save their jobs, Foron agreed to work with Greenpeace as a last resort to save its manufacturing operation.

After extensive talks in July 1992, Greenpeace granted Foron $17,000 to produce ten prototype hydrocarbon refrigerators. Racing against a liquidation timetable, Greenpeace and Foron fought a war of nerves with Treuhand who tried to block the project. At a stormy press conference showdown, Treuhand reluctantly allowed the Foron–Greenpeace project to proceed. The successful prototype, branded the 'Clean Cooler', not only won over Treuhand, but resulted in them providing Foron with substantial financial assistance and support in securing private investors. In March 1993, Foron's 'Clean Cooler', using Greenfreeze CFC-free technology, made its market debut (Walsh 1995).

To assist in developing the market for this product Greenpeace pre-sold the refrigerator to German consumers. This alarmed Western German refrigerator makers, who feared a substantial loss of market share. In an attempt to slow pre-production orders, these competitors launched a disinformation campaign through the trade press, warning retailers that Foron's Clean Cooler was 'an unacceptable danger in the home' and 'a potential bomb in the kitchen' and that Greenfreeze was 'energy-inefficient' (Vidal 1992). Greenpeace was charged with being irresponsible and obstructing constructive efforts to find feasible environmental solutions (*Air Conditioning, Heating and Refrigeration News* 1993). Greenpeace's grass-roots publicity and product endorsement, however, generated over 100,000 orders in less than a year in Germany alone (Vidal 1992). One by one, the negative charges were reduced or dropped as Greenpeace's advocacy motivated the scientific community to align with Foron and the Clean Cooler against the chemical lobby. Later, Foron's Clean Cooler won the German Environment Ministry's prestigious 'Blue Angel' award. By 1994, all German refrigerator manufacturers had either switched to Greenfreeze CFC-free technology or were planning to convert, fulfilling Greenpeace's environmental goal of eliminating CFCs in German refrigerators.

Greenpeace directly assisted Foron, by providing both technical and financial support for its activities. While the direct aid for Foron is noteworthy, it was

Greenpeace's indirect assistance, or strategic bridging, that proved indispensable to Foron. For the vulnerable Foron, Greenpeace brokered and negotiated a political *bridge* by influencing the Treuhand privatisation agency and a credibility *bridge* by influencing customers and the scientific community to support the Clean Cooler. Without Greenpeace's strategic bridging efforts, Foron would not have acquired critical stakeholder support and would have most likely gone bankrupt. This case demonstrates that environmentalist partners can become valuable bridging agents, providing firms the necessary sociopolitical linkages and credibility for green marketing programmes that appeal to consumers and society at large (Mendleson and Polonsky 1995; Stafford and Hartman 1996).

Discussing green alliances, Lober (1997) describes environmental groups as 'collaboration entrepreneurs', entities who play key roles in orchestrating co-operative environmental initiatives, selecting participants, and lobbying for participation. Environmental groups act as bridging agents by forwarding their own ends *while at the same time* serving as links between the marketer and other stakeholders (Westley and Vredenburg 1991). As such, strategic bridging enables the development of environmental solutions especially when diverse stakeholders are unable to negotiate or co-operate freely due to mistrust, tradition, logistic problems or when there is a need for a third party to restore a balance of power, resources and expertise (Sharma *et al.* 1994). In particular, green alliances facilitate opportunities for marketers to address or satisfy demands of multiple stakeholders simultaneously through the crafting of *enviropreneurial strategies*—entrepreneurial innovations that simultaneously integrate and satisfy environmental, economic and social objectives for corporate benefit (Menon and Menon 1997). Foron's Clean Cooler is an example of an enviropreneurial product that benefited the environment, the company and society.

This chapter describes how marketers can use green alliances to establish strategic bridges to relevant stakeholders and bolster green marketing strategy. The analysis begins by delineating how increasing global environmental demands on marketers have necessitated an expanded, multiple-stakeholder view of green marketing called 'enviropreneurism'. Green alliances are described as catalysts of enviropreneurship; in addition to offering marketers expertise and scientific technology (Milne *et al.* 1996), environmental groups can become strategic bridging agents, brokering and negotiating marketer linkages to relevant stakeholders to obtain resources, political support and market credibility (Westley and Vredenburg 1991). Freeman's (1984) stakeholder management process is extended to strategic bridging to uncover relevant stakeholders and critical 'gaps' between stakeholder and marketer interests that must be bridged for the attainment of enviropreneurial objectives. Analysis of the Foron–Greenpeace green alliance exemplifies the strategic bridging process, and the set of stakeholder strategies Greenpeace employed on behalf of Foron are described to illustrate how environmentalist partners can engage diverse stakeholders for marketer benefit. Despite their strategic opportunities, green alliances and strategic bridging pose challenges inherent in the collaboration of partners who traditionally share conflicting values (profit versus ecology), and emerging managerial, social and research implications for green alliances are discussed.

Enviropreneurial Marketing: A Multiple-Stakeholder View of Green Marketing

An expanded, multiple-stakeholder view of green marketing has emerged called 'enviropreneurial marketing strategy', defined as 'formulating and implementing entrepreneurial and environmentally beneficial marketing activities with the goal of creating revenue by providing exchanges that satisfy a firm's economic and social performance objectives' (Menon and Menon 1997: 54). While traditional 'green marketing'—the production, promotion and reclamation of environmentally sensitive products (cf. Boone and Kurtz 1998)—has been the primary marketing response to consumers' concerns about ecological issues, global pressures from other sectors of society have escalated in the 1990s and are now also motivating shifts in firms' environmental behaviour (Roberts 1996).

Poor corporate environmental practices, for example, commonly lead to activist demonstrations, consumer boycotts and embarrassing media coverage (McCloskey 1992). Social and environmental performance have a strong bearing on the desirability of a corporation's stock and its access to credit (Gallarotti 1995; Sarkin and Schelkin 1991). Some corporate polluters are encountering difficulties recruiting employees from an increasingly socially aware workforce (Fischer and Schot 1993) and, frequently, it is a firm's employees who are 'blowing the whistle' on their employer's environmental violations (Nixon 1993). Growing 'stakeholder activism' has made environmentalism a critical strategic issue for marketers in the global economy (Dechant and Altman 1994; Porter and van der Linde 1995), and improving stakeholder relationships and management are necessary for effective organisational performance across markets. Enviropreneurship provides an appropriate stakeholder approach for today's global environmental demands.

The term 'stakeholder' is being used increasingly in the literature (Donaldson and Preston 1995) and, while there is no universal consensus, one of the most widely accepted definitions is suggested by Freeman (1984: 25): 'any group or individual who can affect or is affected by the achievement of the firm's objectives'. Stakeholder theory posits that marketers should adjust their priorities to bring them in line with their stakeholders' interests (Hosseini and Brenner 1992; Roberts and King 1989; Rowley 1997). In order to develop effective enviropreneurial marketing strategy, a marketer must consider all of its stakeholders' interests and design environmental strategies that minimise stakeholders' potential to disrupt and maximise stakeholders' potential to support green marketing activities (cf. Atkinson *et al.* 1997; Harrison and St John 1996; Hosseini and Brenner 1992; Miller and Lewis 1991; Morris 1997; Savage *et al.* 1991; Polonsky 1995b, 1996).

Enviropreneurship adopts a proactive, entrepreneurial innovation and technological approach to environmental problems that accommodates or capitalises on other stakeholder needs to meet economic objectives. Emerging theoretical and empirical evidence indicates that environmental initiatives can lead to cost savings, increased profits and competitive advantages (Porter and van der Linde 1995; Russo and Fouts 1997). Menon and Menon (1997) assert that enviropreneurial activities can meet stakeholder interests at three levels:

- **Strategic level.** Strategic enviropreneurial initiatives reflect social responsibility and a desire to bring marketing activities into line with the expectations of current and future stakeholders. This may involve re-engineering corporate-wide activities for environmental sustainability. Marketers attempt to create a long-term, entrepreneurial prerogative through the development of new technologies, markets and products that create change within the industry or market.
- **Quasi-strategic level.** Quasi-strategic enviropreneurial initiatives are manifest in a desire to make corporate behaviour compatible with prevailing norms and expectations of critical stakeholders. Activities are modified in response to some immediate environmental need, and marketers attempt to integrate environmental concerns with strategy goals to achieve advantages within current businesses and markets.
- **Tactical level.** Tactical enviropreneurial programmes reflect social obligations to meet the minimum requirement to satisfy immediate market forces or legal constraints. Marketing mix variables are manipulated to position marketers/products as 'environmentally responsible', but they may provide limited environmental benefit.

While the specific approach (strategic, quasi-strategic and tactical) varies according to marketer, enviropreneurship is implemented so that the marketer considers how it can maximise the ability to address stakeholder interests (Menon and Menon 1997). Green alliances, such as the collaborative Foron–Greenpeace partnership, are catalysts of enviropreneurship.

'Collaborative Windows' and Green Alliances

Stakeholder collaboration is part of a new trend in environmental problem-solving that has unfolded in the 1990s (e.g. Long and Arnold 1995). Lober (1997) posits that this is the result of a 'collaborative window' that has opened in the global environmental arena where emerging problem recognition, public policy, organisational and social forces are fostering stakeholder collaborations to address ecological problems. 'Partnerships' between nations, government agencies and the public and private sectors are becoming commonplace (Milne *et al.* 1996). This is due to the realisation that environmental problems are complex, transcending governmental boundaries. Moreover, the disparity of power and expertise among international stakeholders requires collaboration in order to deal with problems effectively.

Global concerns, such as ozone depletion and climate change, have encouraged co-ordinated international plans of action from political bodies (cf. National Association of Attorneys-General 1990), international trade organisations (e.g. ISO 14000 series) and trade treaties (Levy 1997). Voluntary government programmes (e.g. the US Environmental Protection Agency's 33/50 programme) are stimulating over-compliance with environmental standards (Arora and Carson 1995) and flexible regulations (e.g. the 1990 US Clean Air Act's acid rain tradable credits programme) are encouraging firms to ally in order to develop and experiment with innovative enviropreneurial initiatives (Reitman 1997; Stewart 1993). The traditional 'command-and-control' character of environmental laws and regulations is shifting to market-based incentives and encouraging multiple-stakeholder collaborations (Hartman and Stafford 1997; Hemphill 1996). Further, private-sector social responsibility investment forums (e.g. Coalition of Environmentally Responsibility Economies [CERES]) are raising marketer accountability through environmental performance disclosure and institutional investor–environmental group collaboration (Wasik 1996). Thus, international political, social and market forces have created a 'collaborative window' for stakeholders to address global environmental problems.

Green alliances between marketers and environmentalists are perhaps the most unconventional outcomes of the 'collaborative window' because they involve formal co-operation between traditional adversaries (Hartman and Stafford 1997). Environmental groups commonly have scientific, legal and environmental expertise as well as public support for their activities (Lober 1997), and, as strategic partners, environmental groups can enhance the legitimacy of a marketer's enviropreneurial strategies. Greenpeace, for example, championed Foron's CFC-free Clean Coolers and used their network of support among retailers and final consumers to build pre-production market demand for the refrigerators. Green alliances also present opportunities for 'self-regulation' (Hemphill 1996), and environmentalists can facilitate negotiations with regulators and government agencies concerning corporate environmental activities, as demonstrated by Greenpeace's appeals to Treuhand on behalf of Foron. Thus, marketers are recognising that greater benefits can be achieved by working with environmentalists, rather than adopting a confrontational approach. Green alliances may bring about 'first-mover' benefits, both in markets and in addressing public policy concerns (Murphy and Bendell 1997). Likewise, environmental groups are increasingly turning away from traditional anti-business tactics for addressing environmental problems (Dowie 1995; Lober 1997; Mendleson and Polonsky 1995), although this shift in emphasis is not always embraced by all parties on both sides.

Murphy and Bendell (1997) characterise this broader collaborative trend as part of a more extensive shift away from a 'blame-and-protest' culture towards a 'solutions' culture by environmentalists. The Foron–Greenpeace alliance marked the first time Greenpeace had backed a commercial product/technology and experimented with market-based principles to promote its agenda (Kalke 1994). Shortly after Greenfreeze's success in Germany, Greenpeace International's director Paul Gilding described the group's evolving environmental strategy, declaring, 'There's certainly a stronger predilection now for interfering in markets' (Levene 1994). In another statement, Greenpeace spokesperson Richard Titchen explained:

> We won't stop the actions that get much attention in the press and that have made Greenpeace famous, but now that people and companies have become more conscious of environmental problems, we consider it more effective to demonstrate solutions that are actually viable to industry (*Business and the Environment* 1994).

Greenpeace has proclaimed it will 'create new alliances with sectors such as businesses and industries' (*Business and the Environment* 1994), advocating technological solutions to environmental problems (Corder 1997). Using the market system, corporate collaboration, green technology development and stakeholder relations, Greenpeace and many other major environmental groups are recognising that the success of their activities is dependent on how well they satisfy or accommodate the needs of various stakeholders in society, including businesses (Hartman and Stafford 1997; Lober 1997; Mendleson and Polonsky 1995). Green alliances allow firms to become more involved in environmental solutions. While some criticise the use of green alliances as a 'sell-out' to business interests (e.g. Dowie 1995), Jay Dee Hair of the National Wildlife Federation has framed his organisation's industry collaborations by saying, 'We're not selling out, we're buying in!' (Dowie 1995: 75).

Menon and Menon's (1997) enviropreneurism framework is useful for understanding how different types of green alliance can address stakeholder concerns. For example, green alliances can be created to identify and implement corporate-wide programmes on the *strategic* level to achieve environmental sustainability of the firm's practices and policies; 'cradle-to-grave' product stewardship and total quality environmental management programmes can address the expectations of current stakeholders and future generations. On the *quasi-strategic* level, green alliances can align corporate practices with current stakeholder expectations as in the Foron–Greenpeace alliance's efforts to market an ozone-safe refrigerator in accordance with the Montreal Protocol. On the *tactical* level, loose environmentalist–business linkages can be formed, such as corporate sponsorships, to meet immediate corporate citizenship obligations.

Strategic Bridging and Freeman's Stakeholder Management Process

For marketers, one of the key advantages of strategic and quasi-strategic green alliances is the bridging potential of their environmental partners to other relevant stakeholders. Sharma, *et al.* describe strategic bridging as:

> . . . characterized by the presence of a third party as a stakeholder, which is separate and distinct in terms of resources and personnel from the 'island' organizations it serves to link . . . Unlike mediators, bridges enter collaborative negotiations to further their own ends as well as to serve as links among domain stakeholders (1994: 461).

In green alliances, an environmental group works on behalf of its corporate partner, leveraging its environmental credibility and advocacy image to bridge other society stakeholders and advance its corporate partner's enviropreneurial activities along with its own ecological agenda. Bridging organisations, while diverse, hold a common vision toward solving problems in contexts characterised by high interdependence and turbulence (Brown 1991). Additionally, because bridging organisations maintain their independence, they can negotiate bilaterally with relevant stakeholders. This freedom allows bridging participants the flexibility and opportunity to develop interpersonal familiarity that may eventually break down social and institutional barriers that typically separate stakeholders. Brown observes:

> As a central actor among diverse constituencies, the bridging organization potentially has great influence over events. It can be a conduit for ideas and innovations, a source of information, a broker of resources, a negotiator of deals, a conceptualizer of strategies, a mediator of conflicts (1991: 812).

Through their strategic bridging capabilities, environmental groups can wield extraordinary influence among sociopolitical and economic constituents. Green alliances offer a form of marketer collaboration that is uniquely suited to the development of enviropreneurial strategies that take into consideration diverse stakeholders. However, similar to other types of alliance, it is essential that firms find the most appropriate environmental partner. Forming alliances with 'radical' environmental groups might fail to bring about the desired outcomes (Mendleson and Polonsky 1995).

Understanding who enviropreneurial stakeholders are, their stakes, and their perceptions of marketer activities is central to enviropreneurism and strategic bridging. Freeman's (1984) four-step stakeholder management process is a useful framework for evaluating domain stakeholders and bridging opportunities. The first step in the process is to identify stakeholder groups relevant to the firm's enviropreneurial activities. In undertaking this activity, it is important not only to consider traditional marketing stakeholders, (i.e. consumers, suppliers, employees, etc.), but also the broader range of external groups who may, in fact, have essential environmental information, expertise or motives that may assist or hamper the firm in achieving its wider enviropreneurial objectives (Polonsky and Ottman 1997; Starik 1995). Although specific stakeholders vary depending on the context, groups that typically need to be considered for enviropreneurial strategy include environmental groups, the local community, regulators and policy-makers, the media, shareholders and investors, employees, the academic/scientific community, and trade and industry (e.g. Azzone *et al.* 1997). As in the case of the Foron–Greenpeace situation, the network of relationships between marketers and their stakeholders *and* between various stakeholders may be extremely complex. Figure 1 illustrates the primary stakeholder relationships, their influences, and strategic bridges enacted during the Foron–Greenpeace alliance.

Foron had a broad range of stakeholders who had direct and indirect effects on marketing activities. Key stakeholders who directly affected Foron included the Treuhand privatisation agency and private investors, who controlled financial resources; Greenpeace, who held scientific expertise and some financial resources; and the scientific community, who wanted environmentally responsible refrigerators. There were also a number of important indirect stakeholders, including policy-makers and supporters of the Montreal Protocol, who gave Greenpeace the impetus to collaborate with Foron; competitors, who undertook a media campaign of disinformation and scare tactics to negatively influence appliance dealers and consumers; and the media, who actively promoted both sides of the debate from information

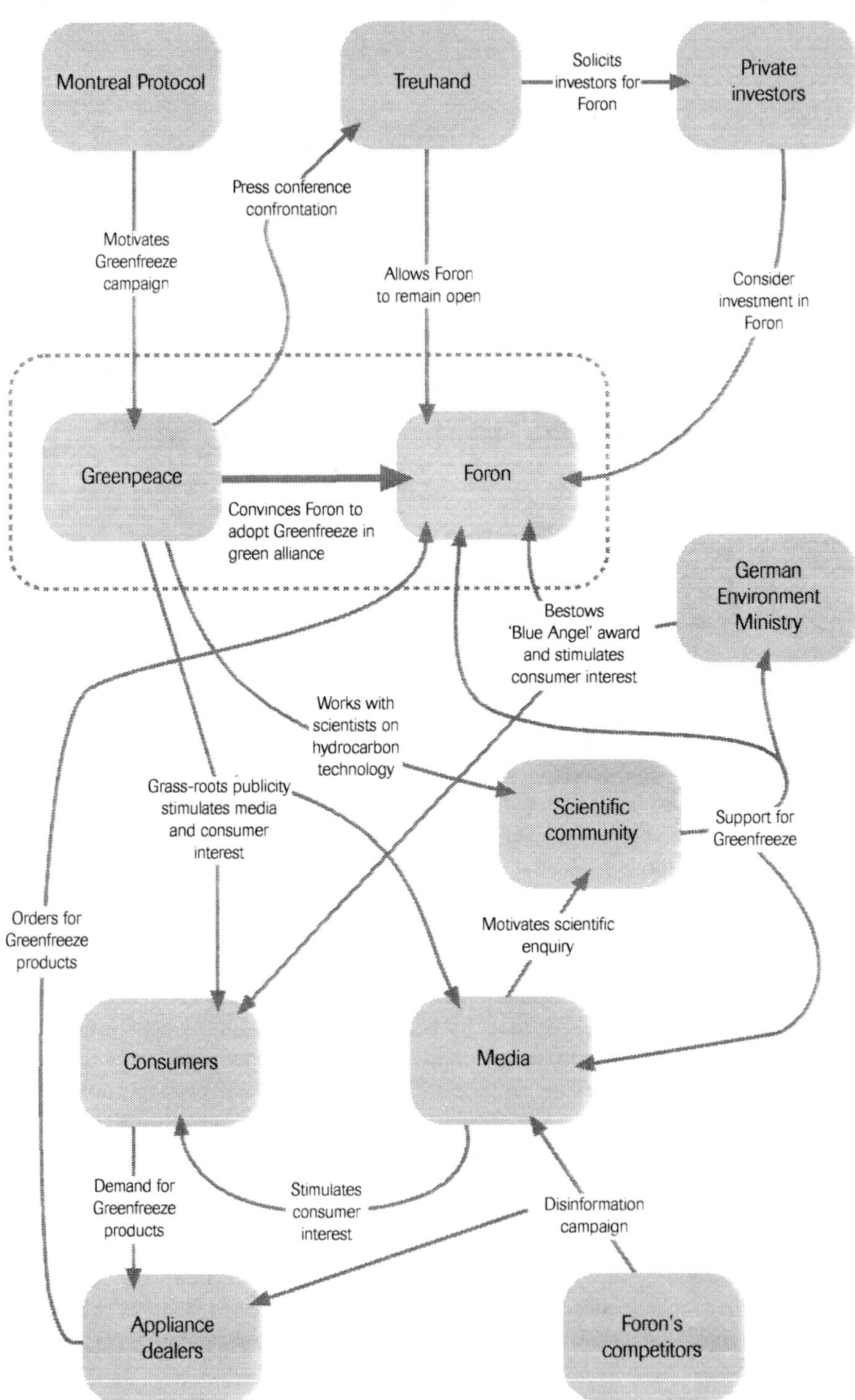

Figure 1: KEY STAKEHOLDER RELATIONSHIPS, INFLUENCES AND STRATEGIC BRIDGES

provided by various sources. As illustrated in Figure 1, the complex interactions among the marketer and its stakeholders highlight that there is an interconnected network of stakeholder relationships and potential influences in enviropreneurial activities (cf. Rowley 1997).

The second step in the process is to recognise stakeholders' interests and their ability to co-operate or threaten marketing outcomes. Thus, the firm must determine each group's stake and its importance to the success of its enviropreneurial marketing activities. In the Foron–Greenpeace alliance, the government had a high stake, for, without its state-funded support, even grudgingly given, the firm would have ceased to exist. Other stakeholders had significant indirect effects. Competitors, through the media, tried to scare off potential distributors and consumers, claiming that the new technology was dangerous. However, it should be noted that stakeholders' interests are dynamic. In the Foron–Greenpeace case, the media went from generally opposing Greenfreeze to supporting it. This shift only arose because of various indirect influencing forces, many orchestrated by Greenpeace. Thus, changes in stakeholders' behaviours or perceptions require monitoring and managing by the marketer and/or its bridging partner.

The third step of the stakeholder management process is to determine how well each stakeholder's needs or expectations are being met and its consequences (i.e. a gap analysis of marketer behaviour and stakeholder expectations). Such gaps represent stakeholder 'dissatisfaction'. Within the traditional marketing context, dissatisfaction by customers (one stakeholder) would result in a loss of sales or decline in performance. In enviropreneurship, there is greater potential for dissatisfaction because diverse stakeholders are considered in marketing programmes, and their interests and perceptions are likely to conflict. For the Foron–Greenpeace alliance, Greenpeace's actions to encourage Treuhand to extend state support to Foron conflicted with competitors' interests, and the threat of Foron's Clean Cooler sparked a disinformation media campaign in retaliation. Focusing on any one stakeholder may alienate another. Thus, enviropreneurship requires that marketers and bridgers understand and address complex stakeholder interests.

The last stage of the process is to modify marketer objectives, priorities and actions to consider stakeholders' interests. This activity is most critical, for it is where enviropreneurial marketing activities (strategic, quasi-strategic or tactical) can be used either to align green alliance activities with stakeholder interests or to constrain the stakeholder's ability to hamper marketing objectives. Various researchers have advanced a variety of strategies for firms to engage stakeholders directly, ranging from outright collaboration to confrontation to isolation (see Freeman 1984; Harrison and St John 1996; Polonsky 1996; Savage *et al.* 1991). In strategic bridging, however, partnering environmental groups may employ these strategies to engage relevant stakeholders on behalf of their marketer partners. Some of the stakeholder strategies employed by Greenpeace to support Foron are discussed in turn.

For example, Greenpeace used an adaptive strategy in working with Foron. The adaptive approach is where a strategic bridging agent attempts to modify its own behaviour or that of its marketer ally to meet another stakeholder's interests (Polonsky 1996). The financially distressed Foron adapted its product to meet both

Greenpeace's technological recommendations and the Montreal Protocol standard because it perceived this technology as a way of ensuring its viability. At the time, Foron was no more environmentally committed than other refrigerator manufacturers, but Foron engineers perceived the CFC-free Greenfreeze technology as a potential market opportunity. Eventually, after the Clean Cooler's success, Foron committed itself to a long-term enviropreneurial market position. In late 1994, Eberhard Gunther of Foron announced, 'We want to stand our ground with intelligent, innovative and, above all, ecological appliances' (Kalke 1994: 24). Energy-efficiency became Foron's enviropreneurial focus and, later, another innovative refrigerator design was developed which further reduced the appliance's energy consumption, thus making it even less environmentally harmful (Kalke 1994).

Greenpeace also used an aggressive strategy among some stakeholders. This is where a strategic bridging agent attempts to change other stakeholder's views directly or the other stakeholder's ability to influence the firm's marketing outcomes (Polonsky 1996). Examples of activities that could be used in an aggressive strategy include: public advocacy, educating target groups, or organising protests. Each of these may have varying levels of effectiveness in changing a stakeholder's beliefs and/or behaviours and will depend on the stakeholders' interests in relation to the specific situation under consideration.

When Treuhand attempted to block the production of the ten prototype hydrocarbon refrigerators, Greenpeace orchestrated a press conference showdown. This aggressive action forced Treuhand to allow Foron the opportunity to continue its operation. Greenpeace's aggressive action directly challenged the authority of the privatisation agency. While Greenpeace's aggressive approach was successful in making Treuhand rescind its dissolution decision, it may *not* have facilitated a change in Treuhand's mind-set. That is, it is unclear if Treuhand believed the Foron–Greenpeace partnership would result in desirable outcomes for the firm or environment. The use of an aggressive approach to change stakeholders' beliefs may be more successful when bridging groups assume a formal public advocacy role rather than an overtly antagonistic one. Changing a stakeholder's beliefs may take more time than changing an immediate behaviour.

When a bridging group decides a stakeholder belief change is warranted, it may use an aggressive approach in an indirect manner (i.e. by influencing other stakeholders to exert pressure) rather than a direct manner. For example, prior to the Foron–Greenpeace alliance, Greenpeace appealed directly to a number of German appliance manufacturers to consider Greenfreeze. After they refused, Greenpeace allied with Foron, demonstrating that Greenfreeze could be developed into a viable product. Greenpeace then attempted to change the industry's perceptions about Greenfreeze by appealing to consumers and the media, who, in turn, engaged the scientific community to support the technology. These actions indirectly influenced the industry eventually to accept Greenfreeze. Thus, Greenpeace's aggressive strategy was enacted indirectly, stimulating consumer demand and changing media disclosure, which, in turn, pressured industry to modify its beliefs and behaviour.

Before the Foron–Greenpeace alliance, Greenpeace had employed a co-operative strategy with scientists at the Hygiene Institute to develop the CFC-free Greenfreeze technology (Beste 1994). A co-operative approach is where the strategic bridging agent works with a stakeholder to achieve a desired set of outcomes (in this case, an alternative refrigeration technology to meet the Montreal Protocol). Co-operation insures that the interactive nature of the marketer and its stakeholders is recognised (Polonsky *et al.* 1997; Rowley 1997), and solutions can be designed as long-term strategic adjustments. Greenpeace's Greenfreeze campaign co-ordinator, Wolfgang Lohbeck, had worked as a department head in Lower Saxony's Ministry of Science before joining Greenpeace (Beste 1994). His personal connections within the scientific community ultimately helped to leverage its support for Greenfreeze, confirming its environmental superiority and safety. This bridge, in turn, encouraged consumer demand and government endorsement for Foron's Clean Cooler.

Risks of Green Alliances and Strategic Bridging

Despite advantages, environmental groups engaged in green alliances as strategic bridging agents present many challenges both to themselves and to their stakeholders. Sharma *et al.* note that groups serving as strategic bridging agents between the firm and its other stakeholders need 'to obtain "back-home" commitment from its constituents—because it remains at all times an independent entity with its own agenda' (1994: 461). That is, the bridging group's stakeholders must support the bridging group's activities. In this way, environmental groups must find mechanisms to integrate other stakeholders who may possess divergent values, interests, wealth, power, culture, language and structural characteristics (Brown 1991). The more divergent stakeholders' interests are, the more difficult the bridging problem may be. Further, members of the environmental group must understand the diverse stakeholder perspectives that its leaders are trying to integrate; any internal group contentions that arise from enviropreneurial programmes which represent 'compromised' solutions to broader environmental problems will seriously weaken an environmental group's strategic bridging ability (Westley and Vredenburg 1991). For a bridging agent to be successful, it must effectively market its partner's cause and become an advocate for the specific enviropreneurial marketing activity to its members, without tarnishing its own reputation. Westley and Vredenburg report: 'Agreements between environmentalists and business are often rejected by the environmental groups themselves, thereby reducing confidence in the process of collaboration' (1991: 72).

This bridging challenge is highlighted in a Canadian green alliance, initiated in 1989 between Pollution Probe and grocery retailer Loblaws. The environmental partner attempted to link environmental stakeholders and consumers to the retailer via a green product endorsement. Many of Pollution Probe's staff and membership disagreed with the endorsement, and Greenpeace questioned publicly the 'environmental soundness' of the endorsed products. The media frenzy that followed damaged Pollution Probe's credibility and its ability to bridge necessary stakeholders (see Stafford and Hartman 1996; Westley and Vredenburg 1991). In sum, strategic bridging agents face extraordinary leadership and management challenges (Brown 1991).

Marketers also face risks. Collaboration with environmental groups can offend core customer segments if they view environmentalists negatively as extremists (cf. Rohrschneider 1991). Moreover, green alliances can propel marketers and their enviropreneurial programmes into the public spotlight, encouraging scrutiny from other critical or threatened stakeholders, as exemplified by Foron's competitors' reaction to Greenfreeze's technological challenge. Central to the success of strategic bridging is the compatibility of co-operative objectives and vision between the marketer and the environmental group (cf. Brown 1991). Over time, as partners' objectives are met, or as they change or diverge, the strategic bridging partner may become less willing to broker and negotiate linkages between the marketer and other domain stakeholders, potentially jeopardising the marketer's competitive advantage (Westley and Vredenburg 1991). Consider the events following the successful product launch of Foron's Clean Cooler.

By 1994, Foron and Greenpeace were not surprised when West German appliance manufacturers switched to the hydrocarbon technology (Kalke 1994). For Greenpeace, the industry's adoption of its Greenfreeze technology was the realisation of its primary campaign objective of eliminating CFCs in German refrigerators. With the marketing experience gained from the Foron alliance, Greenpeace introduced the hydrocarbon technology to China, India and other developing countries (Beste 1994). Greenpeace literally gave the technology to willing enviropreneurs, convinced that Greenfreeze, if readily available, would be adopted widely in the developing world.

For Foron, however, the German industry's adoption of Greenfreeze presented a grimmer marketing reality. Being the first to develop the technology commercially did not lead to Foron developing a long-term competitive advantage, as the technology was not exclusively held by them. This resulted in the other, better-resourced firms marketing their versions of environmentally responsible refrigerators more effectively. Thus, despite its initial success, Foron's line of Clean Coolers did not rescue the firm from its financial crisis, and Foron's market share eroded as more sophisticated, rival hydrocarbon refrigerators appeared on the market. In 1995, Samsung entered into negotiations to buy the company, but bowed out after six months, citing that Foron did not fit with its planned European strategy (*Handelsblatt* 1995). Shortly after, Koc of Turkey began acquisition negotiations, only to withdraw the following year due to Foron's poor sales and financial situation (*Handelsblatt* 1996). Greenpeace had already abandoned the company to concentrate on its global Greenfreeze campaign, and Foron lacked the financial resources and marketing know-how to establish itself independently. In March 1996, Foron declared bankruptcy (*Die Welt* 1996).

The aftermath of the Foron–Greenpeace alliance illustrates a key risk associated with strategic bridging. Because strategic bridgers are motivated to collaborate and engage other stakeholders on behalf of their corporate partners through forwarding their own agenda (Westley and Vredenburg 1991), once that agenda is met or is no longer being served through collaboration, the bridger will no longer be committed to its corporate partner. Understanding the nature of strategic bridging through green alliances warrants consideration of partners' values and goals (cf. Rokeach 1973). Fundamentally, environmentalists and businesses hold diverging, if not conflicting, goals and/or even values. For environmentalists, ecological goals are foremost, whereas, for businesses, profit and market objectives are paramount. In general, green alliances represent lower-level instrumental (or means) values for their partners—potentially desirable mechanisms for achieving each individual partners' goals (cf. Lober 1997). Thus, there is frequently little long-term commitment by the either partner to the other or the other's agenda. In these situations, the environmentalist partner is likely to exit the relationship when the alliance no longer meets its ecological goals. If this breakdown occurs before the corporate partner has achieved its objectives, linkages with critical stakeholders may be placed at risk, as the firm no longer has an effective bridge with them.

Via the Foron–Greenpeace alliance, Greenpeace had introduced an alternative refrigerator technology throughout the German refrigerator industry, which had been opposed by the industry. Therefore, it no longer needed to continue supporting Foron, even though Foron had not resolved its financial dilemma. Though speculative, it is conceivable that Greenpeace *would* have assisted Foron in bridging and procuring investors if establishing consumer acceptance for Clean Coolers had taken longer *or* had the industry delayed its adoption of Greenfreeze; either scenario would have required Greenpeace to continue helping its cash-strapped partner by bridging necessary stakeholders, including investors, for product success. Wolfgang Lohbeck of Greenpeace observed, 'It was a piece of luck that we could win one company over to our way of thinking and that this firm could turn facts quickly into marketable realities' (Beste 1994: 29). Perhaps Foron's products were too successful in that their immediate market demand signalled Greenfreeze's market opportunities to competitors and constrained the time Foron needed to leverage Greenpeace's bridging capabilities to investors to remain competitive. After the industry-wide conversion to Greenfreeze, assisting Foron was no longer instrumental to Greenpeace's agenda.

Strategic bridging in green alliances, therefore, creates challenges both for environmental groups and for marketers. As a strategic bridging agent for a marketer, an environmental group must broker linkages among other diverse stakeholders to support its corporate partner's enviropreneurial strategies. At the same time, the environmental group must ensure that its bridging activities do not jeopardise its credibility and thus its support from its stakeholder network (Westley and Vredenburg 1991). Because green alliances represent only instrumental mechanisms for partners to achieve their dissimilar terminal goals (ecology versus profits), a marketer needs to monitor its own versus its environmental partner's progress toward goal fulfilment. Should an environmental partner meet its objective and exit the relationship, it could leave the marketer unprepared and vulnerable.

Managerial Implications

Green alliances are a new strategic domain for most marketers and environmental groups. Two key issues appear to warrant managerial consideration, market positioning and inter-partner relationship management (Mendleson and Polonsky 1995; Stafford and Hartman 1996). Market positioning of a green alliance requires attention to building legitimacy of what society may view as an unorthodox relationship. Environmental groups are traditionally viewed as non-profit 'watchdogs' of the corporate world, and close business ties can undermine public trust of the group's advocacy role (Murphy and Bendell 1997). Thus, preserving the integrity of the environmental partner's reputation is critical for bridging relevant-domain stakeholders (cf. Stafford and Hartman 1996; Mendleson and Polonsky 1995). Green alliance partners need to bridge the media, disclosing negotiations, precepts for deliberations, designs of co-operative programmes and updates on progress. Engaging the media and allowing stakeholder scrutiny contributed to the Clean Cooler's success.

Partners need to be prepared to deal with external threats (Brown 1991). To marshal support, enviropreneurial strategies need to be founded on specific ecological objectives respected by both the environmental and scientific communities. Bridging the scientific community was a significant success factor for the Foron–Greenpeace alliance and, later, Foron committed to produce 'energy-efficient' appliances as its enviropreneurial objective (Kalke 1994). Further, partners need to maintain their 'independence', taking individual responsibility for their own expenses and maintaining the right to disagree publicly with one another (Stafford and Hartman 1996). Working to maintain the environmental partner's credibility throughout the collaborative process is critical to the environmental group's ability to broker and negotiate linkages to relevant stakeholders and support the alliance's social legitimacy.

Relationship management between the environmental group and marketer also requires attention. Organisational culture differences between partners commonly disrupt the cultivation of inter-partner trust and relations (Milne *et al.* 1996). Milliman, *et al.* assert that both partners need to be adaptive and flexible.

> For corporate officials, this means trying to understand the emotional and spiritual views of environmental groups. For environmental groups, this means having the willingness to temper their idealism and moral convictions to pursue cooperative, often compromised, solutions (1994: 43).

The willingness to share information and communicate fosters mutual understanding and inter-partner respect. Personal relationships across partners facilitate cohesive linkages and commitment for continued dialogue and collaboration (Stafford and Hartman 1996).

Social Concerns with Green Alliances

Beyond the managerial implications involved in green alliances, a number of broader social issues warrant consideration. Foremost is the need to ensure that environmental groups are not perceived as 'caving in' to business interests (e.g. Dowie 1995) and thus seen no longer to be objective in relation to given environmental issues. Environmental group initiatives with business partners are typically expensive and time-consuming, and the results for the environment usually represent compromised, incremental progress rather than significant changes in corporate environmental behaviour (Murphy and Bendell 1997). Adversarial protests and tactics that rally stakeholders against environmentally detrimental corporate practices can frequently instigate sweeping changes, as exemplified by the Dolphin Coalition's consumer boycott of tuna that catalysed industry-wide 'dolphin-safe' fishing practices (Stafford and Hartman 1996). Are the traditional campaigns of protest becoming obscured by closer business ties? Environmentalist enthusiasm for green alliances may be sidelining other priorities aimed at broader society, such as consumption reduction, product re-use and recycling.

Another concern centres on whether a partnering environmental group should be playing the role of a 'free' environmental consultant. As noted earlier, the incentive of an enviropreneurial green alliance strategy for a marketer is either to improve internal efficiencies or to develop differential advantages over competitors (cf. Lober 1997; Menon and Menon 1997; Porter and van der Linde 1995). Is it appropriate for a non-profit environmental group, who may rely on grass-roots donations from the general public (consumers), to support just *one* marketer in a green alliance? Are the broader environment/societal benefits resulting from green alliances outweighing the potential marketer advantages (e.g. price premiums) that may be gained over the general public?

Lastly, there is the issue of transparency (Murphy and Bendell 1997). The co-operative relationship between a marketer and its environmentalist partner requires a degree of confidentiality. Many firms have misgivings about opening their operations and marketing processes to environmentalists for fear that information may be used against them if the green alliance fails. Agreements on confidentiality are necessary to elicit marketer participation, but they may conflict with the environmental group's accountability to its membership and its 'watchdog' role in society. With increased emphasis on green alliances, 'there is a danger that the environmental group may begin to adopt the kind of paternalism normally associated with "establishment" institutions' (Murphy and Bendell 1997: 129).

Conclusions and Future Research Directions

This chapter has examined the stakeholder dynamics and strategic bridging strategies Greenpeace employed on behalf of its green alliance partner, Foron, for the marketing of an enviropreneurial product. Green alliances between environmental groups and marketers are outcomes of a 'collaborative window' that has opened in the global environmental arena where emerging problem recognition, public policy, organisational and social forces are encouraging stakeholder collaborations for solving complex ecological problems (Lober 1997). As demonstrated in the Foron–Greenpeace alliance, environmental group partners provide marketers with (1) scientific, legal and environmental expertise and (2) strategic bridges by negotiating and brokering linkages to relevant stakeholders to procure necessary resources, legal support and market credibility. Strategic bridging, however, poses significant challenges for green alliance partners that warrant careful monitoring of goals, strategic positioning and relationship management.

While stakeholder paradigms are still emerging in the marketing literature (cf. Polonsky 1996), the increasing practice of enviropreneurship and green alliances warrant further consideration of stakeholder frameworks and concepts in marketing research (cf. Menon and Menon 1997). Polonsky asserts:

> Marketers have long realized the importance of monitoring the business environment in an attempt to identify shifts or gaps in the market. Stakeholder theory extends this proactive approach, suggesting that a broader set of influences should not only be monitored, but also included in strategy formation (1996: 228).

Bridging strategies (e.g. aggressive versus co-operative) warrant further investigation, delineating when they may be enacted, tactics for implementation and contingencies for effectiveness. While this chapter has limited itself to strategic bridging, future research needs to explore other aspects of green alliances, including their motivations, forms, success factors and environmental, economic and social outcomes (cf. Schot *et al.* 1997). Because green alliances are new phenomena, involving complex stakeholder, relationship and marketing processes, case research may be the most appropriate means for initial investigations of these issues (Lober 1997; Yin 1994). Considering the increasing acceptance among marketers and environmentalists to collaborate, researchers need to examine more fully collaborative processes if the enviropreneurial potential of green alliances is to be maximised.

11

How to Select Good Alliance Partners

Easwar S. Iyer and Sara Gooding-Williams*

BASED ON ABUNDANT scientific evidence and literature, our understanding of the ecological imperative has increased dramatically (Meadows *et al.* 1972; Carson 1962; Commoner 1972). However, the expertise required to promote these imperatives through specific strategies have not always been available within a firm and have had to be acquired from the outside. One approach to building these capabilities is to form partnerships and alliances with other organisations. Such alliances, while providing the much-needed technological or managerial expertise, also require learning new behaviours (Milne *et al.* 1996). Since there have been no models to guide the formation and management of alliances, they are often unsuccessful. To increase the odds of forming and managing successful alliances, a new framework or typology is needed.

Environmental organisations have been partnering other non-profits, for-profit businesses or governmental agencies in pursuit of their organisational mission. For that reason alone, we felt it was important to understand how these organisations conducted themselves and the manner in which they formed alliances. In this research, our goal is to create a typology of environmental organisations and describe an exemplar in each type so that potential partners understand the range of partners that existed. Our sample was drawn from the membership directory of the National Wildlife Federation (Gordon 1993).

The purpose of our research was to develop a typology of non-profit organisations that could be used to guide the search for alliance partners. To strengthen the

* The authors specially thank George R. Milne for his help at various stages of this project. They also acknowledge the assistance received from Bobby Banerjee, Mary Ellen Gordon, Rajiv Kashyap and Laurie Selvaggio. Partial funding for this project provided by the Department of Marketing at the Isenberg School of Management is also gratefully acknowledged.

proposed typology and its usefulness, this research combined two studies dovetailing into one another. The first study used a survey design; its primary purpose was to develop categories of environmental organisations. In the second study, key informants at selected organisations were contacted for in-depth telephone interviews.

The rest of this chapter will be divided into three parts. First, we will describe our survey and present the results of the data analysis that helped generate the typology of environmental organisations. Second, we will describe procedures involved in conducting in-depth interviews with key respondents from each organisational type. The verbal protocols derived from such interviews will be used to highlight key characteristics of each organisational type. Finally, we will discuss issues of managerial relevance and describe how our results could help in selecting appropriate alliance partners.

First Study

Survey Instrument and Procedure

The first study adopted a survey methodology since it allowed us to gather information from the entire universe of non-profits promoting environmental issues. Our final survey instrument was the result of numerous pre-tests and included items beyond those reported in this study. In general, we followed procedures recommended by Dillman (1978) and accepted as standard in the field. In the interest of brevity, we are not presenting the details of the entire procedure here, but the interested reader can obtain it elsewhere (Iyer and Gooding-Williams 1998).

We mailed our survey to 547 environmental organisations listed in the US National Wildlife Federation (Gordon 1993) and received 197 usable responses. Thirty questionnaires were returned by the post office and 22 respondents indicated their unwillingness to participate. We received five responses too late to be included in our analysis. This resulted in a response rate of over 40% (197/490), well over acceptable standards. Moreover, the high response rate allows us to generalise about the population with much greater reliability.

Data Analysis

We used a median split on the three organisational characteristics, i.e. its **age**, **size** and **number of alliances**, to create the eight-way categorisation.[1] The **age** of organisations in our sample ranged from four to 143 years old; the median age was 23 years, resulting in 104 organisations being categorised as young and 93 being categorised as old. **Size** of an organisation was measured in terms of its annual budget. Our sample included organisations with an annual budget as small as $500 and others as large as $725 million; the median was $500,000. Using a median split

1 The model was significant and explained between 30% and 57% of the variance. Details can be obtained from the authors.

Age of organisation	Old (>23 years)				Young (<23 years)			
Size of organisation	Large (>$500 000)		Small (<$500,000)		Large (>$500,000)		Small (<$500 000)	
No. of alliances	Many (>7)	Few (<7)	Many (>7)	Few (<7)	Many (>7)	Few (<7)	Many (>7)	Few (<7)
N	21	25	14	33	21	18	22	43
Mean no. of alliances (#)	36.4	3.8	385.7	6.1	97.2	3.7	37.6	3.1
Mean size ($'000)	20588.1	42284.6	169.5	165.1	32727.6	5558.3	128.6	115.9
Mean age (years)	54.1	61.6	34.8	39.8	11.8	14.7	11.9	14.6

Overall model significant: d.f. = 21,374; F = 9.79; $p < 0.0001$

Table 1: RESPONDING ORGANISATIONS' PROFILES ON INDEPENDENT VARIABLES

resulted in 112 organisations being categorised as small and 85 being categorised as large. The **number of alliances** formed by the organisation was measured through self-report. This measure had a wide range, with some organisations having participated in no alliances at all and some others that had participated in over 5,000 alliances. The median number of alliances in our sample was seven; based on this we categorised 119 organisations as having participated in few alliances and 78 as having participated in many alliances. A profile of the eight types of environmental organisation is provided in Table 1. It is important to note that cell frequencies across all the eight categories were quite even. At the low end, there were 14 organisations categorised as old, small and having many alliances and, at the high end, there were 43 organisations categorised as young, small and having few alliances. Across the eight categories, mean age ranged from 11.8 years to 61.6 years and mean size ranged from $115,900 to over $42.2 million. The mean number of alliances ranged from 3.1 to 385.7 across the eight categories. Thus, we identified eight types of organisation that varied on the three characteristics of their age, size and number of alliances.

Typology of Environmental Organisations

Our next step was to identify organisational characteristics that were unique to each type but disparate between different types. In other words, organisations would have to be very similar on certain characteristics to be grouped within one category, while simultaneously they would have to be very dissimilar from organisations in other categories. After careful consideration, we identified the following ten additional characteristics:

- Number of members in the organisation
- Strategic goals of the organisation
- Ability of organisation to attain these goals

- Workforce in the organisation
- Advertising options used by the organisation
- Cost of administration
- Sources funding the organisation
- Recipients of funds from the organisation
- Range of activities engaged in by the organisation
- Scope of activities engaged in by the organisation

We examined each of the eight types of organisation for the characteristics that were unique to that type. For instance, one type of an organisation may have had a large cost of administration, whereas another may have had a much lower cost of administration. As another example, one type of organisation may have engaged in heavy advertising, while another type may have spent little. Such differences are captured in our description of prototypical member of each type. We caution the reader that such typologies enhance recall, comprehension and communication, but also tend to mask differences (see Table 2 for a summary).

Type 1: Old Solicitors. This type of organisation is large and has many alliances. It has a very large base of paid employees as well as unpaid volunteers. It tends to have an international focus and more than one-half of its revenues are generated from membership dues or government grants. What really sets this type of organisation apart from the rest is the extensive attention devoted to advertising. It has, by far, the largest advertising budget, spending over ten times as much as the next-heaviest spender. This organisation is also the heaviest user of direct-mail advertising. It would also seem that it is very efficient in its resource utilisation in general, and advertising in particular. For example, its administration cost is clearly below average and its advertising–income and advertising–expenditure ratios are fairly close to average.

Type 2: Old Intellectuals. These are old and large organisations that have only a few alliances. Their primary goal is to increase public awareness about environmental issues. The most striking feature of this type of organisation is its largesse to universities; it is the single-largest contributor to university-based research. We think this tendency is a sign of the organisation's intellectual orientation. Moreover, this type of an organisation perceives itself as not attaining its lofty global goals very often or very satisfactorily, although it does accomplish many things at the local level. This, we believe, is quite symptomatic of an intellectual tradition, in which local goals may be attained but global goals may remain elusive. In other words, the organisation believes it wins many battles but the war is still being fought on many fronts. It is quite likely that this organisation may be finding global goals elusive and, in the intellectual tradition, displays dissatisfaction in not attaining them. All these suggested to us an intellectual orientation, and hence the title.

Type 3: Old Cosmopolitans. These are older organisations with many alliances, but are relatively small. They employ very few people and depend more on unpaid volunteers. In fact, the ratio of paid employees to volunteers is the smallest of any type of organisation. Although its advertising budget itself is relatively small, it represents a large share both of its income and of total expenditures. In fact, advertising–income and advertising–expenditure ratios for this organisation are the highest and are over twice the group mean values. This type of organisation also has the most international focus and the least regional focus. It does not depend on government grants at all; instead, most of its revenues come from membership dues and charitable trusts. At the other end, these organisations primarily donate to communities and charitable trusts. These observations lead us to name this type as 'cosmopolitan'.

Type 4: Old-Fashioned. These are older and smaller organisations that have few alliances. The mere fact that they engage in a low number of alliances ought to be sufficient to view them as old-fashioned, although they are not alone in that respect. However, there are many other characteristics that truly make them old-fashioned: for instance, their advertising budget is the lowest by far, and the little they spend on advertising is devoted largely to publications. Even more striking is the fact that they depend on membership dues almost exclusively; almost one-half of their revenue is generated through dues. At the other end, they give most of their money to individuals. This combination of factors—i.e. not having many alliances, not advertising, relying on publications for disseminating information, and primarily relying on membership dues as a primary source of revenue—typifies these as being old-fashioned organisations.

Type 5: Young Achievers. These are young organisations that are also large and have many alliances. They have the lowest cost of administration of all types of organisation. Moreover, they have the highest scores on goal attainment. They are highly satisfied with their achievement, mostly surpassing their expectations, and falling short of their goals least often. These noteworthy features make them 'young achievers'. They have a reasonable advertising budget and engage in a fair amount of direct advertising; in that sense, they are modern as well. The even spread in the various sources of their income tends to insulate them from sudden shocks. All in all, these are organisations that have clear goals, and they attain those goals in a relatively cost-effective manner.

Type 6: Young Entrepreneurs. These organisations are also young in age and large in size, but have formed very few alliances. In some ways, they are the opposite of the young achievers. They have the highest ratio of paid employees to volunteers; in fact, they are the only type to have more paid employees than unpaid volunteers. Probably as a result of that, their cost of administration is the highest among all groups. They exhibit very few other notable characteristics, but it is clear that these organisations are spending a great deal of money on salaries and administration. More likely than not, the bulk of their administrative expenditure supports their employees.

	Type of organisation							
Primary characteristics	Old Solicitors	Old Intellectuals	Old Cosmopolitans	Old-Fashioned	Young Achievers	Young Entrepreneurs	Young Grass Roots	Young Preachers
Number of members		Large membership		Small membership				Small membership
Strategic goals			Strong goal articulation		Influence business practice and consumers		Increase public awareness	
Goal attainment	Low alliance goal attainment				Alliances surpass goal expectations			
Workforce profile						High ratio of employees to volunteers		
Advertising profile	High advertising and direct mail expenditures		High advertising and in-house publications				High advertising expenditures	
Administration cost						High administrative costs		
Funding sources	Government grants		Individual contributions and charitable trusts	Membership dues	Government grants	Charitable trusts		
Recipients	Individuals and communities	Universities				Individuals		
Range of activities				Plant and animal life	Land and energy issues	Animal life	Water issues	
Scope of activities		National	International				Regional	

Table 2: TYPOLOGY OF ENVIRONMENTAL ORGANISATIONS: DISTINGUISHING CHARACTERISTICS

Type 7: Young Locals. These organisations are young and small and engage in many alliances. Their goals are largely to increase public awareness and to preserve natural resources. They have a very large base of unpaid volunteers that work to promote their agenda. This is reflected in a relatively small ratio of paid employees to unpaid volunteers. These organisations are among the least international and most regional in their focus: in fact, most of the issues addressed by these organisations tend to have a local flavour and interest. In very many cases, once the issue is redressed, the organisation either disbands or refocuses on other local issues. Our label reflects these organisations' tendency to be active primarily in the local community.

Type 8: Young Preachers. This type of organisation tends to be young and small with few alliances. Their goals are also to increase public awareness and preserve natural resources. However, they do not have such a large base of unpaid volunteers; rather, they tend to get their message across through advertising and publications. They invest a fair amount on advertising, largely devoted to publications. In fact, they are the second-largest users of publications as a means of disseminating their message. Their operational style and substance earns them their label as 'young preachers'.

Second Study

The first study resulted in the development of a robust typology of environmental organisations. To enhance an understanding of each organisation's self-perception, we conducted the second study. The purpose of this study was to get qualitative information from key members[2] of each type of organisation. To this end, we conducted three in-depth telephone interviews with members from each of the eight types of organisation for a total of 24 interviews.

Interview Procedures

We randomly telephoned organisations from our sample until three interviews from each of the eight types (T1–T8) were completed. Respondents were guaranteed anonymity for themselves and their organisation. The interviews ranged from 15 minutes to over one hour in length, averaging approximately 40 minutes. The telephone interview procedure was based on a loosely structured, open-ended format that allowed respondents to speak freely about their experiences with alliances over the past three years. Respondents were asked to describe how alliances were initiated, what their organisations looked for in an alliance partner, whether and why they found alliances useful or problematic, and the effectiveness of the alliances in which they participated. Though informant responses were often coloured by personal opinions, most were very familiar with their organisation's experiences with alliances

2 Following standard research practices, these key members are henceforth referred to as 'informants'.

and able to recount detailed stories of alliance successes and failures. What follows is a detailed description[3] of how and why different environmental organisations act within alliances.

Type 1: Old Solicitors. These types of environmental organisation were older, with large budgets and extensive alliance experience and included such well-recognised names as the Nature Conservancy, Keep America Beautiful and the Cornell Laboratory of Ornithology. These organisations often seemed to have broad conservationist goals that were pursued through a wide range of projects. Even though members of this group were quite experienced with alliances, their overall satisfaction with alliances was the lowest reported by any group as the following informants noted:

> Alliances make life much more complicated. It keeps our lawyers busy writing contracts—and that takes time and money (T1.3).
>
> At the foundation of all this is that, if there is a degree of openness and trust, alliances can work well. Unfortunately, we've worked with other organisations that have tried alliances, but come in with a great deal of distrust. They feel we have ulterior motives, that we want to steal their members or take away their donors (T1.3).

These problems, in part, reflected the large size of these organisations; they appeared threatening to their smaller alliance partners. T1 organisations were most comfortable partnering with large national corporations that had a broad-based appeal.

Pursuing alliances was a crucial strategy for Type 1 organisations, since their goals were to have a broad impact through many diverse projects. Their funding patterns of widely spreading individual and community grants across many projects reflected their interest in serving as a catalyst in many different settings, and extending their national network of influence. As one informant commented:

> We work with many, many organisations, and we work with volunteers all over the country. We actively seek out all sorts of partners to extend our mission (T1.3).
>
> Partnerships help us not reinvent the wheel. We often adopt programmes that others have developed. Working with partnerships means we can do more; it means we can teach more people. Alliances open doors and lines of communications so we know more of what's out there and more about how best to work with others (T1.192).

Type 2: Old Intellectuals. This group of environmental organisations was older, with large budgets and limited alliance experience and included organisations such as the American Petroleum Institute, the American Farm Bureau Federation and the Population Institute. These organisations tended to have highly focused goals and a well-defined set of concerns, often with an emphasis on plant life and air pollution issues. According to our informants, their organisational goals were to conduct

3 Researchers often use the term 'thick and rich' to describe these kinds of detailed notes.

research and frame policy; this explained their strong ties to the university community. However, these goals often hampered alliance formation, which may in part explain why these organisations had relatively few alliances:

> We work through universities to provide fellowships and scholarships related to conservation activities. We generally don't approach other organisations to form alliances. We are a private family foundation, and our trustees don't allow any contracts with the government. And, unless there is a bequest, we don't huckster for support. We will occasionally work with other organisations on boards or commissions, but we have no formal relationships with any businesses (T2.81).

> We don't like to formally have alliances. We find we give up our identity within an alliance and then we have to start budgeting for the alliance. We are constantly in the eye of the storm in terms of coalitions approaching us, since we have such a large grass-roots advocacy base. Numbers do matter in public policy, and emerging groups that care about a single issue don't have the base we do, but we can't work with them (T2.177).

> We very rarely work with for-profit businesses. Usually, we'll only work with them on a research or educational effort, and usually we try to stick with a trade association for these projects (T2.177).

> We try to educate allied professionals and community volunteers about our issues . . . You can think of it as a kind of technology transfer, where we know what to do and want to tell others. Extending our reach is the main goal of our organisation, so these alliances really support these goals (T2.90).

Type 3: Old Cosmopolitans. This group of environmental organisations was older, with extensive alliance experience but relatively limited budgets, and is comprised of organisations such as the International Council of Environmental Law, the National Trappers Association and the American Association of Zookeepers. Type 3 organisations consisted of loosely structured networks of local chapters, each run relatively independently. Thus, most alliances were formed at the local level and were distributed unevenly among the local chapters:

> Our organisation has chapters spread out across the country. By and large, the chapters are on their own. The better chapters have several alliances each, but some chapters would benefit from information on other alliances: what successes there have been, what has been accomplished. This information is not well known (T3.95).

It was also clear from our interviews that budget limitations were of concern to this group. In particular, not having as much funding as other alliance partners was perceived to interfere with their ability to participate as equal partners in the alliance:

> Generally, groups that are better funded than their partners are difficult to work with. They are just not as open to other ideas and suggestion. The more money you have, the better your chances of winning the debate. This holds on a state, federal and international basis. Having money gives you the ability to get your message across (T3.161).

Type 4: Old-Fashioned. This group of environmental organisations was older, having limited budgets and alliance experience and includes many smaller local and regional conservation organisations, such as the Upper Mississippi River Conservation Committee, the Rye Nature Center and Association and the Society for Range Management. Type 4 organisations tended to emphasise local membership and conservation efforts. A typical goal for these organisations was to monitor local habitats and develop grass-root support for conservation efforts, as described by this informant:

> Say we have an endangered snail in a particular habitat. Our group would identify the existence of this endangerment and notify the government to take action to protect this creature. Our organisation alerts the government to issues. Within our group we have about a dozen 'watchdogs.' Each one stays alert to identify issues and alert leadership so they can take appropriate action—such as a letter-writing campaign (T4.91).

Another important goal of Type 4 organisations was education, and this often required forming alliances with local school systems:

> Environmental education is another area where we work with many other organisations. Our alliances are basically with other environmental organisations, but we need others (like the local school system) to meet our goals. We work with 1,000 teachers in our watershed, and send a newsletter to them, helping them with their teaching techniques. This year we're writing about professional profiles of biologists, etc. as role models for students (T4.124).

To combat the limited budget and heavy dependence on membership dues for revenues, Type 4 organisations were perpetually looking for new funding sources in a sparse and competitive milieu. They often formed alliances with like-minded organisations on well-defined issues in order to compete effectively with larger organisations, as stated by this informant:

> We have been forming alliances to pursue fundraising. The foundations and corporations that will fund—they fund programmes which need to have consistent messages and, goals. They are more likely to fund you if you are working with other organisations on specific projects (T4.124).
>
> Alliances are useful to multiply knowledge of the few staff that we have and to leverage the few resources we have to reach a critical mass that can attract other funding sources (T4.149).

Type 5: Young Achievers. This group of environmental organisations enjoyed large budgets and had extensive alliance experience, yet was relatively young. Type 5 included organisations such as the Rainforest Action Network, the American Council for an Energy Efficient Economy and Wetlands for America. These organisations pursued strategies to influence business practice and consumer behaviour and relied heavily on government support. Most importantly for our study, these organisations reported a much clearer pattern of alliance success than any other group. These organisations approached the challenge of changing behaviour in innovative ways, using novel incentives for consumers and businesses, as the next two examples of alliance activity reveal:

> We work with two yacht manufacturers in supporting a fish release certificate programme. It's a voluntary programme sponsored by the yacht companies to get anglers to release certain types of fish. The angler fills out a release form and sends it in, and then gets a parchment certificate, suitable for framing, recognising their assistance in fish conservation (T5.86).

> We were part of developing the Super-Efficient Refrigerator Programme, which was developed in collaboration with two dozen electric utilities, two non-profits, US EPA, US Department of Energy and major appliance manufacturers. These manufacturers competed to make super-efficient refrigerators—without CFCs. The utilities put up a bounty for the winner. The programme is working very well (T5.33).

Type 5 interviewees also highlighted the importance of institutional goal alignment, integrity and commitment of resources:

> The alliances are most effective when you're working with people with similar ideas and goals . . . But if people don't fulfil their end it becomes very difficult. You end up delaying or the other organisation ends up doing its work (T5.87).

> What makes for an effective alliance? It comes back to what's internal to the organisation. It's important to have institutional continuity. If we were a flighty crew, the possibility of alliances would be far less likely. Institutional consistency is a must for having an alliance. Also, we are unlikely to pursue alliances with groups we find shaky (T5.33).

Type 6: Young Entrepreneurs. This group of environmental organisation had relatively large budgets, but was younger and had limited alliance experience. Typical organisations were the Pacific Whale Foundation, the International Crane Foundation and the Committee for the National Institute for the Environment. Type 6 organisations tended to represent specific interests in animal conservation and depended heavily on paid employees rather than on volunteers to achieve their objectives.

This group was highly selective in choosing partners, only entering into alliances under exceptional circumstances. Often these opportunities were revenue-enhancing, providing opportunities to partner with business in a way that helped the environment. Informants were very current in their knowledge and often mentioned opportunities in marine eco-tourism, foreign agricultural development, international distribution services for eco-friendly products and database management that provided revenue for their organisations. One of the reasons for their caution in forming alliances was their belief that alliances represented a drain on their resources.

> It takes a great deal of effort to make alliances work. This effort causes a decrease in research and education programmes. We have to balance how

> much effort to devote to the business aspects of our work so we can do our non-profit projects (T6.51).

> We work with other NGOs on an informal basis, many NGOs, but we don't have formal relationships. Our problem with alliances is that we have a small staff. Our approach is to be catalytic in getting others to identify problems and follow through with the work (T6.39).

> Its too much work to be part of these groups. It takes energy and takes away from our primary mission. We don't have the staff to join these alliances, but more, even if we had the staff we believe you don't build a democracy from the top down, you need grass-roots organising to re-engage people in democracy (T6.142).

Type 6 organisations faced an odd dilemma caused by their relatively large budgets: they had to put on the appearance of a struggle in order to continue raising funds at the high levels they did. So, although Type 6 informants spoke like savvy business people, they felt pressured to cultivate an image that differed from standard business practice:

> The public expects non-profits to be wearing sackcloth outfits and to survive on rotting mangoes. If your office is too fancy, there is a suspicion of your credibility (T6.51).

Type 7: Young Locals. This group had extensive alliance experience, despite its relative youth and small budgets, and included organisations such as the Wildlife Center of Virginia, Great Lakes United and the Desert Turtle Preserve Committee, Inc. Our informants most often represented regionally focused organisations, with concerns for conservation and heightening public awareness about their chosen conservation issues. One way these organisations attempted to increase public awareness was through educational efforts; such efforts often required forming alliances. For example:

> It's easy to pick our best alliance. We developed a new type of education programme with the local National Forest. They have no environmental educators on staff nationwide, but they have a mandate to do education. We have staff, but no land. It's been a local process to work together. Around here we talk about it as a marriage made in heaven. We both get to fulfil our mission much more effectively than if we weren't working together (T7.13).

Besides direct public education, these organisations partnered with government agencies to raise awareness on important environmental issues within the government itself:

> Most of our partners are government agencies. We approach an agency that has some involvement with a particular species of animal on an ongoing basis. Sometimes they approach us for 'public input' that they need as part of their mandate. Sometimes we work with them on an ongoing basis, and have formal agreements to do so (T7.107).

Type 8: Young Preachers. These organisations were younger, with relatively limited budgets and alliance experience and included a wide assortment that was the most difficult to characterise as a group. Organisations such as the American League of Anglers and Boaters, the Raptor Education Foundation, Inc., the Jackson Hole Land Trust, the Population Environment Balance Group, the International Bicycle Fund and the American Canal Association make up this diverse group. We found it particularly difficult to contact these organisations, since several did not even have full-time staff to provide phone coverage. This deficiency seemed to account for their lack of involvement in alliances. As one respondent put it:

> The size of the organisation makes a difference. The number and quality of the staff make a difference in what you can do. A good example of an alliance is the Environmental Defense Fund. They worked with McDonald's Corporation to get some important things done. But you need staff—trained people to make that happen. They have PhDs and lawyers, and we don't. You need the staff to provide the follow-through (T8.15).

In addition, this type included very small grass-roots organisations committed solely to representing their constituencies, rather than furthering a specific environmental agenda. These goals precluded formation of many alliances, not only with business, but also with larger, more 'mainstream' environmental organisations, as the following informant revealed:

> We don't want to be the sole deciders of the environmental and social justice goals. We would like to impart skills and information back to the people affected. We have this general negative attitude toward corporate sponsorship and getting involved with business. We came to the conclusion that even liberal guilt money from the foundation world is not totally clean. Its just the wrong side of the issue. We don't want to contribute to growing corporate power. Increasingly this is a mark of significant division between the big national environmental organisations and local groups that work for social justice and environmental justice (T8.16).

These organisations tended to spread their message through advertising and publications. They engaged in advertising as and when it was needed.

Discussion

The environmental organisations that we surveyed and interviewed represent a vast heterogeneous set, making the task of creating a typology very difficult. Within those limits, our two studies represent an attempt to generate a meaningful typology of what inherently is a heterogeneous collection of environmental organisations. Thus, we do not want to aver that these eight types exhaustively and accurately represent the entire universe of environmental organisations. However, we do believe that managers seeking partnerships with environmental organisations would do well to ascertain many, if not all, of the characteristics we have focused on in our studies. With knowledge of these characteristics, the odds of forging a successful alliance should increase considerably.

Any firm with a relatively large budget and an environmental agenda of animal preservation or such would do well to look closely at Old Cosmopolitans (T3). This alliance is most likely to succeed, since T3 organisations have a lot of energy but limited funds. As another example, if a business wanted to deal with a partner organisation espousing a professional culture, Young Entrepreneurs (T6) would be the obvious choice. However, given T6's guarded approach to forming alliances, the odds of forming a partnership may be small, but the odds of success once a partnership was formed would be very high. These are only two examples and, rather than list every possible alliance opportunity, we highlight the key characteristics of each type below.

Old Solicitors

- Impressive in their size, scope and number of alliances
- Potential problems related to their large organisational size
- Possible barriers to establishing trusting relationships with partners
- Perceive themselves as catalysts for change in the national scene

Old Intellectuals

- Can provide important resources to government, businesses and other environmental organisations
- Primarily concern themselves with clean water and wildlife issues
- Prefer alliances oriented towards educational and research projects

Old Cosmopolitans

- Revenues derived from individuals and charitable trusts
- Not much funding available for alliances
- Limited budget interferes with ability to participate equally in alliance
- Alliances formed at local level rather than national level

Old-Fashioned

- Monitor local habitats
- Develop grass-roots support for conservation efforts
- Form alliances to compete effectively with larger organisations for funding
- Form alliances only with organisations that share its culture and promote its issues

Young Achievers

- Pursue strategies that influence business practice and consumer behaviour
- Rely heavily on government support

- Highest rate of success in forming alliances
- Offer novel incentives for consumers and businesses to achieve behaviour change

Young Entrepreneurs

- Speak and act like savvy for-profit business organisations
- Enjoy relatively large operating budgets, but perceive that as mixed blessing
- Feel pressured to appear struggling in order to acquire high-volume fund raising

Young Locals

- Generally focused on well-defined geographic regions
- Goals include conservation and increasing public awareness
- Involved in direct public education
- Partner government agencies to increase awareness within the government itself

Young Preachers

- Very small grass-roots organisations committed solely to representing their constituencies rather than furthering a specific environmental agenda
- Goals preclude forming alliances with business and other 'mainstream' environmental organisations
- Not very involved in forming alliances
- General lack of trained staff and personnel

This summary emphasises the key characteristics of the eight types of environmental organisation. Those wanting to find suitable partners would be well advised to define their own goals, strengths and limitations based on which they could select the type of organisation with which to form an alliance. More detailed review of the potential partner must be done, but at least the vast majority can be eliminated during the first review.

12

Growing Credibility through Dialogue

Experiences in Germany and the USA*

Katharina Zöller

COMPANIES AND THEIR ROLE in society are subject to more intense public scrutiny than ever before. Since the 1970s, companies have been increasingly required to justify their activities, not only to their shareholders, but also to society as a whole. Stakeholders such as citizen, consumer and environmental groups are now able to exert an influence on companies' performance through public protests and boycotts. Trust and credibility have become important factors in a company's performance (Renn and Levine 1991).

Public mistrust of corporations has grown partly as a consequence of a raft of environmental problems; the cases are well known: the proposed sinking of a Shell oil storage platform in the North Sea (Löfstedt and Renn 1997), the *Exxon Valdez* oil tanker disaster in Alaska, the chemical hazards of Bhopal and Sandoz (Brand *et al.* 1997). In reaction to this loss of faith, many companies initiated environmental communications measures. At first, predominantly one-way PR tools included:

- Issuing mission statements about environmental and ethical goals
- Publishing information brochures (environmental declarations, etc.)
- Organising open houses, where anyone interested could visit the production plants and discuss environmental and other issues with employees
- Participating in public debates, etc.

* Thanks are due to Martin Charter, Luise Jaki, Ortwin Renn and Sabine Mücke, who all reviewed this chapter and contributed numerous helpful remarks.

At the same time, these one-way tools have been enriched by forms of dialogue (round tables, etc.; see below) encompassing 'symmetrical', i.e. two-way, communication (Grunig and Hunt 1984) which makes for trust and consensus building. This means that some companies are not only sending out messages (one way), but are ready to listen and to take into account feedback from their social environment.

A company dialogue is a meeting where companies discuss all relevant public interest issues with stakeholders or individuals searching for consensus or agreement when assessing actions or options for action. The goal is mutual understanding and—wherever possible—reaching agreement on controversial points in a consensus-oriented manner. True dialogue therefore implies the serious consideration of the results of company policy (Hansen *et al.* 1996: 311). The benefits for the organisation are more accurate planning (i.e. products design, etc.) and greater acceptability in the eyes of the public.

A dialogue that obeys certain principles such as interactivity, openness, timeliness and neutrality (see below) should result in: more and better information for everyone; greater transparency; room for new arguments; clear differentiation between consensual and contentious points; an (at least partly) better problem-solving capacity; and more adequate risk assessment (Fietkau 1994: 7).

Dialogue and Greener Marketing

What has dialogue communication to do with greener marketing?

First, dialogue is in some cases a precondition of marketing. If a company lacks credibility, it will not be successful in selling products or marketing services. Not only do politicians need to legitimise their actions and policies; companies too depend on credibility and trust. Cases such as Brent Spar or Hoechst (see below) demonstrate that companies are very vulnerable if they neglect to communicate not only the benefits but also the risks of their actions. In some cases, dialogue processes provide an ideal forum in which to explain why and how things have happened and to respond adequately to public concerns, fears and criticism. Open communication demonstrates the company's willingness to learn from mistakes and to initiate change.

Second, most dialogues cover 'green' aspects such as emissions, waste, traffic problems, energy supply, etc. In some cases, even products or their residues (for example, chlorine, ozone-depleting substances or enzymes in detergents) have been dialogue topics. Companies can make use of the effects of dialogues in their marketing.

Third, sustainable development includes the search for better solutions for resource management and environmental policy to ensure quality of life for the present population and future generations. Agenda 21 demands participation in order to develop more sustainable lifestyles by learning from each other's experiences. And the best way to do that is through dialogue.

Characteristics of Dialogues on Environmental Issues

Dialogues vary in their duration, their mandate or goal, their spatial range and their organisational forms.

Duration. Some dialogues are single events: for example, to inform about environmental policy changes or to gather information on the environmental consequences of a new production unit. In a continued dialogue, topics that need longer discussion time are raised, i.e. the consequences of an accident involving release of emissions.

Dialogue mission statement. The mission statement of a dialogue, or mandate, can vary between general recommendations on a project to detailed feedback that influences decisions. In most cases, a mandate demands recommendations rather than decisions. This means that the company is not obliged to act on all recommendations, but is required to comment on those that cannot be realised.

Spatial range. Depending on who is potentially concerned, participants come either from the local or a wider environment. Most cases described below deal with local issues, apart from the Procter & Gamble Germany dialogues, which were organised at a national level (see below).

Organisational forms. Dialogues can be organised in various forms. These differ mainly in the way in which participants are selected. Three different forms are presented here: round tables, citizens' panels and consensus conferences.

Round Tables

For as long as there have been conflicts, round tables have been organised in order to resolve disputes (Claus and Wiedemann 1994; Fietkau 1994). They follow the simple principle that people or groups of people who disagree on a proposed action convene with a neutral facilitator or mediator in order to search for equitable solutions that everyone can live with. At a round table, participants work on a constructive solution to their problems. The facilitator structures the process and ensures compliance with certain rules.

Participants at a round table are generally delegates from interest groups (i.e. industry federations, unions, sports and other leisure clubs, environmental and citizens' groups, etc.) who are affected by the conflict or problem. The delegates must be authorised to speak on behalf of the group they represent. They must also report the results back to their group. Ideally, the participants choose the facilitator themselves; he or she must at least be trusted by all parties. The participants have the right to determine the agenda in co-operation with the facilitator. When initialising a dialogue, the organiser is obliged to ensure that all relevant groups and people are represented in the process so that the results of the round table can include all crucial aspects. If this is not the case, groups who were excluded from the dialogue will probably not accept the results.

Round tables are best suited for dialogues where interest groups (and not individuals) fairly represent the interests of those who are affected. Since delegates of interest groups are mostly multipliers, the dialogue should produce far-reaching results. When working with interest groups, one faces the problem that they often adopt rigid positions and are less flexible. The above-mentioned principle of delegation is often precarious, as support from and transfer into the groups is sometimes ambiguous. For example, many environmental groups have a certain position they have to defend and risk losing credibility if they negotiate these positions.

Examples of the use of round tables are described below.

Citizens' Panels

Citizens' panels (Dienel 1990; Stiftung Mitarbeit 1990) are groups of 20–25 people who work on a topic[1] for a certain time under the guidance of a facilitator. The participants are selected randomly (similar to opinion polls) and invited to participate. Depending on the subject and the amount of work involved, between 3% and 10% of the invited citizens accept the invitation to participate. The incentive for participation is to exert more influence on policy-making, to participate in decision-making and to acquire publicity for one's own interests. The participants, generally lay people, as they are selected randomly, obtain information from different, often controversial sources and experts. The outcome of a citizens' panel is a so-called 'citizens' report', which is presented to the organisation that set the panel up and to the public by speakers of the panel or by the facilitator.

The random selection has the following advantages: every citizen has an equal opportunity to participate; and people are invited who would not normally participate in public discussions. Participants do not represent specific interests, but bring in their everyday experiences. They are more flexible than stakeholder delegates, as they do not have to defend the missions of an organisation. As a random group of citizens will generally have less expert knowledge than stakeholder delegates, the panellists will need to be furnished with extensive information. A major disadvantage of the model is that citizens do not function as multipliers in the same way as representatives of interest groups (the latter will subsequently report results of the dialogue back to their group and to other interested parties).

Citizens' panels have—up to now—been organised only in the political environment of local and regional governments. But such panels could also produce interesting and useful results in the context of industrial politics, corporate identity or product marketing. Since panellists become very well informed about the options,

1 Topics at the Akademie für Technikfolgenabschätzung (Centre of Technology Assessment) in Baden-Württemberg were 'Biotechnology/Gene Technology', 'Waste Management Plan for a German County' and 'Climate-Compatible Energy Supply'.

they can offer constructive feedback with respect to company policies. The difference between a citizens' panel and a consumer focus group for market research is that a citizens' panel gets more deeply involved in the organisation's policy and in problem-solving. As the panels work for at least three to four days or for a set period of time (for example, eight evenings in six months), they become 'experts' on the issue and are able to offer useful evaluations.

Consensus Conferences

Consensus conferences (Joss and Durant 1995; TAB n.d.) have been organised for some years in the political arena in Denmark, the Netherlands and the UK, and lately also in Switzerland and Germany. Similar to citizens' panels, they operate with lay people who are invited to participate through newspaper advertisements. The organiser chooses the participants based on age, sex, profession, education, etc. The 10–20 people chosen from the pool of volunteers prepare the agenda for the consensus conference over two weekends.[2] Before the meetings begin, participants receive information material. During the two preparatory weekends, they put questions, chosen by themselves, to experts. These experts are invited to the actual conference to respond to the questions; this is when the conference is open to the public. In the end, a recommendation by the panellists is drafted and presented to the press and the public. The process is accompanied by a 'steering committee' consisting of stakeholder representatives. The committee helps the organiser prepare and conduct the conference and also serves as a multiplier. Consensus conferences can be used by businesses—just like citizens' panels—to generate well-grounded feedback on company policy or for marketing purposes. The citizens in the conference come up with a 'position of reference' and can act as 'governors of common sense'.

Preconditions and Principles of Dialogues

To call a process a 'dialogue', the following preconditions have to be met (Hansen *et al.* 1996: 316; cf. also Burkart 1996):

Willingness to learn from each other. Within the company, it is important that management and the staff are willing to learn from the dialogue with their stakeholders. Structures within the company need to be flexible enough to implement proposed measures resulting from the dialogue. If the public relations manager is the only one who is enthusiastic about the new communication instrument and the general manager is unwilling to listen to the participants in the dialogue, the process will fail.

2 Topics in Denmark were 'Air Pollution', 'Future of Private Traffic' and 'Clean Agriculture'. In Switzerland, a so-called 'Publiforum' was organised in early 1998 on the subject of 'Electricity and Society'.

A bias for action. A successful dialogue needs a bias for action. For example, if a company needs to build a water treatment plant that is in the final state of planning, 'dialogue' can only be aimed at confirming the plans of the company. This will not be accepted by pressure groups. In an earlier planning stage, there are still choices that can be made (i.e. size, sewage measures, operation, etc.). Public consultation can help in designing new options and in evaluating existing options.

Personal, financial and timely resources. Dialogue with citizens needs time as well as personal and financial resources. All parties must be able and willing to commit those resources to a dialogue. It is important to remember that not all participants have access to the same amount of resources: citizens' groups and individuals participate on a voluntary basis in their free time. It is useful for a company or another sponsor to finance excursions and lectures and reports by experts for the benefit of all participants of the dialogue. If information is generated solely by the company that organises the dialogue, it will be generally presumed to be biased.

Dialogues should obey the following principles:

Interactivity. Communication between participants should be symmetrical, i.e. two-way. Mutual understanding is the goal. There is a difference between panel discussions and dialogues: although panel discussions allow participants to ask questions, a real dialogue implies a mutual exchange of arguments and feedback.

Openness. The focus of the dialogue should leave open the possibility of introducing new aspects at any time. Not only benefits and advantages, but also risks and disadvantages should be discussed.

Start at an early stage. If decision-making is at an advanced stage, there is usually insufficient room for change. Dialogue should start early. Once deep conflicts arise, it becomes increasingly difficult to initiate a fruitful dialogue.

Neutrality. Neutral facilitation is essential. It is not recommended to entrust an employee of the company with the process because he or she is not considered neutral. A facilitator should be accepted and trusted by all participants.

Inclusiveness. All interests concerning the topic must be included in the dialogue through an interest group or a single person.

Case Studies

The following case studies represent applications of the round table model and were initiated by private chemical companies (Ahrens and Behrent 1996). In Germany, dialogues by means of citizens' panels or consensus conferences have only been conducted in the political arena, although they have the potential to be useful for companies.

Chemical companies have a high risk potential compared to many other industries. Due to several severe accidents, i.e. Bhopal (1984) and Seveso (1986), the industry has increasingly lost public trust. 'The public is concerned, largely sceptical and often fearful of chemical manufacturing' (SOCMA n.d.). In reaction to this, the chemical industry has initiated its 'Responsible Care' programme,[3] which is designed to improve the industry's environmental performance based on ten guiding principles and six codes of management practice. It includes dialogue with concerned and interested citizens (see CAP below).

Hoechst Neighbourhood Circle

Hoechst is a conglomerate of chemical companies producing pharmaceuticals, agricultural products and commodities for industrial chemistry. Hoechst has 160,000 employees in 120 countries in Europe, America and Asia. The headquarters of the company is at the biggest site, Frankfurt am Main.

In 1993, the company suffered several severe accidents where toxic emissions escaped into the environment, a worker was killed and some people injured. Due to the severity of the problem and additional communication errors, the public became highly alarmed and the company received very bad press coverage for several weeks. Even now, years later, Hoechst has a worse reputation than other chemical companies in Germany (Kepplinger and Hartung 1995).

Reacting to the accidents, the company set up a round table among affected neighbours of the Frankfurt site, known as the 'Hoechst Neighbourhood Circle'. The circle meets regularly, about four times a year. Participants include local club members (sports clubs, etc.), mothers' groups and citizens' groups, clergy, environmental groups, children and youth groups, and the local ex-pats' association. Local government employees act as advisers if desired. The general manager of the site, as well as other managers and staff members, listen to the discussions and offer feedback. A retired local politician serves as facilitator. The mandate of the group is to discuss all relevant questions concerning the relationship between the company and the neighbourhood. Topics have included air pollution and emissions, transport of hazardous material, energy supply, and the environmental auditing system. In the aftermath of emergency situations, extra sessions were held.

What are the goals that the company and citizens pursue? The company wishes (Henschel 1997):

- To increase credibility and acceptability
- To improve its public image
- To accelerate processes of permits for licensing plants and technical facilities

3 The Canadian chemical industry started the programme in 1985, followed by the US in 1988 and Germany in 1995.

The citizens expect:

- To minimise environmental pollution and health burdens
- To improve transparency on what happens 'behind closed doors'
- To derive better publicity for their interests
- To gain access to and influence some of the company's decision-making processes

Hoechst managers believe that company decisions have gained greater public acceptability since the group was established (Schönefeld 1996). Through dialogue, a new forum for discussing different perspectives has been created with new views that can be voiced and exchanged. Members of the public, local politicians and the media have accepted the Hoechst Neighbourhood Circle as an important organisation and started to feed new topics into the group. The reasons for an accident in 1996 could be better explained by the company and did not create as much upheaval as in 1993. Other benefits refer to the general improvement of relations between the company and its neighbours. The environmental groups, for example, succeeded in initiating a series of epidemiological studies (health examinations of neighbours, especially children) after the 1993 accidents. In addition, some other long-standing conflicts could be resolved: the company now publishes detailed emissions data of its Frankfurt plant. This has been demanded by neighbourhood initiatives for a long time. Hoechst has also finally accepted the need to install early-warning sirens in case of an accident.

Procter & Gamble Germany: WAGE and HAGE Dialogues

Procter & Gamble (P&G) is the second-largest producer of cosmetics in the world. In Germany, P&G has recently conducted two dialogues. The first was the WAGE dialogue ('WAschen und GEwässerschutz': washing and protection of rivers and seas); then came the HAGE I and II dialogues ('HAar/HAut und GEsundheit': hair/skin care and health), both with scientists, politicians, administrators, companies and representatives from federations and environmental organisations from all over Germany (Hansen and Schoenheit 1994; Hansen *et al.* 1997). The HAGE II dialogue, a two-day workshop in February 1996, dealt with skin care and health. P&G wanted to create a forum for discussing important aspects and problems related to skin care with its stakeholders in order to identify potential issues and to establish and deepen contacts. Goals were:

- Identification and analysis of opportunities and risk as well as causes of risk in skin care
- Working out solutions for handling risk
- Assessing agreement and disagreement
- Establishing better information and communication structures

- Finding out about potential co-operation

Participants in the dialogue were representatives from environmental and consumer organisations, industry, retailers, journalists, toxicologists, experts in cosmetics, cosmetic chemists, dermatologists and specialists in allergies. In three working groups, the participants worked out recommendations on questions such as:

- What are the environmental and health risks of natural ingredients, of preservatives and perfumes? What is environmentally sound packaging?
- Information, communication, manipulation: what kind of information does the consumer want and need? What makes for credibility in information about skin care? Can an independent data bank help inform the consumer?
- Products of the future: sustainability aspects of target group cosmetics such as children's cosmetics or cosmetics for the elderly, acceptability of product innovation.

An investigation by the organisers of the workshop from the University of Hannover shows that P&G, as well as the participants, are quite satisfied with the results of the dialogue. Of the participants, 77% believe that the results of the dialogue will have an influence on future P&G product policy.

Neighbourhood Circle Riedel-de Hän

The 'Neighbourhood Circle Riedel-de Hän' was founded in 1995. Riedel-de Hän is a chemical company with about 1,400 employees which produces intermediate chemical products. Its main site, the city of Seelze, has about 32,000 inhabitants. The circle meets three to four times a year, its task being communication with its neighbours and other local stakeholders as well as with delegates from surrounding communities who all are given an opportunity to comment on or to criticise company policy (Hammerbacher 1995). The benefit to the company is the opportunity to react immediately to issues and problems. The circle is organised and facilitated by an experienced mediator. The plant manager, the environmental manager, the PR manager and the leader of the works council are present to listen and to respond to feedback from the group. Topics for discussion have included the planning of an incinerator for hazardous waste, emissions monitoring in the vicinity of the site and questions about site expansion and developments within the company.

Community Advisory Panels at Dow Chemical Germany

Dow Chemical is a US company with almost 43,000 employees in 33 countries which manufactures performance plastics and chemicals, plastics, hydrocarbons and energy. The headquarters is situated in Midland, Michigan. Dow has three German sites where it produces plastics and intermediate chemical products. Being a US company, Dow has transferred elements of its American public relations programme to foreign sites, which includes the organisation of Community Advisory Panels (CAPs; see below) at 22 sites around the world, three of them in Germany. The 'Contact Group Citizens—Dow' at the Rheinmünster site (south-west Germany) was established in 1991 and meets several times a year. Topics have included the shift of transportation from road to rail, waste reduction, incineration and offensive smells.

The North American Chemical Industry and Community Advisory Panels

The chemical industry in the US and Canada has established a pioneering role in dialogue communication (Meister 1996). The initial reaction to the US Emergency Response and Community Right-to-Know Act of 1986, which obliged companies to publish emissions data was that the industry was unwilling to release raw data without the opportunity of explaining it. The best way of communicating with (not merely informing) the public about emissions seemed to be through dialogue. Thus the CAP idea was born. Subsequently, CAPs have become a standard approach in the chemical industries' communication programme.

A CAP is a group of opinion leaders from the community surrounding the manufacturing plant. It allows the management the opportunity to conduct a direct, continued, personal and open dialogue with its neighbourhood in order to improve the acceptability of company policy and to obtain feedback evaluating the company's activities. In most cases, an independent facilitator organises the CAP. Members include neighbours such as teachers, clergy, local politicians, environmentalists, police officers and others, altogether between 10 and 15 people. The CAP is independent of the company, sets its own goals, chooses its own topics and produces a mission statement. The style of communication is open and driven by mutual respect. Company managers have had to learn to take account of the needs and concerns of CAP members and to be ready to react point by point to the interventions and concerns. Openness and fairness, as explained before, are the main criteria for CAPs.

In 1994, there were 244 CAPs in the US, most of which are single-company CAPs, i.e. they refer to one company only. However, there are also CAPs that report to several companies in the same area. In addition to these multi-purpose CAPs, some ad hoc CAPs for special topics have been organised: for example, when a major investment was launched. The topics discussed in the CAPs include emissions data, emergency plans, traffic problems, environmental problems, investment projects, but also dismissals or the relevance for the site of national political issues.

CAPs will usually work successfully, if the following conditions are met:

- Before initiating a CAP, the company conducts research into the attitudes, expectations and interests of the citizens in the local community.
- Those chosen to be involved in a CAP have good communication abilities, and act as multipliers in the community, which is not to say that only industry-friendly citizens should be invited; on the contrary, constructive critics are welcomed to give the CAP its dynamism and legitimacy.
- The goals, mandates and statutes are clearly articulated.

- The CAP meets regularly.
- The topics focus on problems at a local level, because that is where the company can best influence and change the situation.
- The plant manager participates regularly in CAP meetings.
- Companies demonstrate openness for change. If participants feel that their recommendations are not going to be taken seriously, they will cease to participate; or, in the worst case, formerly well-meaning members might become adversaries. If they experience the opposite, their motivation will increase, as they realise that their commitment can move things within the company.

CAPs are useful tools for all (not only chemical) companies who wish to improve relations with the community and its neighbours. They could be especially helpful for companies whose manufacturing plants are situated close to residential areas. Companies in the same area should co-operate and initiate a multi-company CAP.

Conclusions

The main lessons to be learned from failed communication strategies (i.e. Shell Brent Spar, Sandoz Schweizerhalle, Hoechst) is that one-way information based only on risk assessment by experts (often presumed biased) will in many cases fail if the fears and concerns of the majority of people and especially of the local community are not taken into account. Any action carried out by a company that could concern and affect other people should be discussed with and reviewed by an independent panel. The panel should always include some representatives of pressure groups in order to ensure openness, be it at a local, regional, national or international level (depending on the people concerned). Industry would benefit from a greater understanding of the fears and concerns of the public (see Löfstedt and Renn 1997: 135).

Dialogue is an efficient means of incorporating these concerns and for growing trust and credibility. Dialogue approaches 'help to see the logic and rationale of other people's perceptions that affect one's own perceptions and they reduce the so-called "surprise factor" and the potential for conflict' (Löfstedt and Renn 1997: 135). But they are not a means of crisis communication: dialogues have to be established before the cause escalates in order to increase credibility, or after the crisis in order to learn from the mistakes.

13
Building Environmental Credibility

From Action to Words

Lassi Linnanen, Elina Markkanen and Leena Ilmola

SEVERAL TRENDS are shaping the communication practices of those companies willing to adopt sustainable development strategies. These include, among others, the emergence of a global marketplace, the increasing importance of public discourse, the need for environmental accountability and disclosure, a strong movement towards environmental management systems, as well as several forms of stakeholder and supply chain co-operation (Frankel 1998). Generally speaking, the field of environmental communication is becoming more complex. In addition, the boundaries between businesses and their stakeholders are becoming less clear, and stakeholders will more clearly be seen as a power base and a catalyst for innovations in businesses. Therefore, an increasing amount of transparency is needed to ensure the legitimacy of corporate actions.

Environmentally sound business practices and environmental achievements will bring the business and its products **environmental credibility**. Even though numerous companies have achieved technological and cost benefits by applying advanced environmental innovations (see e.g. von Weizsäcker *et al.* 1997), there are several communication-related barriers on the way. Environmental excellence will often turn into competitive advantage only after its benefits have been communicated to customers and other stakeholders.

Environmental credibility will be earned through years of hard work. Despite companies being engaged in image-building exercises, their credibility must be based on water-tight facts, even if the public are only interested in environmental results, not in the facts behind them. Seeking to gain a competitive advantage through environmental credibility is more difficult than, for example, using price as a marketing tool. The difficulty is aggravated by:

- Emotions. Each person tends to have a personal relationship with the environment.
- Long chains of argument. Environmental cases must be thoroughly reasoned.
- Rapidly changing trends. Customers shift the focus of their attention unexpectedly.
- The number of stakeholders. Everyone considers themselves environmental experts.
- The sender effect. Communication by industry is not regarded as credible.
- Risks and crises. These can easily obliterate the results of years of hard work.

In this chapter, we will concentrate on three central themes that shed light on the environmental communication challenge. First, we introduce the **cycle of inclusion** as a process for shifting the focus of stakeholder co-operation from manipulation to dialogue. Second, we analyse and conceptualise the **environmental communication agenda** of successful companies. Third, we focus on **risk and crisis communication** as a phenomenon peculiar to environmental issues.[1]

From Manipulation to Dialogue: The Cycle of Inclusion

Public environmental debate shifts its focus swiftly and unpredictably, and businesses are constantly required to react to new concerns. Interpreting weak market signals that arise from the results of contemporary research or from initiatives by various non-governmental organisations has become more and more crucial. The trap is that, precisely at the time when businesses should be accelerating their speed of reaction, the time needed to solve complex problems grows longer (Bleicher 1992).

An environmentally astute management has to deal with many different groups with many different views. Where environmental issues have considerable implications for a business's activity, management is compelled to follow the climate of opinion. Successful co-operation with stakeholders calls for a careful analysis of relevant markets, enabling dialogue and continuous co-operation (see Fig. 1).

The aim of the cycle of inclusion is to secure the ability of businesses to continue their activities in all situations (see also Wheeler and Sillanpää 1997). This does not mean, however, that businesses should seek to manipulate their stakeholders. Businesses should rather influence environmental debate in such a manner that their own ability to do business will not be curtailed in the future. It is the businesses' responsibility to see that all the stakeholders are provided with sufficient informa-

1 The material in this chapter is based largely on an environmental management textbook originally published in Finland (Linnanen *et al.* 1997).

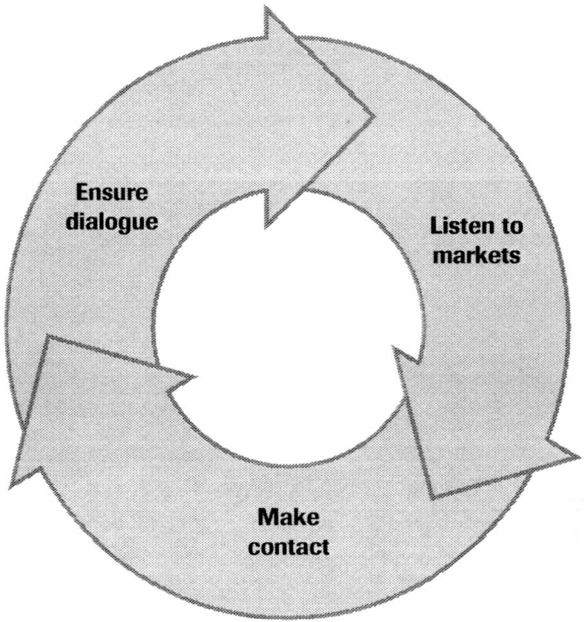

Figure 1: CYCLE OF INCLUSION

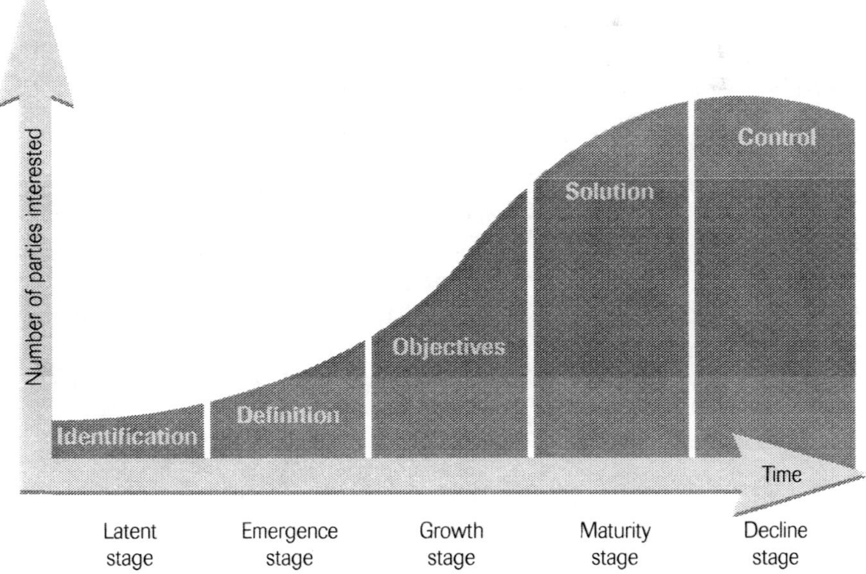

Figure 2: THE LIFE-CYCLE OF AN ENVIRONMENTAL DEBATE

Source: Dyllick 1989

tion on the activities of the company. At their best, businesses can actively strengthen their own environmental credibility during a debate.

Listening to Markets

By listening to markets, businesses will collect information on the aims and activities of important stakeholders and the extent of their knowledge about the business's activity. Building up environmental credibility requires close observation of environmental debates (see Fig. 2).

Each environmentally related discussion has a life-cycle of its own. The parties engaged in the discussion, the forum of discussion and the content of discussion are different for each of the five stages. During the first, **latent stage**, the environmental issue or problem will be identified. The issue will often be raised by environmental activists or researchers, and the debate itself will be conducted by environmental experts in scientific publications. The forums for this stage will be journals, action programmes and environmental organisations' manifestos.

During the latent period, the task is to position oneself in relation to the expected issue. If the environmental issue raised could form a competitive advantage for the company, or it can quickly be developed into such, it is worth becoming active at this stage. The latent period will afford the time to develop internal preparedness.

In the second, **emergence stage**, the issue to be raised, the concepts to be used and the forum for discussion will be defined. Environmental organisations and experts will take an active part in this issue-definition process and the debate will now be conducted in professional magazines and on the podiums of international conferences.

To a great extent, this stage will set the tone for subsequent stages. If the subject is of central importance to a company's strategies and objectives, there is every reason for it to be active in initiating the debate. The purpose of communication when issues emerge is to gain control of the 'battlefield' of discussion, that is, the subject being debated, the method of approach, the choice of forum, and the act of initiating and taking active part in the debate.

In the third, **growth stage**, the environmental debate moves into the domain of the mass media. Environmental organisations are particularly active at this stage, as measurable targets are defined and mechanisms to evaluate the importance of the issue are created. If, during the earlier stages of the debate, channels of co-operation have been created with other main players, the focus of communication can be cast wider.

Large stakeholder groups, such as customers and consumers, now become interested in the subject. This is the stage where the company can cut a competitive edge and make a visible difference. Direct customer communication will be initiated, informing customers on the benefits derived from the company's expertise. Customer magazines, business-to-business sales deals, brochures and trade fairs will function as channels of communication. The company's activities and products will also gain media exposure and the use of advertising should be considered. Should the company be exposed to an attack at this stage, it can only defend itself in a reactive manner, the problem here being that reactive communication is no help in earning environmental credibility—quite the contrary.

In the fourth, **maturity stage**, a vigorous debate about the content of regulations and control mechanisms will take place. Unless the company has previously been active in the course of the debate, it will have to accept both the chosen emphasis of the debate as well as the forum in which it is being conducted. Means of wielding influence at this stage are few.

In the final, **decline stage**, communication is characterised by follow-up and control mechanisms. According to established practices, the company will report on its environmental efforts and results: for example, by publishing an annual environmental report. Environmental activists will already have lost interest in the subject by the maturity stage, and, by the decline stage, the general public will likewise leave the matter to the public authorities. This does not mean, however, that an issue that has reached decline stage could not rise again.

An example of a multi-layered environmental debate can be drawn from the pulp and paper industry (Linnanen 1995). The forestry issue has gradually interlinked several seemingly independent themes, which include: depletion of tropical rainforest through fires and deforestation, tropical timber logging, questions related to landscape and multiple use of forests, biodiversity discussions, northern clear-cuttings, and the fate of primaeval forests. The debate has recently focused on the origins of timber used in paper and other wood products, and related certification procedures. The forestry issue has continued to intensify and broaden the geographical scope of environmental discussion in the pulp and paper industry. Forests first came into focus when discussion broadened from the point-source pollution of paper mills to include regional issues, such as forest depletion due to acid rain. The biodiversity discussion has globalised the theme, and the forestry issue has also been highly influenced by the global warming debate, as forests are major stocks of carbon. However, local issues such as the landscape effects of logging have also become more important.

◻ *Influencing the Debate*

A debate can be influenced but it cannot be manipulated: it is not a monologue. Active participation in a public debate means that the company itself takes part in the debate honestly and actively. Defending adamantly one's own point of view is not the best tactic during an environmental debate: co-operation and listening to markets are liable to yield far better results.

As an expert in its own field, the company is responsible for bringing into the discussion scientific data, the opinions of international experts and comparable statistics. The company can also invite other experts, organisations that are thoroughly familiar with the subject, or consumers with knowledge of the issue to take part in the debate. If the subject is particularly important to the company, it is worth making an agenda for the discussion and covering all the areas that are likely to crop up.

Making contact with stakeholders requires motivation on both sides. The first step is to arouse the interest of stakeholder groups: stakeholders are more likely to want to spend time in discussion if the company can provide relevant and up-to-date information. Co-operation with experts, a local community group or a leading professional publication can create an avenue for stakeholder contact.

Ensuring Dialogue

Making contact also lays the ground for useful, interactive communication. Successful dialogue calls for a constructive approach from businesses, laying aside confrontational 'us-against-them' attitudes. The function of communication will extend, becoming more diverse than the previous one-way channel from business to stakeholder. Instead of engaging in defensive or obstructive behaviour, a business can secure its own environmental competitiveness by continuing to be open, interactive and co-operative. Various advisory bodies, such as a panel of scientific experts or a local community group established in the vicinity of a factory, can act as forums for co-operation. Indeed, several businesses have successfully set up co-operative relationships with environmental and other NGOs (Elkington 1997).

Those businesses that respond most swiftly to altered market conditions accrue considerable advantages in marketing and image improvement. Arguing one's case on purely scientific grounds alone is inadvisable, as, however accurately they put across their respective viewpoints, industry and environmental organisations may still disagree on the respective weight to be attached to different matters. What is essential, however, is to learn to live in the world of diversity and unpredictable markets. Even though proactiveness always entails risk-taking, acting as a forerunner in environmental matters is not a bad strategic solution in the long run.

From Glossy Pictures to Reality

Especially in cases where environmental issues influence the purchasing decisions of customers, it pays off for businesses with strong environmental performance to invest in building up environmental credibility by means of communication. This way a new opportunity will be created for bringing innovations relating to products and manufacturing methods effectively to market.

For the purposes of building up environmental credibility, the recipient of the message should be determined, i.e. the target group of communication should be chosen, and the target group will partly determine the content of the message. When considering the message, it should also be decided whether corporate or product image should be chosen as the basis of environmental marketing, and which material should be used. Having analysed the communication strategy with respect to these factors, a practical communication plan can be drawn up and the opportunities of exploiting environmental achievements can be considered.

Choosing the Target Groups

In building up environmental credibility, the most important stakeholder group will often be customers, but several other groups will also need informing of the aims, content, plans and results connected with the company's environmental efforts. In the following, we focus on communication to final consumers, but most of the analysis is also applicable to business-to-business environmental communication.

Communication is rendered efficient by targeting. Especially in relation to environmental matters, different stakeholder groups are interested in very different issues. All too often, communication is tuned to operate in terms of the sender of the message; in other words, only those issues are dealt with that are of interest or importance to the sender. The content of the message must relate to the expectations and attitudes of the recipient, for communication that has no relevance to the wider community will be rejected. Furthermore, the message must be tailored to the recipients' level of knowledge. Different communication channels, such as the Internet, professional journals or retail advertising, reach different audiences. The probability of the message actually reaching its target is increased by focusing on those themes that are of interest to the receiver, and expressing the message by using language familiar to the chosen audience.

In choosing the target group, it is wise to divide each separate audience into the smallest target groups possible. The more precisely defined the target group, the more effective the communication. The significance of the target group to the company's business dictates how finely the divisions should be drawn. Nevertheless, the realities of time and budget need to be taken into account when defining these groups.

In choosing the target groups for environmental communication, posing the following questions will prove helpful:

- Who is going to benefit from the environmental improvements?
- Who is interested in environmental issues at present?
- Who has the power to influence the company in environmental matters?
- Who is going to be interested in the environmental matters of the company in the future or affect its operational environment?

Making Environmental Credibility Part of Corporate or Brand Image

Environmental excellence can be linked with either corporate or brand image. International operations call for investments of such magnitude that businesses will usually be compelled to choose a strategy based *either* on corporate image *or* on brand image, and building up an image requires long-term commitment. Superficial solutions, aimed at catching the latest change in public mood, form no part of carefully planned communication.

The following considerations affect the choice between corporate and brand images:

- Whether the company operates in consumer or business markets. In business-to-business markets, corporate image is usually of higher importance, whereas in consumer markets brand image is more decisive.
- Which is better known at present, the company or its product?
- With which stages of the product's life-cycle are environmental benefits connected? (see Fig. 3)

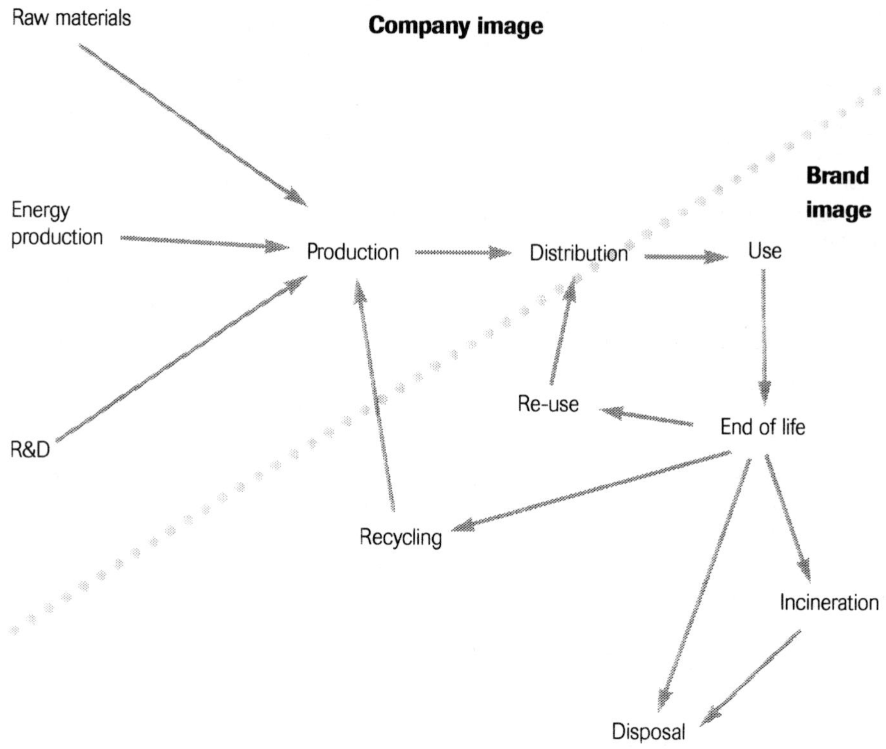

Figure 3: ENVIRONMENTAL CREDIBILITY AS PART OF THE CORPORATE OR BRAND IMAGE ON THE BASIS OF A PRODUCT'S LIFE-CYCLE

If environmental benefits are mainly achieved in the early stages of the life-cycle of a product, as is the case when emissions are cut at production stage, the benefits are felt mainly by the local community or society at large. In these cases, emphasising corporate image will usually be the best option. If, on the other hand, the environmental benefits are reaped by consumers, as they relate to the qualities or use of the product, the benefits are connected to the product and can naturally be made part of the brand image.

However, it is quite possible to add environmental value to a product even in the case of bulk goods. In the pulp and paper industry, for instance, the Swedish cellulose producer Södra Cell succeeded with the aid of intense marketing in creating a brand image with its ozone-bleached TCF (Totally Chlorine Free) cellulose in the early 1990s. Environmental issues can be beneficially exploited in a situation in which all competing products fulfil the other demands required by consumers.

Building up corporate image cannot be neglected entirely even if a brand-image-based marketing strategy is chosen. Stakeholders usually judge the whole company, not just its products. The significant point about using environmental issues in marketing is that communication requires that the sender commands a high level of credibility. Corporate environmental reporting is a trend that supports the corporate image approach.

The background with which the company is associated, such as industry sector, also has a bearing on the credibility of the company's environmental message. A business with a considerably greener corporate image than its own industry sector may encounter a surprising amount of opposition when attempting to highlight its own environmental achievements. It is difficult for a company to be greener than its own industry sector. Many chemical companies, previously regarded as the worst environmental offenders, are among the few large corporations to have taken on the challenge of sustainable development seriously (Hart 1997), but they still face credibility problems due their past sins, as the intense argument about genetically modified food indicates.

Presenting a Persuasive Environmental Case

Communication is made effective by crystallising and repeating the message. By 'the message', we mean the core information that the sender wishes the recipient to retain after contact. This answers the question: How is the customer going to benefit from using our product or choosing us as their partner? The message is not usually expressly stated in the marketing material: rather, it is the conclusion a recipient is constrained to draw having read the preliminary text or heard the sales promotion speech.

The process by which effective environmental communication approaches are crystallised can be divided into four different stages. The first step is to identify all possible benefits arising from the company's environmental excellence in products, services or operations (see Fig. 4). In the second stage, the list of possibilities is narrowed down by identifying all the purchasing criteria applied by the customer in making a decision. If there are several factors, they should be listed in order of priority. A good question to ask is: If the prices of two products are the same, which is the one chosen by the customer? Or, in sifting through factors affecting corporate image: If the prices quoted by two different companies are the same, on what basis does the customer choose a supplier?

Leading on from this, the third stage is where the environmental strengths and weaknesses of competitors are analysed, and only those strengths that competitors lack or possess in a lesser degree are left on the list. A customer is going to choose a product or supplier that stands out in a positive way from its competitors.

The final selection is where the most practicable cases are chosen from the final shortlist. The environmentally friendly corporate or product image is built around the strengths that fulfil the following criteria:

- They are of an enduring nature.
- They are easily harnessed (data can be obtained directly from an environmental management system).

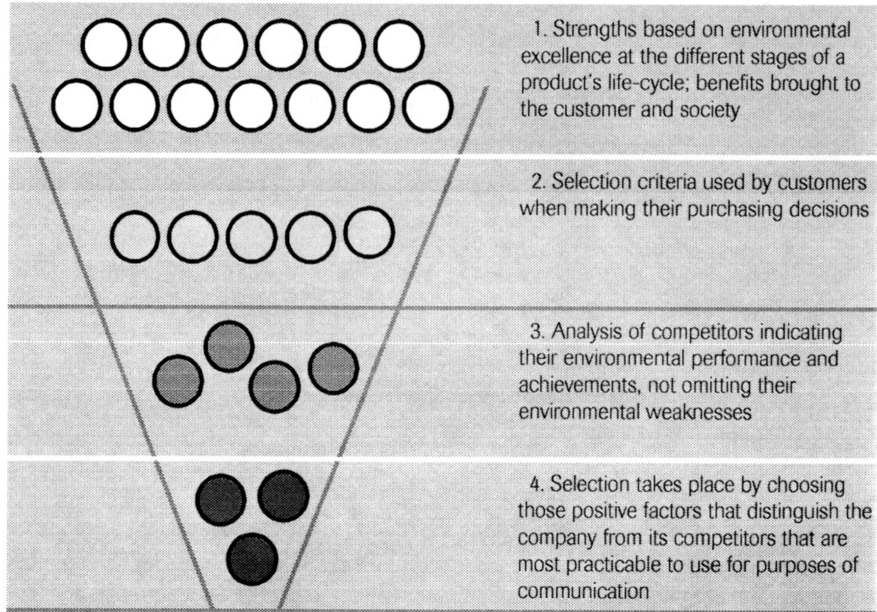

Figure 4: THE CRYSTALLISATION PROCESS OF ENVIRONMENTAL COMMUNICATION STRATEGY

- They are communicable (taking into account how they attract attention, how easily they are understood, the receptiveness of public opinion, the level of knowledge of the target audience)

Devising a Communication Programme

The establishment of a communication programme designed to improve environmental credibility is guided by selected corporate or product image factors. Once the benefits yielded by environmental performance have been crystallised and the messages that are of interest to the target groups have been defined, the foundations will be in place. Naturally, this should be aligned with overall corporate strategies and objectives.

In devising a communication programme, the first step is to prioritise the most important target groups. Resources are often limited, and accurate targeting determines the efficacy of communication. When setting the priority order of the target groups, the following factors are worth considering:

- Who is expected to manifest the most significant change in their behaviour?
- Who is going to make the actual decisions?
- Who is going to make a decision based on environmental factors?

- Who are the opinion leaders in any particular environmental debate?
- Who will have the most decisive impact on the operational environment of the company during the planning period?

Communication channels should also be selected carefully. Of all these, the mass media will reach by far the widest audience. Editorial departments may be provided with information on the business's operation and background, or advertising space may be bought. And marketing mailshots and customer magazines are ways of communicating to customers directly. When used properly, the Internet is also among the most effective environmental communication channels. Personal communication is one of the best methods, especially if the message does not correspond to recipients' prevailing attitudes and expectations. The more difficult the message, the more important it is to approach the target group via opinion leaders or experts.

One of the most neglected communication channels is the local network, i.e. the employees of the company and their neighbours, relatives and friends. The neighbouring communities of a company's production sites will also act as communication channels similar to the local network, and their message carries much weight. To meet this challenge, early environmental reports, such as those from Novo Nordisk and other companies, were initially targeted at employees.

A third communication channel based on personal contact is the company's own distribution channel, which is often neglected. Advertising, sales promotion and information can all arouse interest in a product, but the final decision to purchase will often be made only during sales negotiations. Bearing this in mind, it is important that sales personnel are able to articulate the company's environmental achievements and the facts behind them to the customers with sufficient credibility. If the chain of communication is long, the importance of training increases dramatically. It is not difficult to imagine, for example, that a summer sales assistant in a speciality store in a small Portuguese town will not automatically be able to explain to customers why an environmentally friendly German product is 30% more expensive than its Portuguese alternative.

Profiting from Environmental Achievements

The nature of communication of environmental issues differs from other types of communication in that behind each claim there must be a credible chain of argument and supporting material. Communication of environmental issues should be based on fact, and there is a clear trend towards more quantitative data. Environmental management systems actually produce documents on a continuous basis which can be used in support: a print-out from an environmental database produced the very same morning in which negotiations are taking place provides more credibility in a sale situation than a glossy brochure covering identical issues.

The credibility of environmental material varies in the eyes of external stakeholder groups. Surveys and inspections carried out by the business itself have the lowest credibility. In contrast, evaluations carried out by impartial, third-party inspectors, or eco-labels granted according to strict criteria, do provide excellent marketing material. Programmes or labels whose criteria have been established with the full participation of opinion leaders count as the most credible.

Sponsoring environmental organisations and projects can at times offer a more effective and less troublesome way of signalling the company's commitment to environmental issues than advertising. Co-operation with environmental organisations is to be recommended for several reasons: first, it is a way of acquiring valuable information and expertise, which the company itself may lack; second, despite full-time environmental managers or board-level responsibilities, few companies can afford to employ a broad array of functional environmental experts; third, it may also lend credibility to a company's environmental credentials by association—in the eyes of the recipient, a message is only as reliable as its sender.

Publicity is a good means of building up environmental credibility. Companies should aim to get the media to publish positive articles and news stories by organising events, through press releases, by granting interviews, giving presentations or writing articles for the press. The best thing about media publicity is that it carries more credibility than advertising, but the downside is that it is uncontrollable. A well-known example of using of public relations and sponsorship for marketing purposes instead of advertising is The Body Shop, an international retail chain selling natural cosmetics. One of its guiding principles used to be the rejection of conventional advertising, which was exceptional, especially in a sector such as the cosmetics industry which is one of the most dependent on advertising.

The Fragile Nature of Environmental Credibility

Developing environmental credibility takes years, but can be lost in an instant. With an environmental management system in place, accidents can still happen at manufacturing sites, or a faulty product can generate an unexpected environmental hazard when used. Building up environmental credibility requires realistic communication by companies that does not omit information about risks. Communicating risks along with the company's environmental excellence generates credibility capital that ensures that an environmental crisis is not going to destroy the company's image.

Internationalisation will present companies and their environmental communication with additional risks. While environmental credibility needs to be earned locally, it can be lost globally. In an era of Internet communication and efficient mass media, an environmental crisis will be international news in no time, impacting on the international corporate image of the business concerned, even when the effects of the crisis are of a local nature. Environmental communication must therefore succeed in operating in many different arenas, even though at the same time the target is not a geographically homogenous market with shared attitudes. A recent example of this global communication challenge was the one faced by Shell. The public row over the environmental and social effects of oil drilling in Nigeria in 1995 and its tragic culmination in the execution of nine dissidents had an impact on Shell's image in Europe as well as in Africa. Earlier, the wave of negative public attention towards the attempt to sink the Brent Spar oil rig in the Atlantic generated another movement that led to an international boycott (see Chapter 17 and Elkington 1997). Both of these serve as concrete examples of the global nature of the environmental and social responsibility debate.

Communicating Risks

Environmental credibility is earned by providing recipients with diverse and thorough information. Environmental credibility cannot survive on positive messages alone. Precise presentation of risks coupled with a description of the measures taken by the business to minimise those risks is of interest not only to experts, but to shareholders, investors and financiers as well.

Large-scale accidents are not the only things that fall into the category of environmental risk. All stakeholders, especially public authorities and the distribution channel, should be provided with regular up-to-date information on risk potential and issues. As the environmental responsibility of a company extends to cover a product's entire life-cycle, logistics, distribution, the use of a product, its possible recycling and disposal all create their own risks. As far as reputation is concerned, it is not important who is actually responsible for the emergence of a crisis: the crucial issue is the product or the company with which the problem is associated (see Fig. 5).

Provided that a company has established a working communication relationship with its most important stakeholders, one of the most important issues to be discussed is risk. The risks connected with a manufacturing site, the security systems designed to prevent them and the measures to be taken once an accident has taken place all need to be discussed with the local community. One of the tasks of scientific staff will be to comment on the company's risk analysis and to assess the effectiveness of preventative measures. Environmental activists will be asked to give their views on the nature of risks and their prevention.

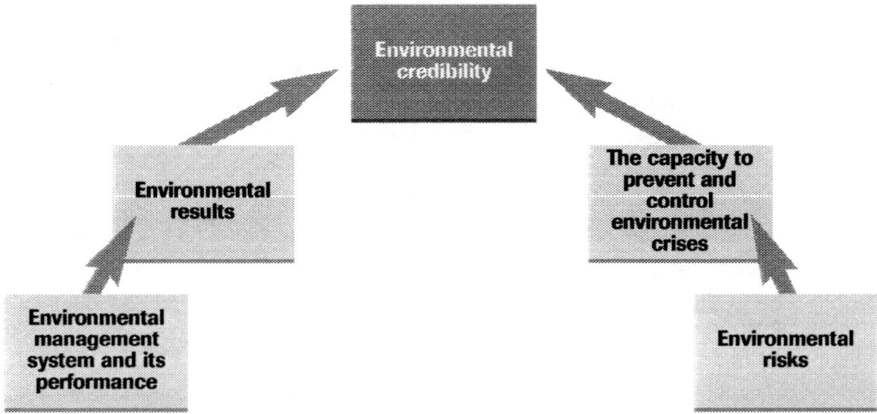

Figure 5: THE BUILDING BLOCKS OF ENVIRONMENTAL CREDIBILITY

In addition to other benefits, environmental reports provide a convenient communication framework for dealing with risks. For example, London Electricity chose the risks connected with the seepage of chemicals from its cable network into the soil as the theme of its 1995 annual environmental report. This unconventional move received plenty of positive publicity in England, and the report was awarded the Deloitte & Touche award for environmental reporting.

Crisis Communication

Preparing for environmental crises requires not only an existing action programme for restricting the crisis and minimising its effects, but a detailed communication plan as well. The credibility of a company will be decided on the basis of crisis communication, as news of an environmental disaster will spread around the world in the space of an hour.

It is possible to prepare for environmental crises only after the possible risks have been identified and the most likely crises defined. A crisis can concern raw materials, manufacturing processes, new products, distribution chains or the storage, use, recycling or disposal of a product. It is typical of environmental crises that mistakes made in the company's past will resurface.

The need for preparation for an environmental crisis is determined by the likelihood of it taking place and the gravity of its possible effects. First, by picturing the worst situation possible, the dangers of the most likely crises can be assessed. When making decisions on investments in crisis prevention, it is advisable to compare the cost of those investments with the costs of repairing the damage caused to the brand or the corporate image by a crisis, or with the costs generated by a tightened regulatory framework brought about as a result of a crisis (see Fig. 6).

Preparing efficiently for a crisis calls for sound guidelines on crisis communication and action, with which all key personnel should be thoroughly familiar. Where possible, these guidelines should be formulated together with key personnel, thus ensuring their commitment and capacity to handle the crisis and apply the correct principles of communication.

Whatever happens, a company must be able to carry on its normal business. To this end, it is important that a special task force for crisis control is assigned, who will assume sole responsibility for the handling of the crisis. The rest of the organisation can then continue its activities as usual.

One of the greatest stumbling blocks for crisis communication is inadequate means of connection. In a typical crisis situation, telephone lines are blocked by inquiries from the surrounding community and customers, directors are unavailable at the accident's location, and the media cannot be effectively informed as there are no telefax numbers to hand. The crisis task force therefore needs its own internal telephones and lines that enable its to make outside calls at any time. In addition, it must be possible to set up a recorded message on an answering machine to answer consumers' most immediate concerns. Furthermore, it is vital that the register of company stakeholders is always kept up to date and available for use.

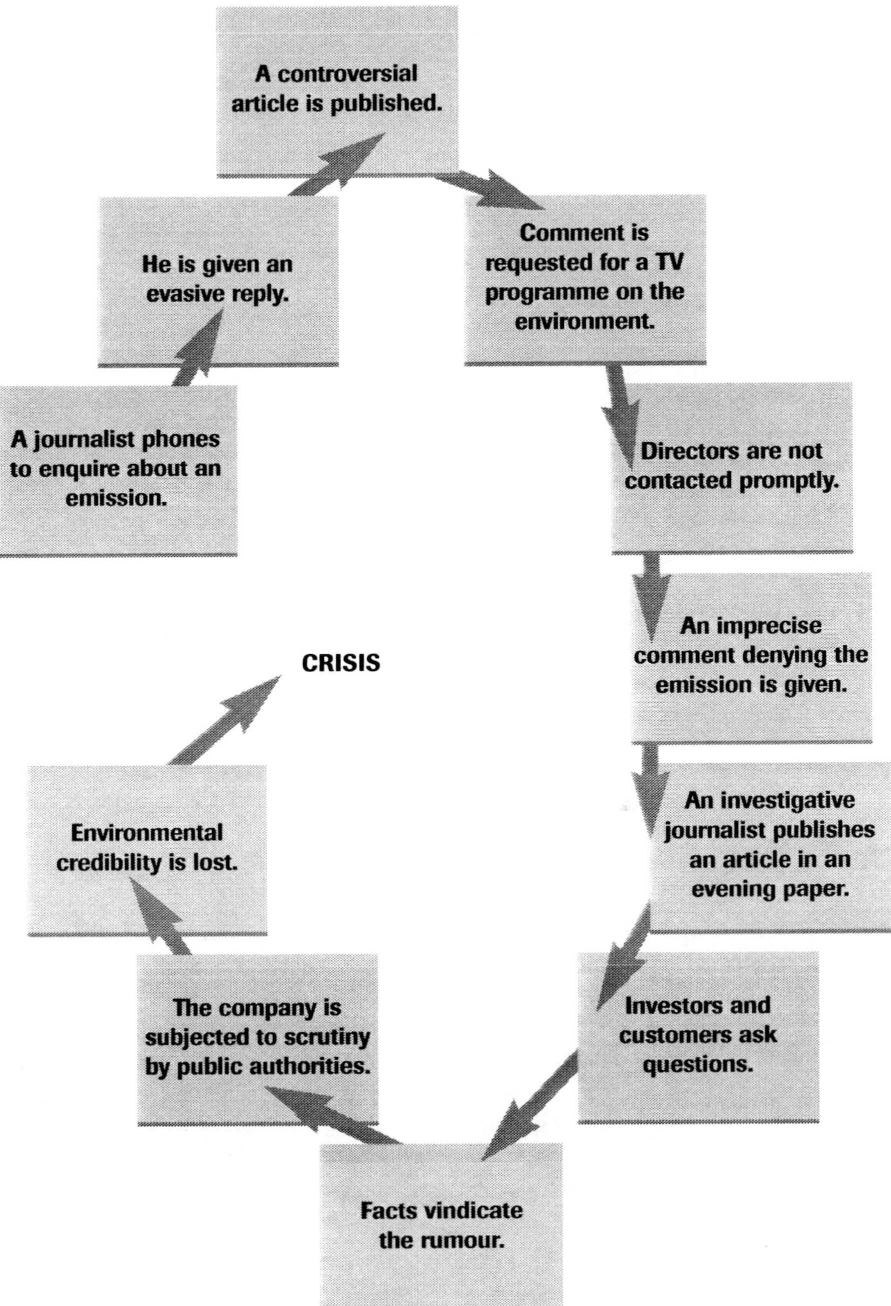

Figure 6: WORST-CASE ENVIRONMENTAL CRISIS SCENARIO

1. Build up environmental credibility and requisite stakeholder connection in advance.

2. Be prepared for a crisis and train your employees on how to communicate in a crisis.

3. Always inform about risks. This will generate credibility and take the sting out of criticism if a crisis occurs.

4. Once a crisis has taken place, admit immediately if you do not know what has happened or how it has happened. Say that the matter is being investigated and say when you are going to provide more information.

5. Accept full responsibility for what has happened. Do not blame your organisation or the circumstances.

6. Do not belittle the situation if stakeholders see it in a more serious light than you do.

7. Be sensitive: if the situation is serious, communicate your own feelings too.

8. Say exactly what is being done and will be done in order to prevent the crisis from reoccurring.

9. Be prepared for a public backlash.

10. Analyse the crisis and the communication measures in retrospect; improve your guidelines.

Box 1: TEN GOLDEN RULES OF CRISIS COMMUNICATION

Crisis communication is the acid test of a company's credibility. In a accident situation, every employee of the company will be regarded by the media as an expert and a source of information, whether this is desirable or not. It is therefore important that all personnel—switchboard, secretarial, sales, etc.—who may come into contact with stakeholders are given training on how to respond to questions and direct inquiries.

Information has to be passed swiftly in a crisis situation, and it must be precise and correct. Effective crisis communication will come to nothing if data on raw materials, quality control of the manufacturing process or the testing of products is collected only once the crisis has emerged. For certain crucial crisis situations, all employees will be provided with their own guidelines regarding co-operation with experts, and these same experts can also be called on to provide statements and assessments in a crisis situation.

Crisis communication does not end when the crisis ends. After the news about the accident itself, it may make the headlines again in connection with coming to grips with the consequences. If the crisis was handled negligently, court decisions will once again bring attention to issues damaging to the company's environmental credibility.

Conclusions

Earning environmental credibility is not an easy task. There are only a few companies that have excelled in environmental communication, which still tends to take the form of tactical ad hoc responses rather than strategic brand building. An integrated approach is needed in situations where different industries, countries and stakeholders are involved. In addition, public concern is gradually shifting not only towards environmental issues but social issues as well, which does not make this complex problem any easier.

However, the need for environmental communication in today's world is paramount. While communication and rhetoric cannot replace the many and different substantive ways of disseminating knowledge, neither can those substantive methods replace rhetoric. According to our analysis, characteristics for successful environmental communication are

- Long-term commitment
- Fact-based information that is of an enduring nature and easily communicable
- Crystallisation of the message
- Willingness to communicate risks as well as positive aspects
- Two-way interaction with stakeholders

Perhaps proponents of corporate environmentalism should learn from direct marketing, which has proved to be an effective way of reaching people. The success of direct marketing arises from a sophisticated form of repetition. People tend to rationalise and simplify messages they receive, which requires the sender's ability to express messages in a compact form which nevertheless gives the right impression.

We should also make a clearer distinction between environmental information and environmental awareness. The discovery of ecological problems has been based on wide, anticipatory scientific studies. Even though this information is technically within the reach of many people in a world of the Internet and mass media, we also need to further the kind of environmental awareness that influences attitudes and consumer behaviour, in order for the fact-based information to fall on fertile soil.

14
Factors Affecting the Acquisition of Energy-Efficient Durable Goods

Hannu Kuusela and Mark T. Spence

> IN THE ABSTRACT, we all love the environment. In the micro-details of our lives, however, it is something to exploit (attributed to Seligan, reported in Geller 1990).

Encouraging consumers to act in socially desirable ways is laudable. In the US, 'Earth Day 1970' set in motion a series of studies, many by behavioural psychologists, which explored methods of curbing environmentally destructive behaviours. 'The research results were encouraging, but large-scale application of . . . practical findings was not to be' (Geller 1990: 269). In hindsight, a catalyst was required to spur pro-environment behaviours. The two oil embargoes in the 1970s, along with other resource constraints such as water shortages in California, served that purpose and behavioural changes ensued. Some of these changes have persisted, such as maintaining a commitment to recycling, but others reverted back in response to the perceived easing of resource shortages. Inflation-adjusted prices for petrol, for example, have fallen in the US since the mid-1980s and, taxes aside, are likely to continue to do so. The response has been to reverse the trend in the US toward purchasing smaller, more fuel-efficient automobiles. The decreased commitment to conserving scarce resources is disconcerting, given findings by Costanzo *et al.* (1986: 522) that 'roughly 85% of the respondents regard the energy crisis as serious, and over half (54%) believe it will worsen.'

In light of the mixed track record by individuals on making pro-environment decisions, this chapter has been produced for three reasons. First, to share with marketers work that addresses how purchasing decisions are made. The focus is on expensive durable goods purchases, specifically household appliances. Household appliances are responsible for approximately one-quarter of total energy consumed by the average household (*Irish Times* 1997). Prior studies that address socially responsible buying behaviours have examined inexpensive products, such as biodegradable laundry detergent (Balderjahn 1988; Ellen *et al.* 1991; Minton and Rose 1997). Second, we provide highlights from a field survey which reflect how Finns make refrigerator (also known as fridge-freezer) purchases. Finally, we suggest how marketers can use the insights concerning how purchases are made when designing promotion campaigns.

We start by presenting three broad methods that can be used to prompt pro-environment behaviours. Attention then focuses on one of these three methods: encouraging consumers to investment in energy-saving durable goods—those that are costly and intended to last multiple years.

Three Approaches to Conserving Energy

Broadly speaking, there are three ways to reduce energy consumption: investment, curtailment and management (Kempton *et al.* 1992). Investments are one-time expenditures in time or money that reduce energy consumption; they can be major or minor. Examples of major investments are insulating one's home, purchasing a more fuel-efficient automobile or replacing a refrigerator; minor investments would include weather-sealing windows, fixing leaky taps or installing toilet dams. Stern (1992) suggests that minor investments are motivated by non-financial factors such as a personal feeling of obligation to conserve energy. Minton and Rose (1997) found that a feeling of moral obligation had a greater effect on inexpensive purchases than did an overarching concern for the environment. One can feel morally obligated to conserve energy without feeling that the environment is in any imminent danger, although the reverse is also possible. Major investments, in contrast, have historically been assumed to be motivated by the expectation of financial gain. Disseminating product-specific information that convinces prospective consumers that benefits exceed costs is therefore thought to be both a necessary and sufficient condition to affect behaviour. Of late, this perspective has been challenged (Black *et al.* 1985; Stern 1986; Costanzo *et al.* 1986; Geller 1992; Finger 1994), an issue to which we will return.

Curtailment, the second of the three categories, is self-sacrificing behaviour: individuals reduce their level of comfort without buying or selling anything. Lowering the temperature in one's home in the wintertime, taking shorter showers and car pooling are good examples. Without prompting, these 'sacrificing behaviours' are unlikely to happen. Stern (1992: 1226) notes,

> [A]dopting new technology is often perceived as an improvement in the quality of life, whereas cutting direct energy use is usually perceived as sacrifice; and improved technologies are inherently long-lived, whereas changes in everyday behavior are easily reversed.

Thus, encouraging investments is more beneficial than curtailment activities because the latter are comparatively easy to discontinue.

Management, the third category, refers to policies that reduce energy consumption. Management activities, whether mandatory or encouraged, prompt investments or curtailment. The government has been proactive in mandating changes to behaviour (e.g. in Finland there are minimum insulation standards for new construction). And both government and industry have offered financial incentives to encourage change (e.g. in the US there are tax rebates for energy-related research and reduced loan rates for energy-saving improvements). In a review by Samuelson (1990) which assessed the effectiveness of offering financial incentives, results varied widely: energy reductions ranged from 3% to 33%. The small reductions may be because the public does not understand the incentives (Costanzo *et al.* 1986). To mitigate misunderstandings, Ireland circulated a lengthy insert in the *Irish Times* entitled 'Energy Awareness Week' (20 September 1997) which described energy-saving technologies/methods with estimates provided for expected payback periods. For example, replacing an older refrigerator with a more efficient one can reduce energy costs by as much as £40 annually. Other countries are providing cost-related information in different formats: for example at the point of purchase, such as providing miles-per-gallon information.

In this chapter we focus on factors affecting energy-saving *investments* in durable goods. To do so, we must understand the psychological processes that affect how individuals make purchasing decisions.

Three Decision-Making Paradigms

Three decision-making paradigms that individuals consciously or unconsciously use which affect energy-saving investments are the rational–economic, attitudinal–behavioural and social norm paradigms. Each has strengths and weaknesses, which are discussed below. Each also implies a different promotion approach.

The rational–economic paradigm has the longest and richest history. An introductory course in microeconomics will start with the assumption that individuals are fully informed utility maximisers: purchases are made that are economically advantageous. Thus, the increase in the price of petrol in the 1970s motivated consumers to purchase more fuel-efficient automobiles. That was rational. Economists would argue that an informed public receiving unambiguous feedback—at the time it was obvious to consumers that energy prices were escalating rapidly—were conducting cost–benefit analysis and therefore purchasing 'optimal' vehicles given their budgets. Hutton *et al.* (1986) found that feedback concerning daily energy consumption does indeed positively affect consumer learning and motivation; and Dennis *et al.* (1990) have suggested ways of disseminating energy-related information for maximum effect. Unfortunately, several arguments discount the rational–economic paradigm.

First, it is often the case that energy-use information is ambiguous. This is best illustrated by an example. Consider a typical utility bill: energy consumption is not itemised by household appliance. As Kempton and Montgomery (1982: 817) note, this is akin to receiving a bill from a store 'without prices on individual items . . . the shopper would have to estimate item price . . . by experimenting with different purchasing patterns '. As a result, it is common for people to overestimate energy consumption of manually operated items such as light fixtures, while underestimating energy used by automatic appliances, such as refrigerators and water heaters (Costanzo *et al.* 1986; Stern 1992). Interestingly, consumers are good at rank-ordering appliances by annual running costs, but not good at articulating *relative* running costs (Crawshaw *et al.* 1985). In sum, ambiguous information concerning relative costs can lead to seemingly irrational actions and purchasing decisions.

Ambiguous information aside, how many people can honestly claim to be knowledgeable appliance purchasers? A benefit of purchasing energy-efficient appliances is reduced future energy bills: hence consumers should be making buying decisions similar to the way corporations do when buying capital equipment—by looking at the benefits and costs over the life of the product. But people tend to use simplistic decision rules. For example, if a person did not factor in rising energy costs (assuming energy costs are rising faster than income), it postpones the break-even time for when cost savings in the form of lower energy bills would justify the upfront purchase cost of a new appliance (Costanzo *et al.* 1986; Kempton *et al.* 1992). An irony is that it is easier to compare the energy efficiency of brands when provided with measures such as miles per gallon or kilowatt usage (Stern 1992), but this increases the likelihood of not considering rising energy costs.

In light of the deficiencies of the rational–economic paradigm in conjunction with the recognition that recent pro-environment actions are often not motivated by financial considerations, the second paradigm, attitude–behaviour, has gained in popularity. Attitudes are defined as 'a learned predisposition to behave in a consistently favorable or unfavorable manner with respect to a given object' (Schiffman and Kanuk 1994). For expensive purchases, attitudes are typically formed prior to making a purchase (Ajzen and Fishbein 1980). The relative effect of attitudes on purchasing behaviour compared to financial incentives is *believed* to be greater for less costly goods, but thus far little effort has been expended on studying expensive purchases. This explains the recent trend toward assessing how much the public is genuinely concerned about the environment, and the extent to which this affects energy use and related decision-making (Black *et al.* 1985; Kempton *et al.* 1992; Stern 1992).

The third paradigm, social norm, takes its lead from innovation diffusion research. Diffusion research examines the speed with which individuals within an environment of interest acquire a new product. People often desire to follow others' leads, particularly when dealing with expensive, risky and/or complicated purchases. For many, the effectiveness of new products needs to be demonstrated, after which purchase of the product is more likely (Rogers 1995). Samuelson (1990) shared findings from two studies highlighting the effectiveness of social diffusion, both of which suggest a strong role for word-of-mouth communication. Communication serves to solidify group norms; it can also make salient the fact that an individual's actions do affect the environment. Ellen *et al.* (1991) found that perceived consumer effectiveness (a belief that individuals acting alone can make a difference) is a distinct construct separate from concern for the environment. People may participate in a recycling programme because they are concerned about the environment, but in their hearts they may feel that, because each is one individual in a world of billions, they will have no significant effect. Another can argue the opposite: because they recycle their daily newspaper, they are directly responsible for saving some trees every year. Perceived consumer effectiveness has been shown to positively affect inexpensive purchases of environmentally safe products; but, again, its effect on expensive acquisitions has not been examined.

Highlights from a Survey on Appliance Purchases

What follow are findings from a national survey of 1,038 households in Finland: 522 of the respondents were women and 516 were men. The study was undertaken: to ascertain where consumers collect energy-related information when considering refrigerator purchases; to determine what criteria are used in evaluating and selecting refrigerators; and to elicit their general attitudes toward energy conservation. By examining these findings, we can gain insight into the extent to which the aforementioned decision-making paradigms are supported.

Respondents were first asked where they would go to find information they considered important and relevant in making a refrigerator purchase. Their responses appear in Table 1. Respondents were able to state more than one information source, hence the total exceeds 100. Furthermore, only categories stated by 5% of the sample or more are shown.

What is striking is that the most biased information sources—sales personnel (who typically receive commissions) and brand-specific brochures—are the information sources that respondents said they would be most likely to use. Conversely, the most impartial information source, construction/trade magazines—magazines produced for building contractors, but usually readily available—were mentioned only 5% of the time. Thus, **consumers may believe they are acting rationally by collecting and comparing brand-specific information, but they are not going to the most objective source.**

Friends were mentioned infrequently (7%), which suggests that **the social norm paradigm is not receiving support**. However, with respect to word-of-mouth communications, these findings may be cultural: perhaps Finns are more reluctant to seek advice from friends than is the case in other countries?

The next section asked what attributes were used by consumers when evaluating and selecting refrigerators. Research on decision-making has found that, when faced with complex decisions, individuals use a multi-stage decision-making process (Payne *et al.* 1993): we first begin with relatively easy-to-execute 'if–then' rules to reduce the total number of options to a smaller, more manageable 'choice set'. For example, if a refrigerator is not of a certain size, it might be eliminated; other attributes cannot compensate for the size deficiency. Similarly, a person may refuse to consider a refrigerator over a certain price. After eliminating some options, compensatory decision rules are then used to select the brand that offers the greatest utility. Thus, if a refrigerator did not perform the best on a certain attribute (e.g. it's not your

Sales personnel	52%
Brand-specific brochures	22%
The press	17%
TV commercials	11%
Friends	7%
Store ads	6%
Construction/trade magazines	5%

Table 1: WHERE RESPONDENTS SAY THEY OBTAIN BRAND-SPECIFIC INFORMATION

	Evaluation criteria (%)*	**Selection criteria (%)***
Energy consumption	35	39
Price	60	38
Outside dimensions	26	30
Warranty/maintenance contract	20	27
Inside dimensions	24	24
Brand name	11	11
Country of origin	10	11

* Respondents could state more than one attribute, hence column totals exceed 100. (n = 1038)

Table 2: FREELY RECALLED EVALUATION AND SELECTION CRITERIA USED WHEN MAKING A REFRIGERATOR PURCHASE

favourite colour), other attributes could compensate (you like the size of the refrigerator). In practice, it is often not a strictly two-stage process. Instead, we go back and forth between using if–then rules and compensatory rules. Furthermore, criteria used to compare brands need not be the same as those used to make a final choice. For example, a person may eliminate brands due to size and style, then make a final selection based on price. This is true both for experts (those familiar with the product class) and novices (Kuusela *et al.* 1998).

Table 2 shows the evaluation and selection criteria respondents could recall using. Only attributes stated by 10% of the sample or more are included in the table and, again, totals exceed 100% because more than one attribute was often recalled. What is striking is that, with one noteworthy exception, there is very little difference between what criteria are used to evaluate brands and what are used to make a final selection: **the exception is price**. Price is by far the most frequently cited criteria used to evaluate/compare brands, stated as important by 60% of the sample. **This does not augur well for a rational total-cost approach to making a purchase**. Energy-efficient refrigerators frequently cost more up-front, but will yield downstream benefits over their life in the form of reduced energy bills. Apparently, consumers are more concerned with immediate losses than future gains. Unfortunately, this suggests that, for many individuals, energy-efficient refrigerators are *not* likely to be purchased. Cost savings over the life of the product must therefore be made more salient to decision-makers. As stated, according to 'Energy Awareness Week' in Ireland, a modern refrigerator can result in savings of up to £40 annually. As further support for the need to stress long-term cost savings, energy consumption (the cost of operating the appliance) was the most frequently stated selection criteria (39%) and the second most commonly stated evaluation criteria (35%). Thus, although energy consumption does not receive the attention price does, over one-third of respondents claim they consider it.

The final section of the survey asked a series of questions that assessed attitudes and opinions about energy conservation. Looking at financial benefits, women and older individuals were more likely to say they will save money by buying a more energy-efficient refrigerator. This lends support for the rational–economic paradigm, at least for these two groups. However, other responses support the social norm. There was relatively strong agreement that saving energy protects the environment and that individuals can have an effect on the environment by acquiring a more efficient refrigerator. Again, the latter was particularly true for women and older individuals; it was also true for those with greater income. These findings support Stern's (1992) contention that appealing to non-financial benefits is advantageous even for major investments. It also shows there is belief in 'perceived consumer effectiveness'. Importantly, across all groups there was strong *disagreement* that 'energy conservation is just a fashionable idea'.

Two other interesting insights concern information source and message. Respondents neither agreed nor disagreed with the statement 'I have discussed electricity savings with friends.' This supports the finding noted previously that friends are rarely reported as a source of brand-related information. The other insight concerns message type. Brand-related information reporting kilowatt-hours of energy consumed is preferred *less* than information on savings in Finnish marks. An argument for not reporting monetary figures is that automobile manufacturers have been successful at getting consumers used to comparing miles per gallon ratings. A Canadian appliance-labelling programme did have the desirable effect of prompting producers to manufacture more energy-efficient models (presumably because making brand comparisons was easier) even though few consumers claimed they read the labels (Stern 1992).

Harnessing Insights from the Three Paradigms

Our objective is not to support or refute these three paradigms, and it is certainly not to criticise how purchasing decisions are actually made. Each of the paradigms has strengths and weaknesses, and therefore can be integrated to yield a more complete picture of how decisions are made. We present the following recommendations for how marketers can harness the insights gleaned thus far when developing promotion campaigns for expensive durable goods purchases. Note that actions that assume the rational–economic paradigm can be undertaken at the brand level; affecting attitudes and social norms, however, is a longer-term strategy and, due to cost considerations, is best done by some form of collective (government?) action.

Rational–Economic Paradigm

1. **Provide unambiguous information.** The public prefers information in monetary units (e.g. energy consumption expressed in Finnish marks), but governments often encourage the use of ratings such as kilowatts used or miles per gallon. Consider providing both an estimate of yearly operating cost and a non-monetary unit of measure. In the event competitors do not disclose this information, your brand, by providing it, would have an information advantage over the competition. It is generally believed that, when information is not disclosed about a brand's attribute, consumers assume the brand performs poorly on that attribute. But beware: avoid the perils of information overload. Consumers cannot be expected to wade through masses of brand-related minutiae. Succinctly organise a few pieces of particularly meaningful information, probably best done in a table format.
2. **Encourage consumers to conduct a total-cost analysis.** Suggest how long the expected payback period is, given normal use. Offer multi-period payment plans so payments occur when benefits are realised.

Attitude–Behaviour Paradigm

1. **Encourage pro-environment attitudes.** Participate and/or sponsor events such as 'Earth Day' to draw attention to the environment. Cities and corporations have worked together to promote and organise a day to clean up streets, parks, beaches, etc. The cities get beautified, corporations get publicity, and the participants may receive free food, prizes, etc.
2. **Disseminate information about the effect that products such as refrigerators have on the environment.** Perhaps this could be done at the events mentioned above. How harmful are the CFCs in older refrigerators? In total, how much energy do refrigerators use?

Social Norms

1. Stress that individuals can affect the environment: in other words, develop 'perceived consumer effectiveness'. Finns believe that their purchasing decisions affect the environment. A variety of publications such as Ireland's 'Energy Awareness Week' discuss what individuals can do to help the environment.

2. Although longer term and probably best undertaken by the government, perhaps in public service announcements, encourage consumers to talk about energy-related issues.

Concluding Remarks

There were three primary objectives for this chapter. First, to share with marketers work concerning how purchasing decisions are made that affect energy consumption, with a focus on durable goods purchases. Second, to provide highlights from a field survey which reflect the extent to which the decision-making paradigms are being supported. And, finally, to suggest how marketers can use these insights when designing promotion campaigns.

Sales personnel and sales brochures, two biased sources of information, were mentioned most frequently as providers of important brand-related information. The least biased source, construction/trade magazine reviews—more of an effort to obtain, but providing objective comparative data—were mentioned by only 5% of the respondents. Thus, people may believe they are acting rationally, but they are probably not. Seven per cent of the respondents mentioned friends as sources of information, which discounts the social norm paradigm. When asked to freely recall criteria used to evaluate brands, price was stated 60% of the time, far more than any other criteria. This suggests that the more costly energy-efficient brands may be removed from people's consideration. In light of the tendency to refer to sales brochures, producers should encourage total-cost analyses. Cost savings over the life of the product must be made more salient to decision-makers, a tactic that is being used in the US to promote refrigerators with expensive add-ons. In this case, the refrigerator featured water purification filters so salespersons stressed the money that could be saved by not having to buy bottled water.[1] Helping people make cost–benefit trade-offs would clearly be beneficial. Multi-period payment plans so that payments are made when benefits are realised would also help.

Respondents did acknowledge that saving energy protects the environment; energy conservation is *not* just a fashionable idea; and that they can have an effect on the environment by acquiring an energy-efficient refrigerator. Thus, there is potential in appealing to non-financial societal gains, even for expensive purchases.

---

1 *New York Times*, 14 September 1997.

15

Greening the Brand

Environmental Marketing Strategies and the American Consumer

Daniel S. Ackerstein and Katherine A. Lemon

'IT'S NOT EASY being green.' Though this phrase has become the time-worn cliché of the 'environmental decade', it is nonetheless true, and perhaps now more so than ever. The tenuous, awkward and often maddening relationship between corporate America, the environment and the consumer has never been more complex. The history of this relationship, and its progeny, 'environmental marketing', is filled with dramatic shifts in consumer attitudes, misdirected and misleading corporate responses to those shifts, and the occasional efforts of government at a variety of levels to intervene. Unfortunately, the following frustrating cycle appears to be emerging. Consumers, seeking environmental performance in the products and services they purchase, select products that communicate environmental performance information or a message of corporate environmental responsibility and stewardship. Corporate America responds, but the voices of companies sincerely committed to such notions are too easily drowned in a sea of often misleading, unethical, or even untrue claims by firms seeking to grasp a share of the environmental market and put environmental-guilt-feeling consumers at ease. Environmental and consumer watchdog groups are happy to see such products arrive, but view them with suspicion. When their in-depth analyses of these products expose vague and misleading performance claims, the alarm is quickly and noisily sounded. Frustrated, confused consumers give up their mission of environmentally responsible purchasing, doubt regarding all environmental claims becomes the norm, and the market for such products recedes again.

Breaking this cycle represents the critical challenge for environmental marketers in the next decade. Environmental marketing presents a bold and exciting opportunity for corporate America to start delivering something many consumers have been seeking for years: products that are environmentally responsible in their production, packaging, performance and disposal. Environmental marketing combines two simple steps. First, companies take responsibility for the environmental effects of producing, selling and disposing of their products. By doing so, producers cut costs and reduce waste—more importantly, they also add value. Second, companies employ marketing strategies to identify customers concerned about the environment, understand their concerns, and deliver to these customers the environmental-value-added products (products that deliver superior environmental performance or are produced in an environmentally conscious or responsible manner) they seek. These strategies must be carefully designed and considered; every effort at environmental marketing represents an opportunity to convince consumers of the sincerity of a company's efforts as well as a risk that jilted environmental buyers will be deceived again.

Survey after survey shows that people are concerned with the health of the natural environment and even willing to pay to see its protection and repair. In the United States, for example, Cambridge Reports/Research International (1992) reported that more than half of Americans believed the government provided insufficient environmental protection. In fact, 75% choose to identify themselves as 'environmentalists' (Mackoy *et al.* 1995). Perhaps more indicative of a budding connection between environmental concern and buying patterns is the finding that 80% of Americans claim to have changed their behaviour due to environmental concerns (Cambridge Reports 1992). Sixty-six per cent say that they have switched product brands in an effort to obtain more environmentally sound products (Ottman 1995). More recently, 70% put 'environment' ahead of 'economic growth' in pursuit of a public balance between the two (Roper-Starch 1997).

Nonetheless, the sincerity of Americans' concern appears to be limited. A recent article in *USA Today* notes that, though many claim to be 'environmentalists', in reality they 'tend to act more brown than green' (Watson 1998). That is, Americans are apt to express intense concern for environmental causes and issues in surveys, conversation and the media, but often fail to manifest the same level of concern in their daily life choices and behaviour. Though many individuals contribute to environmental organisations or support environmental legislation, consumer behaviour suggests a much more complex relationship between environmental feelings and the tangible expressions of those feelings.

One such expression can be found in the behaviour of consumers. The very act of purchasing, choosing one item or service over another, represents an expression of the values and feelings that each of us imbues. Through the lens of environmentalism, the way we consume, the way we purchase, the way we shop, are all expressions of how we feel about the environment and our place therein. As consumers have prioritised the environment in their purchasing decisions, corporations have responded, adding labels, new brand names and environmental performance information on a variety of products in a concerted effort to seize the environmental opportunity.

In fact, for many years, environmentally 'responsible' or 'friendly' products were among the fastest-growing segments of the US consumer products market (*Supermarket News* 1990). As American consumers demanded improved environmental performance, companies answered with a pantheon of environmentally 'conscious' or environmentally 'friendly' products. Third-party 'eco-label' programmes seemed poised to bring order to the confusion surrounding environmental performance descriptions. Academics and professionals spoke of an emerging 'environmental market' and wondered how long it would be before products faced tough environmental evaluations and labelling as to their ecological sensitivity. However, in the 1990s, the domestic environmental market materialised in ways quite different from those many authorities in the field had predicted. Except for a few isolated products ('dolphin-safe' tuna, for example), eco-labelling programmes have fallen far short of projections for relevance in the consumer marketplace. Once touted as the likely final authority on products' 'greenness', labelling programmes in the US have failed to overcome the scientific and technological hurdles necessary to be of genuine use to shoppers. Few products, other than a select group identified by consumers with environmental hazards or prominent issues in the public psyche, currently promote their environmental performance or responsibility. And fewer consumers seem to seek such labels or marketing in making their purchasing decisions.

A company's communication of its environmental performance or responsibility efforts can take a variety of forms. Television and print advertising often generate the most public attention (both potentially positive and negative attention), while other efforts, including corporate environmental reports, public information campaigns and similar methods, can communicate a message of environmental content. In the consumer products sphere, however, the most important means for such communication may be product packaging. The importance of packaging as a communications tool appears to be growing (Belch and Belch 1990) and its role as a vehicle for environmental information may prove to be pivotal.

The specific methods developed by product and brand managers to communicate information via product packaging are countless. But the techniques chosen to communicate environmental information have been somewhat more consistent. Two forms of environmental marketing have become prominent among managers. The '**environmental performance label**' (distinct and wholly different from the third-party 'eco-label') can be defined as *'a simple representative word or icon describing environmental performance'*. Examples include the ubiquitous 'chasing-arrows' recycled/recyclable logo or the CFC-free designation on aerosol canned products. The label usually integrates specific environmental performance information, such as the percentage of product content that is from recycled materials. In contrast, the '**environmental line extension**' *represents a distinct sub-brand of an existing brand-name product; a sub-brand that highlights environmental performance as a key identifying and differentiating factor.* The line extension often makes somewhat less specific or substantiated claims, while implying a more comprehensive environmental commitment on whatever level. Examples might include Ben & Jerry's 'Rainforest Crunch' Ice Cream, Maytag 'Neptune' washers and Procter & Gamble's Cheer 'Enviropak'. Both strategies have been employed in the past with varying degrees of success. But little research exists examining the relative effectiveness of the two strategies in comparison to one another. Might consumers, made cynical by years of deceptive 'greenwashing', prefer the informative, fact-focused specifics of the environmental label, or might they be more likely to select products that suggested a more comprehensive environmental commitment (i.e. the environmental line extension)?

Or might they, in fact, summarily reject both efforts, and perhaps signal the impending end of the environmental market?

The answers to these questions must also consider the nature of the consumer at hand. More environmentally aware or active consumers might respond differently to various packaging-based environmental marketing strategies compared to their less 'green' counterparts. For such information to be useful to marketers and brand managers, identifying which customers responded positively (and negatively) to which environmental marketing strategy would be essential.

The Environmental Market: Just Beginning . . . or Ending?

It is only in the past 25 years that the concepts of 'environment' and 'market' have merged in a manner relevant to the daily lives and decisions of consumers. Since the 1970s, environmental awareness and activism in the United States has grown erratically. In the past quarter-century, American consumers have become better educated on environmental issues and more active stewards of global, local and personal environmental health. Environmental products have evolved on a parallel course: natural and organic grocers and recycled products are hardly new to the consumer products market, but it is only in recent years that the notion of 'environmental marketing' or an 'environmental market' itself took hold.

Environmental marketing is the marketing or promotion of a product based on its environmental performance or an improvement thereof. Environmental, or 'green' marketing, represents both a vast opportunity and a potential minefield. Consumers would undoubtedly prefer their consumption to be less environmentally damaging, but they are also highly suspicious and critical of corporate efforts to use the environment as a sales tool.

The 'environmental market' is a complex notion. This market is made up of consumers who, in some form or fashion, integrate environmental concerns into their purchasing decisions. The complexity arises in the fact that these concerns relate to a wide variety of issues, from local landfills overflowing with diapers to global climate change. They express themselves in a variety of behaviours, from reducing consumption to recycling waste. Compounding the difficulty of analysing the environmental market is the notion that perception is the critical factor: consumers often have no means of evaluating the environmental performance of their purchases; they must make decisions based on perceptions created by a spectrum of conflicting and confusing influences. For example, a product claiming 'recycled content' may be of varying percentages, may be from pre- or post-consumer waste, or may be merely packaged in a form containing recycled materials. The environmental consumer market is also inconsistent: purchasing priorities in one scenario may go unconsidered in another. Though many shoppers consider environmental factors in purchasing paper products or produce, such factors may play a much less prominent role in the purchase of larger durable goods, such as appliances.

The environmental market also includes another, quite different facet. The business-to-business market for products offering environmental performance involves a second set of unique factors and considerations. As such, this chapter focuses on final consumers only, rather than attempting to encompass both of these disparate areas.

To understand more completely the realities behind concepts such as green marketing and the environmental market, it is useful to divide our discussion into four sections: the environmental consumer, the environmental company, environmental products, and marketing the environment.

The Environmental Consumer

Identifying the American 'environmental consumer' has not been difficult. This consumer has most often been described as young, better educated and higher in income, occupational status and socioeconomic status than the average (Cude 1992; Hartman 1992; see also Ellen *et al.* 1991; Schwepker and Cornwell 1991; Granzen and Olsen 1991 for comprehensive reviews of literature). Not surprisingly, liberal political preferences also imply increased environmental consciousness (Cornwell and Schwepker 1992; Van Liere and Dunlap 1980).

A major psychographic study by the Roper Organisation undertaken for the first time in 1990 attempted to identify the different types or levels of environmental consumer in the US. The 1,413 respondents to the survey were divided into five categories based on their responses to a pair of questionnaires describing the level of seriousness with which they viewed certain environmental issues and their engagement in a number of environmental activities (Speer 1997; Cude 1992; Miller 1990; Roper 1990; Schwartz and Miller 1991). The categories outlined in the 1990 study are described below.

'True-Blue Greens' are the most active environmental consumers. They recycle, compost and re-use whenever possible, support environmental groups and regulations and, most importantly, avoid buying products from companies they perceive as not environmentally responsible. Their numbers have remained largely stable: from 11% of the population in 1990 to 12% in 1997. As Tibbett Speer (1997) notes, one True-Blue in every ten Americans is a sizeable share—but what may undermine their influence is their willingness to pay, on average, no more than 7% more for ecologically friendly products.

'Greenback Greens' are more enthusiastic about spending more for environmentally friendly products—up to 20% more. This group tends to be less politically and socially active in environmental causes than their True-Blue comrades. They share many of the ideological and moral viewpoints of the True-Blues, but feel they are too busy to make changes in their lifestyle that would benefit the environment. Greenback Greens are rapidly disappearing, from 11% of the population in 1990 to only 6% in 1997.

Many former Greenback Greens have moved into the category of 'Sprouts'. Sprouts are somewhat concerned and educated on environmental issues and are willing participate in environmentalist behaviour that does not require great effort or initiative: for example, buying recycled products or recycling newspapers. However, they are willing to pay no more than 5% more for such products. In recent years, Sprouts have grown from 26% in 1990 to 37% of the population in 1997.

'Grousers' are reluctant to involve themselves in environmental issues. They are unenthusiastic about spending more for green products and place the responsibility for environmental clean-up on corporations rather than individuals. The Grousers are likely to be less educated and less affluent than the True-Blues, Greenback Greens or Sprouts. Their ranks have fallen from 26% of the population in 1990 to 13% in 1997, a decline that may spell more bad news than good for the environmental market.

Many ex-Grousers have become 'Basic Browns'. This group grew from 28% in 1990 to a dominant 37% of the population in 1996, but dropped back to 1990 levels the following year. The Browns are unlikely to pay more for green products, are uneducated on environmental issues and unimpressed with recycling and other environmental efforts. They are overwhelmingly male, and most often blue-collar workers.

Further research has attempted to fine-tune the volume of data collected by Roper. Bei and Simpson (1995) found that psychological benefits are a significant factor in motivating the environmentally conscious purchasing decision. Their research also noted that consumers expected to pay more for such products, regardless of their willingness to do so.

Today, it seems clear that the environmental consumer both exists and remains identifiable. Though Roper's 1997 report suggests a small decline in environmental consumerism, there is little doubt that an 'environmental market' will continue to exist. Unfortunately, consumers have been more enthusiastic about verbally expressing their concern for the environment than about directing their purchasing habits toward green products (Mayer *et al.* 1993; Shrum *et al.* 1995). The key to reaching this market is twofold: presenting products that are 'green' in a meaningful and relevant sense, and assuaging consumer scepticism that environmental marketing efforts and claims are mere corporate misinformation.

The Environmental Company

The consumer is not alone in the environmental market. Equally important is the role of the companies who design, produce and deliver products to that consumer. The literature on environmental marketing is dominated by tales of companies such as Patagonia, Stonyfield Farms and The Body Shop, whose environmental ethic has become a centrepiece of their corporate strategy and culture.

Corporations undertake environmental initiatives for a wide variety of reasons. Many firms see environmental improvement as an opportunity for cost cutting via waste reduction and increasing the efficiency of their operations. Others attempt to stay ahead of increasingly stringent environmental regulations and enforcement policies. Environmental performance initiatives can also be part of larger corporate responsibility and ethics movements, or designed to improve corporate image and public relations. Most probably, companies are motivated by a combination of these goals. Environmental marketing represents a relatively unique synthesis: often an opportunity for environmental marketing is the by-product of environmental improvements undertaken for other reasons. For example, a diaper company may choose to produce more environmentally responsible disposable diapers (with recycled-content fill, non-bleached fibres or similar features) because they see a market segment of environmentally conscious diaper buyers. On the other hand, that company may produce the same diaper because it wants to improve its image as a responsible corporate citizen, to avoid costly regulatory issues relating to its production process, or to save money by improving operational efficiency. The opportunity for environmental marketing arises in this case as well, even though the motivation for the environmental improvement was unrelated to green marketing potential. (For additional discussion on this topic, see Keogh and Polonsky 1998a.)

Why do many companies choose to improve environmental performance but fail to explore the corresponding opportunities for environmental marketing? To take advantage of these marketing opportunities, companies must carefully evaluate their options from a number of perspectives. First, they must consider the real-world costs associated with improving environmental performance to create an environmental marketing opportunity. Alterations of processes, products, energy demands or waste-streams require a commitment that is both philosophical and financial. Though many environmentally oriented operations improvements save money in the long term, the capital requirement in the short term can dissuade many firms. Management's ability to prioritise environmental considerations is constrained by the spectrum of business priorities (e.g. quality control, cost management) that compete for management and employee resources and mind share.

Second, businesses considering an entry into the environmental market must assess the potential risks associated with doing so. These risks are often intangible. For example, taking full advantage of environmental marketing may require a shift in product positioning (i.e. the perception of the product or brand by the consumer). Any such shift represents a risk for brands or products that have found success with their current positioning or market strategy. Compounding this risk is the persisting consumer perception that environmentally marketed products are inferior in terms of quality. Product managers are reluctant to risk having consumers attach this perception to their brands or products. Finally, some companies fear the 'message' sent to consumers by environmental marketing. Rather than communicating environmental responsibility, these firms fear that green marketing is interpreted by consumers as 'preaching' or a 'guilt-trip'.[1] Few companies are enthusiastic about gambling on their customers' positive reaction to such marketing efforts. The punishment companies face for a failed environmental marketing strategy can be severe. If consumers question the credibility or truthfulness of environmental claims, damage to the brand, in terms of loss of market share or goodwill, can be significant (Ottman 1992; Stisser 1994). In addition, exposure by the media or environmental groups of a misleading, incomplete or genuinely false claim can be devastating to public image, not to mention leading directly to decreased sales (Polonsky 1995a).

---

1 Personal communication with the Vice-President for Environmental Issues of a major beverage company, April 1998.

Third and finally, companies considering environmental marketing need to consider the benefits that such efforts can generate. Improved environmental performance is an obvious positive outcome—not only reducing many long-term costs, input requirements and waste management needs, but also controlling environmental liability and compliance costs. In addition, companies can benefit from an improved brand image in the public eye and a perception among consumers as a more responsible, community-conscious firm. Not least is the potential opportunity to gain a competitive advantage over competing products that do not deliver environmental performance. If consumers respond enthusiastically to environmentally marketed products, the market share gained can be significant.

Environmental Products

During the 1980s, the environmental product market appeared to have a bright future. *Supermarket News* published a report by Marketing Intelligence Service that noted 100% growth in the introduction of green products each year between 1985 and 1990. Though green products still held only a small fraction of the market (4.5%), that number had been a nearly invisible 0.5% just five years before. Articles in business journals and magazines warned corporations to get on the environmental bandwagon or be left behind. A national or international environmental labelling system seemed imminent—a system that would allay the mistrust of consumers already concerned about dishonest or incomplete environmental claims.

Since then, the growth of the 'environmental market' has been difficult to chart. Though 'green' product introductions have grown to represent approximately 10% of the new products introduced each year, they have shown little likelihood of surpassing that level in the near future (Makower 1998). *Natural Foods Merchandiser* magazine claims that sales of green products in the US (including, but not limited to, food) have more than tripled since 1991, and now exceed $100 billion each year (Del Franco 1998). And a 1996 audit of grocery store products across the country found that 66% of the 397 brands examined by the authors made claims, either explicit or implied, or environmental product or packaging performance (Mayer *et al.* 1996). And, yet, opinion exists that, as Malcolm Brown so aptly puts it, finding items marketed and promoted on the basis of their 'greenness' has become 'like a game of hunt the thimble' (Brown 1997).

Price has been, and will continue to be, a critical issue in any discussion of environmental products. Brenda Cude (1993) examined the popularly held notion that the green alternative is more expensive. Her research found that price differences between green and non-green products for a set of various categories were essentially non-existent. Only one product in a multi-category product sample, coffee filters, consistently bore out the notion of an 'environmental premium'. Unfortunately, in the mind of the American consumer, the perception persists.

Marketing the Environment

Though the US Environmental Protection Agency and much of the popular press assumed that the environmental market in consumer and grocery products would flourish rapidly, in reality it stumbled. Quality, convenience and price were the market's Achilles heels. According to Jacquelyn Ottman, a leading authority on environmental marketing and author of *Green Marketing* (1995):

> Recycled paper sometimes looked and felt substandard. An early brand of 'green' fabric softener clogged washing machines. It wasn't just that the products were outdated and low quality, they were overpriced. Consumers soon caught on to the double whammy of inferior goods for ridiculous prices.

Consumers were willing to pay more, but they would not pay more for so much less (Speer 1997).

In recent years, price premiums for environmentally marketed products have largely diminished. Though Cude's research suggests that the price premium for environmental goods is largely mythical, it is the perception of such a price premium that endures. Nonetheless, consumers' willingness to pay more for products they perceive as environmentally sensitive also persists, as documented by Roper's *Green Gauge* report (Roper-Starch 1997). This willingness suggests an important window of opportunity for companies entering the environmental market: the opportunity to increase prices by as much as 5% to cover the (possible) increased costs of greening their products. Consumers today are prepared to *pay more* for such products, as long as they believe they are not simultaneously *sacrificing quality*.

Green advertising also rose and fell in the 1990s. Though two-thirds of American adults recall seeing environmental labels or claims when shopping and 54% recall advertisements with such claims, less than half say they have purchased a product due to those claims (Speer 1997). Part of this discrepancy may have to do with the confusing, often vague nature of much environmental advertising. In fact, even the research on this subject is somewhat contradictory. A 1992 study noted that environmental claims presenting companies as 'environmental good citizens' or which state environmental facts, rather than those that discuss specific virtues of products or processes, are likely to contribute to consumer confusion about environmental advertising (Carlson *et al.* 1992). In contrast, a more recent study by Thorson *et al.* (1995) examining the effectiveness of green television commercials found that not only were consumers positive in their response to green advertising, but that the response was significantly more enthusiastic for commercials that said either 'Look at all the good things we're doing for the environment' or 'Let's all learn how to be better to our environment'. Responses to product- and process-oriented advertisements were less positive. What is clear, however, is that consumers are thoroughly distrustful of green marketing 'hype' and see little connection between purchasing green products and helping the environment (Moore 1993; Mohr *et al.* 1998).

Of course, consumers have every reason to be suspicious of companies' claims of environmental performance and responsibility. Between 1990 and 1994, in the United States alone, more than 60 cases of litigation regarding environmental performance claims were brought to the Federal Trade Commission (FTC), National Advertising Division and state attorneys general. Among the worst offenders were: aerosols, plastic waste disposal and shopping bags, disposable diapers, household paper products, and coffee filters (Scammon and Mayer 1995). The most notorious case involved Mobil Corporation's 'Hefty' brand rubbish bags.

In 1991, Hefty introduced a breakthrough 'biodegradable' plastic rubbish bag. The bags, though indeed biodegradable after exposure to sunlight, wind and rain, were not at all biodegradable in the landfills where such exposure is deliberately avoided. After environmental groups called attention to Mobil's deception, the FTC and a number of states' attorneys general challenged the company in court. As part of the settlement, Mobil agreed to remove the claims from their products (Caswell 1997).

The public nature of this case and similar cases undermined consumer faith in environmental advertising and served to illustrate not only the complex nature of producing an environmentally sensitive product, but the difficulties in making truly environmentally conscious purchasing decisions. Perhaps only now are consumers beginning to view environmental-cause marketing efforts with reduced scepticism. Ropers' 1997 *Green Gauge* report found that 80% of Americans considered such marketing and advertising generally acceptable (Roper 1997).

More recently, a debate has emerged questioning the very existence of an environmental market. Harvey Hartman, president of the Hartman Group research and consulting firm, concludes:

> people talk about the green market, but I don't believe there is one. We've gotten on an emotional bandwagon with this subject. We've developed a sense that the environment is more important in most people's daily lives than it really is (Speer 1997).

Hartman argues that the toughest lesson for green marketers has been that the environment matters much less to consumers than do price, quality, convenience and other factors. Nonetheless, Hartman does note that many consumers (52%, according to the 1998 *Hartman Report*) want to purchase environmentally friendly products if the factors listed above remain constant (Hartman 1998). Karil Kochenderfer, director of environmental affairs at the Grocery Manufacturers of America, agrees: 'Consumers want to buy environmentally sound products.' But only if they are competitive on value and price (Racho 1997).

Recent research by Roper seems to confirm Hartman's argument. Most indicative is the finding that fewer consumers perform environmentally conscious purchasing activities on a regular basis than in previous years. Fewer Americans seek environmental labels, buy recycled material products, use biodegradable cleansers, or avoid styrofoam and aerosols (Roper-Starch 1997). The same research suggests that consumers prioritise price, brand recognition, others' recommendations and convenience more than environmental impact.

Complicating these findings is the inherent difficulty in shopping with an environmental conscience. Joel Makower, editor of *The Green Business Letter*, notes that 'Buying environmental products is hard. The claims are vague, and the science behind them is less than clear-cut' (Watson 1998).

These developments may not signal the impending doom of the environmental market. In fact, they may merely be a sign of the transition of the green market into maturity. One theory argues that the environment is no longer the hot-button issue that it once was; instead, consumers have begun to assume that manufacturers have dealt with environmental issues on their own. Malcolm Brown, in a recent article for *Management Today* (1997), claims that

> the makers of the main brands . . . have made their own standard brands so environmentally sound that concerned consumers now feel they can get the best of both worlds. Previously green shoppers often had to pay a premium for green products which, not infrequently, performed less well than the main brands . . . Now they can buy the traditional brands . . . at competitive prices, safe in the knowledge that they are both effective and environmentally sound.

Green Labels, Line Extensions and the Consumer

In such a critical and challenging arena of marketing, how can brand and product managers hope to reach consumers most effectively with packaging-communicated information about their products' environmental performance? This study asks four important questions in an effort to better understand American consumers' responses to environmental marketing in its various forms:

- First, do consumers prefer products that communicate environmental performance via packaging? Though the answer to this question may seem an obvious 'yes', it is important that this assumption does not go unconfirmed. All other things being equal, most consumers would rather purchase products that are ecologically sound. The real issue is: Do consumers respond to on-packaging environmental information when making purchasing decisions?
- Second, what form of environmental marketing is most effective in communicating environmental value to (a) American consumers in general and (b) major segments of the US consumer population? Both environmental performance labels and environmental line extensions (as discussed and defined previously) have seen success and failure in the past 20 years, but rarely have the two methods been compared directly (i.e. keeping all other product attributes constant). The colourful history of environmental performance claims and labelling efforts suggests that consumers may view a simple, understated environmental performance label (e.g. 50% recycled content) more favourably than an environmentally oriented line extension (e.g. Cheer Enviropak). Alternatively, consumers may be more likely to trust a direct and clear label than they would the more marketing-savvy line extension.
- Third, are individuals who are more environmentally active or aware more likely to purchase products that highlight environmental performance? Specifically, will environmentally active or aware consumers exhibit stronger preferences for environmental labels or environmental line extensions? The Roper

Organisation's identification of the environmental consumer provides a gauge of purchaser environmentalism. There is some question, however, whether the True-Blues, Greenback Greens or even the Sprouts will be likely to seek or trust environmental marketing information on consumer products.

- Finally, how does the effectiveness of environmental marketing relate to a corresponding adjustment in price (for example, a 5% price premium). Can environmental information, communicated either via environmental label or line extension, overcome the negative effects of the price increase in the minds of consumers?

Asking Consumers

The answers to these questions can be found only by asking consumers themselves. Testing the relative effectiveness of environmental labelling and line extensions on consumer purchasing required a forum within which consumers could make meaningful choices among familiar products and brands—'real-world' buying decisions. To provide such a forum, consumers were presented with a simple survey asking them which of two grocery products (identified by brand logos, product descriptions, prices and, in selected cases, environmental marketing information) they would be most likely to purchase. Complicating this simple experiment was the intention to compare multiple marketing strategies with one another, rather than simply to a constant case. To construct an experiment with real-world applicability, it was critical first to identify the consumer 'choice share' commanded by each of the two brands: that is, the percentage of respondents choosing Brand A and the percentage choosing Brand B. For example, if 30% of respondents selected Pampers disposable diapers and 70% chose Huggies when the brands were unmodified, then any alteration in those percentages must be ascribed to the changes in product information, logo or price that have been made. That is, if the addition of an *environmental label* (keeping all other product attributes constant) increased Pampers choice share to 40% and reduced Huggies' share to 60%, then that difference in 'choice share' would be attributed to the addition of an the *environmental label*. By performing the same experiment comparing Huggies to a Pampers *environmental line extension*, it was then possible to compare the relative effectiveness of the two methods in increasing choice share. Additionally, it allowed for a direct comparison, *within a single brand*, between an environmental label and an environmental line extension, in the hope of providing insight into the influence that comparison and relative choice (between brands) might be having on this purchase decision. The experimental design took the form of the example in Table 1, using one of the actual fictional products developed.

To remove any product or subject-specific biases, four different product categories were tested, using two different brand-name products for each category. Additionally, to ensure that subjects were not influenced by the alteration of a single product logo through the course of the survey, no subject was asked to compare products of the same category more than once. To put this more simply: each subject faced four

Experimental condition	Brands seen by consumers		
Comparison condition #1 Control brand versus base brand (no environmental information)	Huggies (Base)	versus	Pampers (Base)
Comparison condition #2 Control brand versus environmental label	Huggies (Base)	versus	Pampers Recycled-Content (Label)
Comparison condition #3 Control brand versus environmental line extension	Huggies (Base)	versus	Pampers Naturfil (Line extension)
Comparison condition #4 Environmental label versus environmental line extension	Pampers Recycled-Content (Label)	versus	Pampers Naturfil (Line extension)

Table 1: CHOICES OF (FICTIONAL) DIAPER PRODUCTS PRESENTED TO CONSUMERS

purchasing decisions; each decision represented both a new product category and a new comparison in terms of marketing strategies.

A number of issues had to be considered in designing the experiment: which product categories to use, what price to charge, which brands to use within each category, how to communicate brand logos and what additional product information to include. Each of these issues is addressed below.

Product categories. Selecting product categories proved to be a challenge in itself. Categories were selected based on four criteria. (1) Each category needed to include products that were perceived by consumers as related to the environment in a logical, intuitive manner. Products that consumers could not connect to environmental issues directly would probably complicate the experiment unnecessarily. (2) Each category needed to represent products that most people either purchase regularly, have purchased in the past, or were likely to purchase in the future. It was essential for the purchase decision not be an arbitrary or unrealistic one. (3) Categories needed to include at least two competitive brand names. Categories dominated by a single brand would not provide a useful comparison for our purposes. (4) Product performance across brands within the category needed to be relatively constant. Again, if the actual performance of the product varied markedly across brands, the influence of that factor on purchasing decisions would probably overshadow other manipulated factors. A simple pre-survey suggested four categories that fit these criteria: laundry detergent, disposable diapers, single-serving juice boxes and toilet paper.

Product pricing. As described earlier, many consumers expect to pay a price premium for environmentally marketed products. Though research suggests that this

Laundry detergent	Tide	Cheer
Disposable diapers	Huggies	Pampers
Single-serving juice boxes	Tropicana	Minute Maid
Toilet paper	Quilted Northern	Charmin

Table 2: PRODUCT CATEGORIES AND BRANDS USED IN THE SURVEY

premium may not be a purchasing reality, to ignore that perception in the consumer purchasing decision could have biased the results of this experiment. Traditional consumer behaviour theory suggests that consumers presented with two similar products are certainly more likely to select the one with an additional positive feature if that feature comes to them at no cost. A 5% price premium was identified by the Roper Organisation's series of reports as the average price increase consumers were willing to pay. The environmentally marketed products in this study were priced accordingly.

Product brands. The survey instruments asked consumers to select products *across* brand names, so it was important that the products be reasonably competitive in terms of market share and consumer preference. A pre-test of brand selections confirmed relatively competitive brands that would not unduly influence respondents. Additionally, a direct comparison of the brands without environmental marketing was included to establish a baseline consumer choice share for comparison of brands with environmental marketing features. A listing of product categories and brands is shown in Table 2:

Product brand logos. Alterations to the brand logos were necessary to create a realistic scenario for the purchasing decision. Each environmental label was designed to be both simple and straightforward, without vague terminology or creative names. Environmental line extensions, however, were more artistic, logo-specific designs using environmental terms and ideas.

Product information. Purchasing decisions integrate more than brand name. To facilitate an informed and complete decision-making experience, a number of product-specific details, including environmental performance factors, were provided. All product performance and feature information were kept constant within product categories, except for the addition of environmental factors when appropriate.

In addition to the four-category product purchasing survey, two sections of the survey were designed to provide a sense of the relative 'environmentalism' of respondents. The two-page Roper survey testing the relative seriousness of environmental issues and levels of environmental activity was used to measure 'environmentalism'. Also, a brief questionnaire was attached to collect demographic information on the participants.

The survey was given to 66 individuals at Raleigh-Durham International Airport.[2] Participants were asked to complete the survey in full and were offered a $2 cash payment upon completion. No participant saw more than one purchasing decision sheet for each product category, each completing a total of four product purchasing choices.

Consumers Respond

In examining the responses to our experiment, we recognised the importance of breaking our analysis into three distinct parts: first, looking at the respondents as a whole, representing the hypothetical 'American consumer market'; second, looking specifically at consumers whom we identified as relatively 'green' or environmentally aware, conscious and active; and third, looking closely at consumers we identified as 'brown' or environmentally inactive and relatively unconcerned with environmental issues. To divide our consumer sample into 'Greens' and 'Browns', we used a simple survey instrument similar to that used by the Roper Organisation. The instrument measured the regularity with which respondents participated in a series of environmental 'activities', including recycling, activism and similar actions (overall mean for this scale: 2.78 on a five-point scale). 'Green Shoppers' were identified as those scoring above the mean (mean for Greens: 3.44), 'Brown Buyers' as those scoring below (mean for Browns: 2.28).

Comparing Brands across Entire Sample: A Green-Seeking America?

Our findings across the entire group of respondents were remarkably consistent. Not only did environmentally marketed products overcome their higher prices to maintain their choice share, they actually increased share in most cases. Choice share for the environmentally marketed product improved marginally when the chosen strategy was an environmental label (from 39.7% to 41.3%) and somewhat more dramatically when the strategy involved an environmental line extension (to 46%). These results suggest consumers in general are willing to pay 5% more for environmentally improved and marketed products; more importantly, it suggests that some consumers will switch from a *less expensive* to a *more expensive* brand to receive the value added by environmental marketing (see Figs. 1–5).

2 Because the experiment was designed to be conducted as a between-subjects design, i.e. each person only saw one condition for each product, the possibility of sample bias is greatly reduced, if not eliminated, because respondents were randomly assigned to each experimental condition. The airport was chosen as a venue to collect data because it represents a relatively homogeneous group of consumers in terms of economic factors. (As this study focused only on differences in response across the experimental conditions, this homogeneity is actually a plus.)

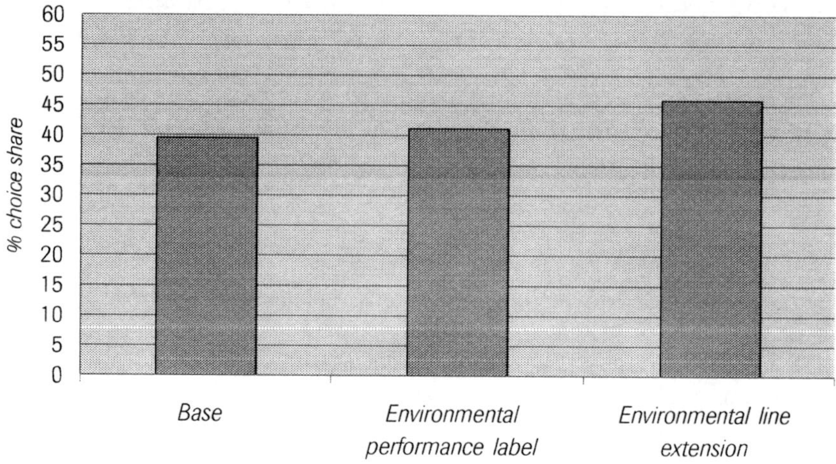

Figure 1: CONSUMER CHOICE SHARE: ALL RESPONDENTS

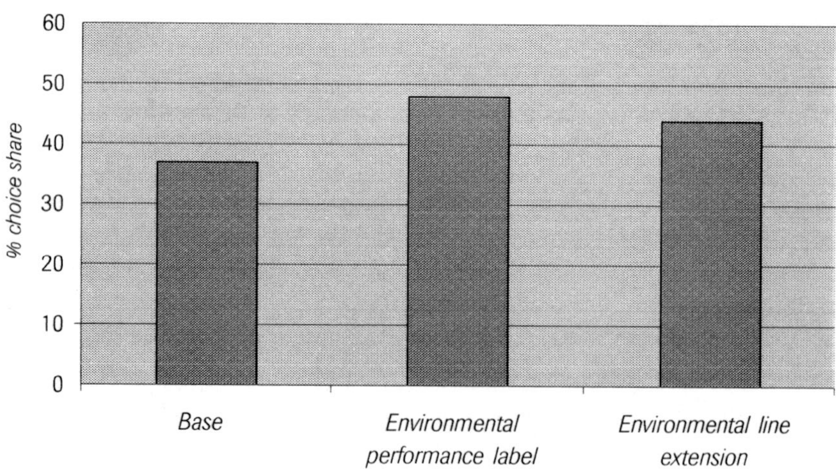

Figure 2: CONSUMER CHOICE SHARE: GREEN SHOPPERS

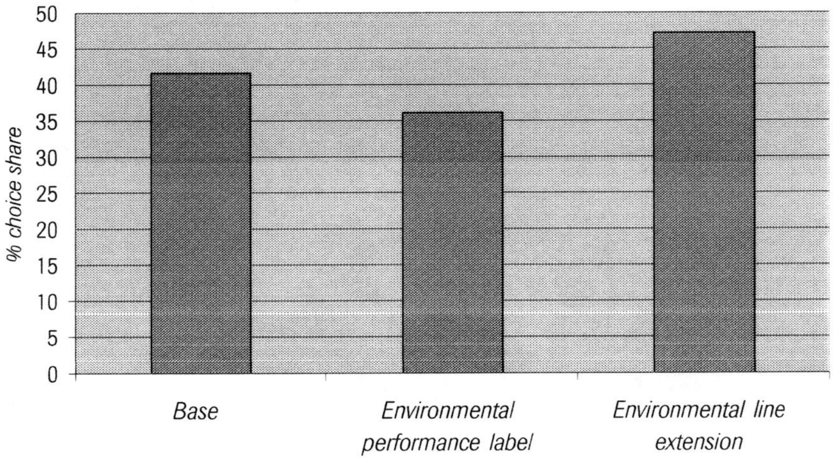

Figure 3: CONSUMER CHOICE SHARE: BROWN BUYERS

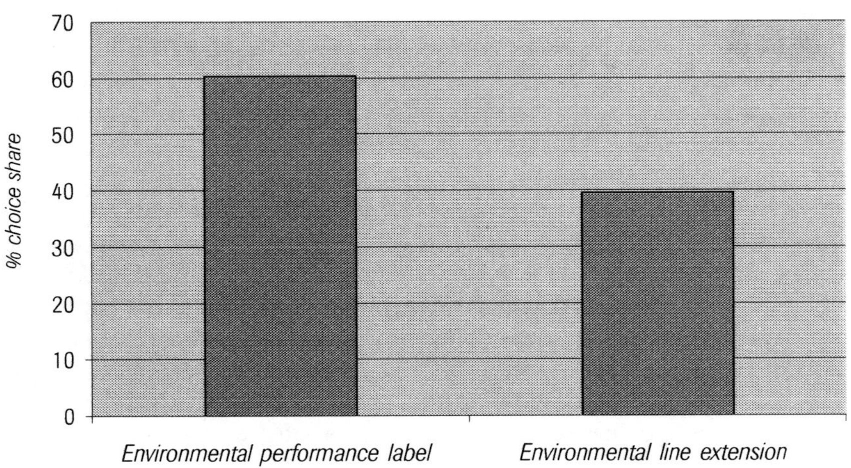

Figure 4: CONSUMER CHOICE SHARE: DIRECT COMPARISON (ALL RESPONDENTS)

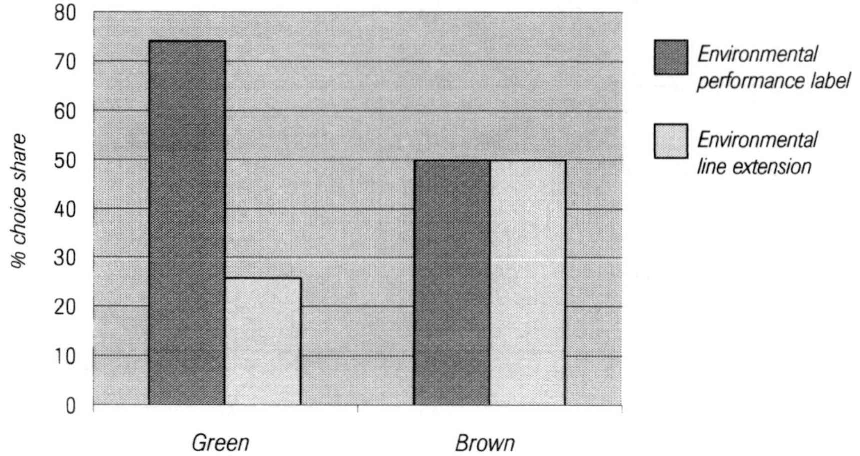

Figure 5: CONSUMER CHOICE SHARE: DIRECT COMPARISON (WITHIN SEGMENTS)

The Greens. The Green Shopper was, predictably, strongly influenced by the addition of environmental marketing to our product set. The Green Shoppers rewarded environmentally labelled products with an impressive 11.1% increase in choice share. Though their enthusiasm for environmental line extensions was somewhat less, a 7% increase, it is still quite encouraging. An interesting question arises here as to why the Green Shopper would be less responsive to one form of environmental marketing than the other. This question will be more aptly dealt with when we examine the direct comparison of these two methods.[3]

The Browns. The Browns' reactions to environmental marketing efforts were more intriguing. Their overall preferences for the base brands was consistent with the greens (58.3%, 41.7%), but Brown Buyers responded *negatively* to the addition of environmental labels (and the corresponding 5% price increase), reducing the choice share of that brand by 5.6%. This is hardly surprising, considering that Brown Buyers theoretically place a lower priority on environmental performance and a higher emphasis on price. What was surprising, however, was the Browns' *positive* response to the environmental line extension. The line extension was able to overcome the negative influence of the increase in price and boost choice share from 41.7% to 47.2%, a 6.1% increase over the base condition (and an 11.7% increase over the environmental label condition). This distinction between the effect of the label and the extension for the Browns is quite interesting, as the results suggest that these consumers may

3 Due to the small size of our sample when divided into market segments, the differences in share described within the 'Green' and 'Brown' markets are not 'statistically significant' and must be viewed as directional in nature.

be exhibiting some risk aversion toward environmental performance in label form, but not when it appears as a line extension.

Comparing within brand. Our final survey instrument attempted to further illuminate the contrasting reactions consumers had to our different environmental marketing strategies. This instrument asked consumers to select from two versions of the same brand: one featuring an environmental label, the other an environmental line extension. The results in this case were quite dramatic. Though Brown Buyers showed complete ambivalence toward the two marketing strategies, awarding each a 50% market share, the Green Shoppers overwhelmingly favoured the environmental label. Their clear preference for the label, with 74% choosing it over the extension, was the clear driver behind our findings for the entire sample, which reflected a 10.3% disadvantage for the line extension.

An alternative 'Green/Brown' scale? Though we elected to base our measurement of respondent 'environmentalism' on their level of environmental activity and/or action, we also measured the extent to which the subjects reported caring about environmental issues, and examined the effect of this factor on their preferences. In contrast to the activity-based measure of environmentalism used previously, this measure related to more general concerns and feelings about environmental issues, rather than actual behaviour (mean = 3.86 on a five-point scale). For consumers who expressed higher than mean levels of concern and/or interest in environmental issues (mean for this group: 4.24), we find a pattern similar to the active Green Shoppers: i.e. they prefer the environmental performance labels over line extensions (61.4% label; 38.6% extension). Among consumers who reported a low level of concern about environmental issues (mean for this group: 2.97), environmental marketing strategies had little positive effect. The addition of an environmental label was enough to overcome the 5% price increase but nothing more—market share remained constant. The addition of the environmental line extension, however, appears to have a negative effect (36.8% base; 31.6% extension). Again, due to small sample sizes, these differences are not significant, but are directionally indicative. Consistent with earlier findings, there was no significant difference in the preferences between the label and extension versions for people with below-average concern for environmental issues.

Sceptical Greens and Benevolent Browns: Discussion

The historical legacy of the environmental marketing movement still bears heavily on the environmental market today. The findings detailed here indicate that different segments of the consumer market reflect this legacy differently. Environmentally concerned shoppers, the Greens, showed an enthusiastic preference for environmental performance labels, suggesting that they remain suspicious of broad, indefinite environmental claims, preferring specific, quantitative performance information. Greens strongly prefer the simplicity and factual nature of environmental labels, as opposed to the more vague, general environmental performance implications of the line extension. Their Brown counterparts, however, are less discriminating. Because they make environmental performance a lower priority, they may not share the feeling of betrayal or suspicion that Green Shoppers feel when environmental marketing is vague and less fact-driven. More probably, they see environmental performance as a simple added value and a 5% price increase as relatively insignificant.

In fact, the issue of shoppers, both Green and Brown, accepting the price–value exchange included in this experiment is remarkable. A significant percentage of consumers appear willing to exchange a 5% price increase for the value added to products by improved environmental performance. Though a 5% price premium is generally accepted in the literature relating to environmental products, one possible explanation remains that, for many consumer products, a 5% increase in price may be insignificant in consumers' minds. Though this explanation merits further research, it is undermined by the fact that respondents were asked to select products across brands—the increase was enough to overcome brand loyalties in many cases.

Getting the Message to Consumers: Action Steps for Managers

The Environmental Market Opportunity

Consumers appear to care about the environment and appear conscious of environmental issues in their lifestyle and behaviour. This makes it all the more surprising that so few companies attempt to leverage environmental performance improvements into their marketing efforts. Our findings consistently suggest that consumers react positively to environmental marketing efforts, and that such efforts represent an added value for which they are willing to pay. Unfortunately, a history of misinformation and deception has tainted the environmental market: consumers remain somewhat suspicious of environmentally marketed products and often question their honesty. The challenge for managers is to communicate environmental performance convincingly, without overstating accomplishments or feeding consumer cynicism.

Do it right.

Communicating environmental performance information is much more complicated than adding a simple logo or clever brand name. Environmental marketing must be an outgrowth of a comprehensive, pragmatic plan to improve the environmental performance of a product, process or company. Environmentally active shoppers and consumer watchdog groups are quick to suspect environmental performance claims, making it all the more essential that such claims be based on solid scientific data, an established environmental management system and a genuine commitment to improving the firm's environmental record. One wise step is to seek partnerships with appropriate environmental organisations. These groups can lend legitimacy to environmental marketing efforts and provide essential guidance and insight in the development of environmental programmes. Think carefully about the types of group that might be able to offer the most valuable assistance, as well as those whose participation will have the most resonance with customers. And seek ways of making those partnerships meaningful and productive, rather than simply an endorsement or name-attachment.

☐ *Know the colours of your customers.*

Chances are, your customers include Greens, Browns and a host of other environmental shades. That doesn't mean that you cannot reach them effectively with environmental information and marketing efforts on product packaging. It merely suggests that it is all the more important to understand which customers purchase your product for what reasons. Intelligent market research will allow you to identify your customers' environmental attitudes and deliver environmental information to them in the form that will have the greatest impact. The study described here suggests that more environmentally aware and active consumers will respond most positively to environmental marketing that is fact-centred and provides specific numbers, terms and information about product performance. On the other hand, less environmentally active consumers are more apt to select a product that promotes its overall 'greenness' and commitment to environmental responsibility with an environmental line extension. It's not hard to see that understanding your customers' environmental attitudes can only enhance your ability to deliver environmental performance and information to them in the form that will be most appealing, effective and appreciated.

☐ *Know the colours of your product—in the eyes of your customer.*

Understanding your customers' environmental attitudes is only one-half of the equation for successful environmental marketing. Managers must also have a firm sense of how their customers view their product in relation to the environment. Products with prominent or obvious connections to environmental issues may be forced to face environmental questions in a very disciplined manner. Products less directly associated with the environment may enjoy greater flexibility in their approach to communicating performance information. Companies who take the lead in this area stand to reap the benefits of establishing themselves as standard-setters, much like Patagonia and The Body Shop have done recently, taking products less directly associated with environmental impacts and making that connection more salient for their customers.

☐ *The right greening may pay for itself.*

Put simply, environmental performance equals customer value. Consumers are willing to dip deeper into their pockets to receive products that communicate environmental performance or responsibility. Obviously, Greens are somewhat more likely and enthusiastic to do so than Browns, but the message is clear: environment = value. Of course, the 5% price increase employed in this study is a relatively arbitrary figure—product managers and marketers may find that environmental improvements pay for themselves in increased efficiency or reduced waste disposal costs, and require no product price increase at all. But our research indicates that, if delivering an environmentally improved product requires a slightly higher price, consumers will pay. Not only are your products likely to maintain market share, but they may actually increase share as 'greener' customers desert brands that fail to communicate environmental responsibility. Perhaps most importantly, your firm will see the larger benefits of improved environmental behaviour: reduced liability and compliance costs, stronger community relations and enhanced product and brand image

Taking the Next Step

Marketers and managers seeking a 'greener' brand have their work cut out for them. The nature of the consumer products market dictates that individual categories, products and brands will face very different and often unique challenges both in improving environmental performance and in communicating that performance to consumers. Marketers, in taking a customer-focused view of their environmental efforts, may elect first to examine how their customers place their product in the environment. What are the perceived environmental impacts? How is its environmental impact perceived relative to competitors' products? To substitutes? By focusing environmental improvements initially on the issues most salient to customers, managers can make their marketing more responsive, build credibility with consumers, and gain a genuine competitive advantage.

Managers may also find it worthwhile to examine the relationship between environmental performance and price. The cost of improving environmental performance on a given product may be significant or may, in fact, be non-existent, thanks to improvements in efficiency and waste reduction. Fortunately, for products that do require a price increase to recoup the costs of environmental performance, consumers have shown a willingness to endorse such improvements by paying higher prices. Again, the flexibility and range of this price–environmental value relationship must be explored on a category-, product- and brand-specific basis. But the history of the environmental marketing movement suggests that, when presented with clearly communicated, credible environmental performance information on a product of competitive quality and relatively competitive price, consumers will respond with enthusiasm.

16

Environmental Performance: What is it Worth?

A Case Study of 'Business-to-Business' Consumers*

Graham Earl and Roland Clift

THE GROWTH in stakeholder[1] interest in industry's environmental performance, especially among consumers, has not gone unnoticed by marketers and manufacturers of electrical and electronic products. Unsurprisingly, this has resulted in environmental performance becoming increasingly emphasised in marketing such products. A graphic illustration of this trend is the burgeoning number of electrical and electronic consumer products which are now being 'badged' with so-called 'green labels'.[2]

In some instances, manufacturers have designed their products to meet green label criteria in direct response to purchaser requirements: for example to meet a public-sector organisation's buying guidelines. However, in many instances, the pursuit of environmental claims has been carried out without independent verification, to try to differentiate a product from its close competitors.

While it may seem advantageous to add environmental features to a product's list of attributes, this may often create obvious trade-offs with other attributes of the

---

* This chapter summarises part of the work carried out by Graham Earl for the degree of Doctor of Engineering at the University of Surrey. Financial support from EPSRC and Paras Ltd is gratefully acknowledged.

The authors also wish to acknowledge Ms Zoë Jackson and Mr Tom Davies of Hewlett Packard, who helped with the design of the conjoint experiments. Special thanks are also extended to all those research engineers on the Engineering Doctorate who assisted with gathering data for the conjoint experiments.

1 Stakeholder groups are defined here as more-or-less organised groups of people who stand to be affected by the decisions and activities of an industry or company. For example, neighbours living near a chemical plant are potentially affected by the way the plant operates.

2 Example 'green labels' used in Europe include Blue Angel, Nordic Swan and the EU Eco-label.

product, most notably its price. As they say, nothing comes for free, and there is always an upper limit to what consumers are willing to pay to improve a product's environmental performance. As such, there is a strong need for both marketers and policy-makers to better understand how environmental issues factor in purchasing decisions.

Policy-makers need this information to help them design effective policies. In this sense, an effective policy is one that is able to efficiently redirect consumer purchasing behaviour towards environmentally 'friendlier' products and hence exert the necessary pressure to ensure responsible environmental corporate action. Marketers are in need of information that indicates how far consumers are actually willing to go to play their part. Knowledge regarding consumer thinking in this regard will have broad-range strategic and tactical ramifications for marketing decision-makers: each element of the marketing mix has the potential to be affected—from redesigning products and packaging to highlighting environmental performance in promotional materials.

More specifically, it is important for marketers to understand where environmental issues stand in relation to more traditional product considerations. Put differently, what types of product attribute trade-offs are consumers prepared to make for the sake of the environment? Potentially, these trade-offs could include, among others, a decrease in convenience, reduced product availability, or increases in price.

Green Consumerism: Fact or Fiction?

There are many sources of information for those seeking to inform themselves about social attitudes towards environmental factors—for example, Social Trends surveys, British Social Attitudes reports, Eurobarometer Surveys and European Values Surveys. Generally, these tend to indicate a growth in environmental consciousness; for example, a study by Social Trends carried out for the UK Department of Environment in 1993 revealed that 85% of adults in England and Wales were either 'quite' or 'very' concerned about the environment. Similarly, a consumer survey by British Social Attitudes on the environment printed in *The Guardian* (16 November 1994) showed that 38% of those interviewed would be 'fairly willing' to pay higher prices to protect the environment.

Looking outside the UK, Drumwright (1994) reported on a US study which indicated that 80% of US consumers are willing to pay more for environmentally friendlier goods. Evans (1990) reported on studies that indicate that 82% of West Germans claimed to take environmental considerations into account when shopping in the supermarket, while Smith (1990) reported that this figure is 67% in the Netherlands and 50% in France.

There is also evidence that consumers not only wish to purchase products that minimise their impact on the natural environment, but are also willing to pay more for them (Coddington 1993). For example, a Mintel survey concluded that 27% of British adults were prepared to pay up to 25% more for green products (Prothero 1990) and, in the USA, Green Market Alert estimate a market growth rate for green products of 10% in (Lawrence 1993).

These impressive figures lead quite naturally to the question: Who are these environmental consumers and how are they best characterised? This is a question that has challenged many researchers over the past 25 years and one that has resulted in many attempts to define the characteristics of the socially or ecologically responsible consumer (e.g. Schwepker and Cornwell 1991; Balderjahn 1988). Other research (e.g. Schlegelmilch *et al.* 1996; Forman and Sriram 1991) has looked at how ecological concern translates into modifications of purchasing behaviour.

These studies have tended to focus on domestic (or household) consumers and have mostly concluded that these consumers are becoming more environmentally sensitive and are becoming more willing to modify their consumption patterns. Significantly, the research listed above has relied predominantly on self-report measures of consumer environmental purchasing behaviour and not on actual purchasing behaviour. Therefore, care must be taken in interpreting this research since there must be a suspicion that these results are reflective of demand effects[3] inherent in the method and measuring instruments.

Indeed, emerging evidence suggests a curious paradox: many green products have not achieved the level of market success that might be expected in a society that claims to be sympathetic to the environment. In many consumer product categories, producers have achieved disappointingly low levels of market share for their green innovations (Aspinwall 1993). A recent survey into green consumption patterns in the UK suggests that the number of green consumers has increased only slightly since 1990, while the proportion of people who have not altered their spending habits despite environmental concerns has increased in the last few years (Mintel 1991, 1995).

The question, then, is which figures should you believe, if any at all, and do all consumers behave alike or do some categories, such as business-to-business consumers—a category of consumer that has not received much research attention—behave differently? If so, what kind of trade-offs does this behaviour imply and what are the implications for marketers?

Project Aims

This study aims to investigate the relative trade-offs company purchasing managers make when purchasing electronic and electrical products. If one assumes it is possible to categorise consumers into three basic categories[4]—domestic, i.e. households; intermediate, i.e. retailers and business-to-business, i.e. purchasing managers in the corporate and central and local government sectors—then this research focuses on the last category. In doing so, it aims to specifically answer the question: Are business-to-business consumers (in this case represented by corporate purchasing managers)

3 For example, if you are asked 'Are you concerned about the environment?', there is an implied demand to answer positively.

4 From this point on, the word 'consumer' will only be used when referring to unspecified consumer types, and can therefore be taken as referring to all three categories; in all other cases, the specific categories defined in the text will used.

willing to forego performance or pay higher prices to improve a product's environmental performance and, if so, by how much?

To meet this aim, the study chose to investigate two closely related products. The first, inkjet printers, are relatively long-lasting and involve an element of investment. The second product, inkjet cartridges, are much more frequently purchased and involve significantly lower per-transaction cost. At the same time, both products belong to a product sector that is subject to rapid technological change and is increasingly being subjected to environmental performance pressures from consumers and regulators. Indeed, both products share attributes that have significant potential to impact the environment, either through using up valuable resources (e.g. energy, materials) or by creating large amounts of waste and potential contamination.

This study has chosen to investigate the business-to-business category, more specifically the behaviour of 'purchasing managers' from the private sector, since they are deemed to be the key informants on purchasing decisions. For example, Jackson *et al.* (1984) have shown purchasing agents to wield the majority of the influence in the purchasing of supplies. Likewise, the purchasing behaviour of the purchasing managers also indicates the company's own attitude towards the 'greenness' of supplier products, which is an important influence on attitudes within the purchasing company itself. Consequently, because of their specific positions and responsibilities, knowledge of the trade-offs purchasing managers are willing to make is critical and powerful data for manufacturers. However, because of the their uniqueness and added purchasing weight, this study has not attempted to achieve the sample rate of other studies that have examined general consumer behaviour.

Conjoint Analysis: A Tool for Measuring Trade-Offs

Conjoint analysis is a market research tool that can be used to measure consumers' trade-offs among products with many attributes. Conjoint analysis relies on the ability of respondents to make judgements about stimuli. For example, conjoint analysis asks the consumer to answer the question, 'Are you prepared to pay £1,000 to upgrade from a Ford to a similar BMW?' This is clearly easier to answer than 'What is the relative importance to you of a car's brand and price?'[5]

In conjoint analysis, the stimuli represent some predetermined combinations of attributes, and respondents are asked to make judgements about their preference for the various combinations of attributes. Conjoint analysis attempts to handle the problem of determining preferred features by systematically estimating how much each attribute is valued on the basis of the respondents' choices between alternative product concepts.

Because questions are framed closely and made concrete, conjoint analysis is distinct from the broad economic approach of contingent valuation. Furthermore, since the conjoint method converts preferences for different performance attributes

5 Green and Srinivasen (1990) provide a good background on the history, theory, development and application of the conjoint method.

to a single variable—utility[6]—it is possible to quantify the relative importance of these to the respondents, which for this study is a sample of purchasing managers.

Methodology and Conjoint Results

In line with the study's aim, a conjoint experiment was designed to measure the relative trade-offs purchasing managers make when choosing inkjet printers and inkjet cartridges. Data from the study was analysed using conjoint analysis software developed by Bretton Clark.[7]

The methodology used covered the following basic stages:

Stage 1: Specification of separate conjoint experiments for each product. Tables 1 and 2 in Appendix 1 summarise the performance attributes and levels used to describe the inkjet printer and inkjet cartridge experiments. Printer casing recycled content and cartridge re-usability are the environmental performance attributes included in each design. The performance levels specified for these attributes were defined so that they did not imply any *direct* financial or operational gain or loss to the purchasing managers interviewed. The idea was that utility values measured for these performance attributes would indicate only business-to-business consumers' preference for environmental performance.

Stage 2: Stimuli design. Conjoint analysis works by asking respondents to **rank** in order of preference a set of product scenarios which have been specified using a common set of performance attributes and performance levels (in this case, those described in Appendix 1). While each product scenario will be specified by the same set of performance attributes, the performance **levels** defined for each attribute will differ on at least one of the attributes. The most common way of displaying the product scenarios to the respondent is through a set of cards. Each card carries a description of the product using the predefined performance attributes and performance levels.

Stage 3: Data gathering. The inkjet printer and inkjet cartridge conjoint experiments were carried out with 22 purchasing managers selected from 13 companies. On average, two individuals were interviewed from each company; in each case, these were chosen for their responsibility for purchasing IT equipment. The companies approached covered a wide spectrum in terms of size (ranging from SMEs to multinationals) and area of operation (consultancy to production).

Stage 4: Produce output results. Tables 3 and 4 in Appendix 2 summarise the **average** utility and attribute importance data calculated for the two experiments.

---

6 Utility is an arbitrary unit used by economists to measure consumer satisfaction, pleasure or need fulfilment derived from consuming some quantity of a good.

7 Other suppliers of conjoint software include Sawtooth Software and Intelligent Marketing Systems. Carmone (1995) offers a useful review of conjoint analysis software.

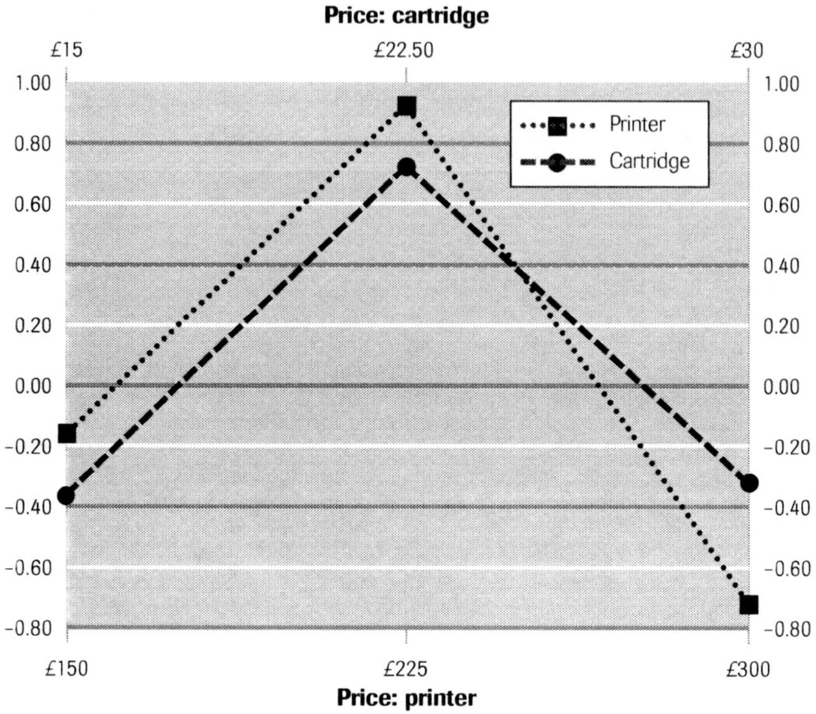

Figure 1: PRICE UTILITY FUNCTIONS FOR INKJET PRINTERS AND CARTRIDGES

A useful representation of this data is achieved by comparing utility levels with performance levels for each attribute. Figure 1 shows the utilities for 'price' and Figure 2 shows the utilities for the 'environmental' attributes for the two product groups.

Reference to Figure 1 shows that, on average, everything else constant, the purchasing managers' utility increases as the price of the inkjet printer and cartridge increases from £150 to £225 and £15 to £22.50 respectively. Their utility then decreases as the price of the inkjet printer and cartridge continues to increase from £225 to £300 and from £22.50 to £30 respectively. It is important to note that, because some utilities are negative, this does not indicate a negative pleasure. In this case, total utility (for any one performance attribute) is constrained to sum to 0, therefore some are negative and some positive.

Reference to Figure 2 shows that utility falls almost linearly as the recycled content of the inkjet printer's casing increases from 0% to 100%, while, in the case of the inkjet cartridge, utility falls off less sharply as the cartridge changes from a disposable to a refillable one and then more sharply as it is changes from a refillable to a recyclable version (see Table 2, Appendix 1, for detailed definition of the cartridge performance levels).

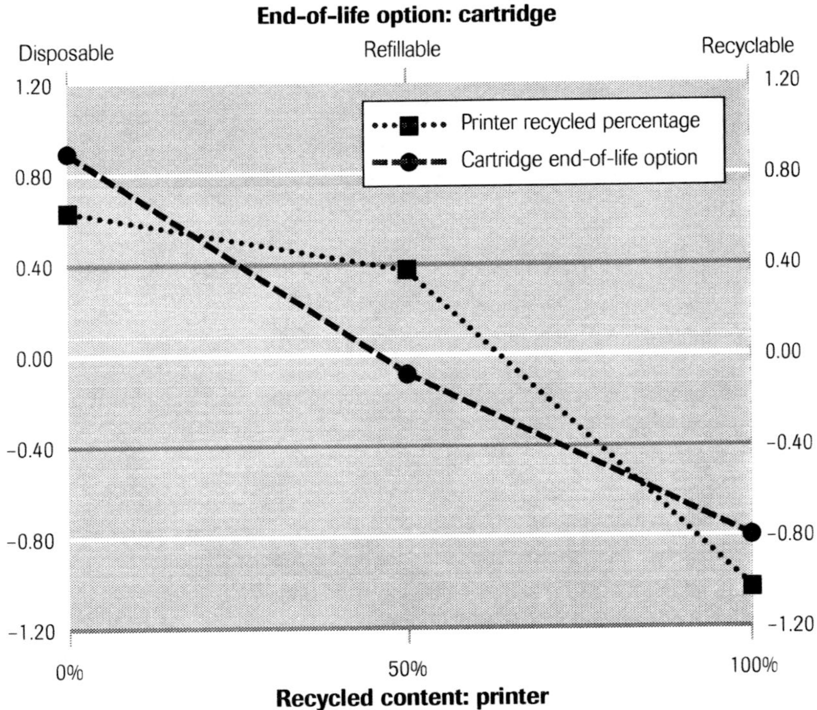

Figure 2: ENVIRONMENTAL PERFORMANCE UTILITY FUNCTION FOR INKJET PRINTERS AND CARTRIDGES

Key Findings

Price is invariably, and not surprisingly, an important attribute. However, reference to Figure 1 shows that, all other things constant, the lowest-priced inkjet printers and cartridges are on average not routinely preferred (do not have higher utilities) over higher price versions. In fact, the detailed individual results show that, for inkjet printers, only 22% of purchasing managers consistently placed higher utilities on lower-priced printers than higher-priced ones, and for inkjet cartridges this figure was 14%. This behaviour suggests that the purchasing managers interviewed are inferring some kind of benefit associated with higher prices which are not defined on the conjoint card, or, in other words, an inferred price–quality trade-off. Alternatively, they doubt the credibility of the lower-priced products described on the conjoint cards.

Recyclate content of the inkjet printer is on average an important negative feature. The utility function (see Fig. 2) shows that, on average, purchasing managers prefer lower recyclate content over higher recyclate content. Indeed, the analysis of the individual data showed that this was true for 85% of the purchasing managers interviewed. This behaviour implies that the interviewed purchasing managers simply do not wish to have inkjet printers made from recycled plastics or that they associate some kind of product performance loss, not defined on the conjoint cards, with printers with higher casing recycled content.

The **spent cartridge option** attribute was deliberately defined so that the possible performance levels would not offer any financial incentive to the respondents. The value of each performance level would therefore relate solely to the importance placed on the cartridge's environmental performance. Reference to Figure 2 shows that, on average, purchasing managers, all other things equal, prefer disposable cartridges over refillable or recyclable ones. Analysis of the detailed individual preference data showed that this was true for over two-thirds of the data sample. The conclusion is that the interviewed purchasing managers prefer disposable inkjet cartridges: although they offer poorer environmental performance, they are easier to use, requiring no refilling or storing for recycling.

Conclusions and Implications

The results analysis has deliberately focused on the spent cartridge option and recyclate content attributes, since these were introduced to 'capture' environmental performance as a selling attribute for the two products studied. In both cases, the environmental attributes were defined so that they do not directly imply financial benefits to the purchasing managers, in order to elicit any preference for environmental performance.

In this study, purchasing managers were required to think about and articulate their trade-offs. Whereas a preference for environmental performance is often assumed, it is only possible to measure real preferences via trade-off decisions that include environmental performance as one decision criterion among several. The results from both conjoint experiments show that price and operational criteria are important for most purchasing managers. Casing recyclate content and cartridge re-usability are also shown to be important product features. To the extent that recyclate content and re-usability represent improved environmental performance, lower rather than higher performance is preferred.

False Perceptions and Misconceptions

For inkjet printers, Figure 2 shows that purchasing managers place lower utilities on (i.e. are less satisfied with) printers with higher recyclate content. This means that, for two printers with equal cost, each offering the same operational performance,[8] the printer made from virgin material offers more utility (i.e. is preferred) over the

8 In this case, 'performance' was measured through the following attributes: print speed, print quality, reliability, service and support, and colour capability.

same printer made from recycled materials. There are a few likely reasons for this behaviour:

1. The purchasing managers misunderstood the experiment's definition of 'recycled' as used to describe the recycled content of the printer casing.
2. The purchasing managers perceive recycled products as inferior to new products.
3. The purchasing managers may have had poor past experiences with products that are recycled or made from recycled material.

The first reason is thought unlikely, since a great deal of care was taken to define and explain each performance attribute fully. It was made very clear that the term 'recycled' referred only to the material used to produce the printer's casing. Respondents were told they were comparing 'printers with different amounts of casing recyclate content' and not 'recycled versus new printers'. It was also made clear that this attribute was totally independent of the printer's other attributes, i.e. the printer's recycled content is not in any way linked with and can therefore not affect any of the other attributes used to describe a printer.

The second reason, driven by the purchasing managers' own perception of what 'recycled' means, appears to be more likely. So, based on the data—which shows purchasing managers (all other things equal) prefer printers whose casing is made from virgin as opposed to recycled material—it is plausible to assume that the purchasing managers did not acknowledge that recycled means 'as good as new'. Or, in other words, printers with casings made from recycled material are perceived as inferior or 'nearly new'. Given that the purchase of a printer is longer term and can be seen as an 'investment decision', it is plausible to think that purchasing managers would be reluctant to invest in products perceived as 'nearly new'.

This reluctance feeds into the last listed point and matches closely with the findings of Polonsky *et al.* (1998a). They found that the future evaluation of newer 'friendlier' products by business-to-business purchasers was clouded due to bad past experiences. In other words, it is not just product perception but poor communication that may be causing the behaviour recorded by this experiment.

The same study also concluded that price was not the most important business-to-business purchasing criterion, being typically preceded in importance by product quality and reliability. Hence, given the arguably frequent and false negative quality perception of recycled products, this may also go some way to explaining the preference by purchasing managers for the printers made from virgin and so, they assume, higher-quality material.

This type of behaviour has created a difficult and very real challenge for many companies. Taking Xerox as an example, quoting Pierre van Coppernolle, director of Rank Xerox 2000 Strategy and Environment from their 1995 Environmental Report:

> Externally, we have a number of barriers to overcome: many of these are caused by misunderstandings about our environmental objectives and perceptions of our re-manufacturing business (Rank Xerox 1995).

Xerox consequently struggles, and continues do so, with the misconception that their remanufactured photocopying machines are 'refurbished', use old components and are in some way inferior to brand-new products. In fact, Xerox has found that the greatest resistance to their remanufactured machines stems from public-sector buyers. In some cases, this is borne out by the organisation's selection protocol, which may stipulate that only brand-new products may be considered for tender. Quoting once again from Xerox's Environmental Report:

> Xerox is currently in discussion with the European Commission to seek its assistance in remedying the exclusions of re-manufactured products by some public authorities in procurement contracts of photocopiers (Rank Xerox 1995).

In other words, Xerox has found that buyers using public money, who are therefore accountable to taxpayers, are reluctant to risk public disapproval by spending money on goods that are not 'brand new'.

Perhaps this partly explains why ICL prefer to market their products as 'second life' rather that 're-whatever', anticipating that the products are less likely to be devalued by the purchaser. However, the analysis reported here suggests that the problem is deep-rooted, not merely semantic. As suggested earlier, it is much more likely to be driven by purchaser perceptions. The answer, therefore, is not simply to change the name of goods or hide the fact that a product is remanufactured; rather, it must address the cause, which in this case seems to be a lack of understanding.

◻ *Greenness is not enough.*

The preference drivers for printer cartridges seem to be slightly different. Because the purchase of an inkjet cartridge is unlikely to be seen as an investment, it is more likely that operational and logistical criteria drive the purchase decision. The conjoint analysis results (see Fig. 2) show that purchasing managers actively prefer disposable cartridges to refillable and recyclable ones.[9] Given no financial benefit, the easiest and most convenient option is then shown to be preferred. If the preference for disposable cartridges is seen as a proxy for convenience, then the analysis shows this is a much more important factor for the purchasing managers than any potential environmental gain.

In fact, very similar behaviour was concluded from a study that interviewed 20 marketing managers from leading UK companies (Wong *et al*. 1995). This study found that the majority of marketing managers indicated that offering green attributes was important, but experience with their own green products suggested that greenness alone was insufficient to sustain consumer demand for the product. According to the managers, the *mass* of consumers attached considerable weight to other product attributes, such as performance, quality, image, safety and so forth. These results

9 There are, of course, schemes that pay users for returned cartridges. However, this study specifically asked respondents to assume no 'direct' financial gain from re-usable or recyclable inkjet cartridges. The aim was to measure whether respondents saw any 'indirect' values associated with the different disposal options studied.

clearly coincide neatly with the essentially similar findings of Polonsky *et al.* (1998a) and serve to reiterate the conclusion that 'greenness' in itself is seldom the overriding determinant of product choice.

Business-to-business consumers are no different

Looking specifically at business-to-business purchasing, Kärnä and Heiskanen (1998) report that some manufacturers of electronic and electrical products claim to have noticed clear differences between business-to-business consumers and domestic consumers with regard to environmental awareness. However, this type of behaviour has generally been limited more to countries such as Germany and Sweden with social pressures that emphasise environmental performance. The implication is that business-to-business consumers in countries such as the UK, where these pressures are weaker or absent, suffer the same misconceptions and lack of awareness as retail and domestic consumers.

Similarly, Business in the Environment, who have carried out research on the level of environmental engagement of the FTSE 100 top companies, found a disappointing level of supply-chain management among the UK's top companies (BiE 1997). In reply to the question, 'Does your company have an environment-focused supplier programme in place?', the survey found that only 38% of the companies interviewed responded with a positive answer.

The solution for manufacturers wishing to specify and sell their products using green credentials is clearly not simple. On the one hand, manufacturers see a diffuse and unspecified demand for environmental solutions; yet, on the other hand, there is very little hard evidence to show reward for their environmental improvement endeavours. In other words, environmental consciousness is evident, yet ecologically conscious decision-making is not.

Obviously, there are exceptions to this rule, as identified by John Carew from Business in the Environment (Carew 1997), for example:

- BT reports that it uses environmental considerations in its purchasing decision-making process.
- IBM carries out risk analyses of strategic suppliers.
- Nortel works with suppliers on specific environmental issues: currently it is trying to tackle packaging issues by working together with Motorola.
- B&Q uses environmental management up the supply chain to increase market share.
- Sainsbury is developing joint ventures in crop management.

Nevertheless, the assumption that business-to-business consumers are going to be the 'forerunners of the environmentally conscious generation of customers of the future' (Kärnä and Heiskanen 1998) appears to be ill-founded. A major challenge suggested by this study is that there is confusion and lack of understanding even among purchasing managers of what some environmental claims actually mean, especially their implications for the product's performance and for the business in general. This problem is not helped by what also appears to be a lack of generally accepted environmental criteria for electrical and electronic products.

Barriers to Green Purchasing

Looking at it from the purchasing company's perspective, John Carew from Business in the Environment has identified two fundamental barriers faced by business-to business purchasers wishing to improve their companies' supply chain management (Carew 1997). The first is gaining policy commitment and the associated mechanisms and procedures to back it up. Assuming that purchasing managers might share similar obstacles to environmental managers, then a recent UK survey by ENDS (1995: 241), which showed that 74% of all environmental managers cite lack of management commitment as a key obstacle to their work, suggests that this is not a straightforward or easy barrier for purchasing managers to cross.

The second barrier is the application of the poorly understood approach of 'whole-life costing'. Carew claims to have found little or no evidence that whole-life costing, which implies including the environmental imperatives of longevity, lower running costs and disposal costs, has been applied properly and as a matter of course in private- and public-sector procurement.

Underlying these barriers is a scarcity of available and reliable information about the environmental characteristics of products and services. In fact, Jean Cinq-Mars, head of the Pollution Prevention and Control Division Environment Directorate, OECD, speaking at the *Greening Government* conference (Cinq-Mars 1997), suggests that lack of information is sometimes considered to be the major obstacle to greener purchasing initiatives as it limits the development of multi-criteria specification of environmental characteristics of products.

Eco-Labels and Regulation

Although third-party labelling schemes may seem to offer a part solution, the role and significance of labelling are still unclear, especially since there seems to be little agreement on an internationally acceptable label. Recent analysis by the OECD (Cinq-Mars 1997) on a few selected eco-labelling schemes concludes that such schemes have had little effect on consumer behaviour, except in those countries where consumers express strong environmental awareness.

Underlying all of this is an evolving regulatory environment. Particularly in the electronics sector, this has been moving towards enforcing producer responsibility, with emphasis seeming to be on 'end-of-life' management (EOLM) rather than eco-design, and is illustrated through the recent European Commission working paper on the management of waste from electrical and electronic equipment (EC 1997a). This European Union initiative has considerable implications for the development and design of electrical and electronic products. In this respect, the paper outlines specific responsibilities for producers of electronic and electrical equipment, which, taken together, aim to:

- Eliminate toxic materials
- Increase recyclability
- Increase dismantlability
- Increase the amount of recycled material
- Improve the reverse logistics associated with these products

For example, a specific measure proposed in these guidelines is a re-use and/or recycling minimum target of 85% by weight[10] for all IT equipment (EC 1997a). The responsibility for achieving this target is placed firmly with the producer. To achieve it, producers will need to provide purchasers of electrical and electronic equipment with the necessary information about the return, collection and recovery systems available to them, and also to emphasise their role in contributing to the recovery, re-use and recycling of end-of-life electrical and electronic equipment.

A Strategy for Closing the Attitude–Behaviour Gap

Overall, the picture for producers seems somewhat paradoxical. On the one hand, the regulatory framework is looking to impose the responsibility for EOLM on the producer through 'take-back' requirements, to encourage the uptake of re-usable and recyclable products and materials. On the other hand, this and other similar studies show that consumers, even the supposedly better-informed business-to-business consumers, show an unwillingness to switch to greener designs and products.

For example, Pegram Walters Associates (1994), in their INCPEN (Industry Committee for Packaging and the Environment) study into green packaging, showed that, while most people regarded packaging as bad for the environment, they still chose packaged goods over unpackaged alternatives when shopping. So, although 88% of the sample considered there to be disadvantages to packaging, not all of them environmental, 68% thought the benefits of packaging, such as hygiene, convenience, product protection and information, outweighed these disbenefits, thus leading to their continued use.

The overriding theme suggested by these and other results is of an 'attitude–behaviour' gap or 'words–deeds inconsistency'. Obviously, green products are selling and this is being helped by government legislation and initiatives; however, there are some fundamental barriers—such as misconceptions, lack of policy commitment, short-termism, etc.—which are currently preventing the closing of this gap in terms of all consumer types. Clearly, it would be unfair to fully associate 'general consumer' behaviour with that of business-to-business consumers; nevertheless, this study suggests that there is a certain level of homogeneity of behaviour.

10 This figure is taken from the European Commission (1997) working paper and consequently may change.

On the positive side, the electronics sector is leading the way in the implementation of the international environmental management standard, ISO 14001. Because this standard aims to push companies to greater understanding of the direct and indirect environmental effects of products throughout their life-cycle, it should help with the marketing of greener designs and closing of the attitude–behaviour gap. It would be foolish, however, to rely solely on ISO 14001 and emerging international eco-labels to solve the perception problems associated with recycled or re-used products.

Form Relations with Stakeholders to Reduce Misconceptions and Build Relationships

It is probably much more sensible for manufacturers to become more proactive and start to develop in-house strategies to help specify, design and market their products. As a starting point, conclusions drawn from this study suggest that this strategy must aim to reassure purchasers of the validity and implications of all green claims. Relating this advice back to potential manufacturers of products made from recycled materials or components, a key message which must be accurately and effectively conveyed is that 'recycled' does not mean 'nearly new' or 'second hand'.

Since consumers, especially householders, often distrust environmental claims because they are perceived to be used to gain competitive advantage and cannot easily be tested by consumers themselves, it is critical that manufacturers are able to demonstrate their credibility and develop an honest and trusting relationship with their stakeholders.[11] To help achieve this, the manufacturing company must aim to identify what kind of information is used and needed by its different stakeholders, and then be proactive in ensuring that this information and the way it relates to their products reaches the stakeholders in a systematic way. Clearly, there are many types of stakeholder, e.g. regulators, household consumers, business-to-business consumers, etc., and, since these groups are unlikely all to have the same needs, they will probably need to be managed differently.

Another key benefit of proactive communication is the opportunity of building stronger and longer-lasting relationships with business-to-business contacts. Drumwright's (1994) study into the use of environmental purchasing as an example of non-economic buying criteria found that, regardless of the motivation to build environmental criteria into the purchasing decision, once companies had started with it, there were no cases in which this practice was subsequently abandoned.

Similarly Business in the Environment (1992) suggest that companies engaging in inter-firm collaboration or partnerships in supply chains can create opportunities for the supplier to embed its business in the customers' value chain. Since each then has a vested interest in the other's success, this then creates a better environment for putting into place effective environmental solutions, resulting in shared environmental benefits, cost savings and/or improved image.

---

11 For example, Toor (1992) reports that 63% of consumers are suspicious of manufacturers' green claims.

Demonstrate the Whole-Life Value of the Product

Secondly, and probably just as importantly, the strategy must ensure that, if green claims are being made, these are linked wherever possible to overall environmental policy and associated financial and operational gains for the business-to-business consumer. Addressing this challenge is given particular importance when viewed in the light of the finding by Drumwright (1994). She found that purchasing managers were generally unreceptive to the integration of environmental purchasing measures, preferring to focus on pricing and other economic criteria. Drumwright speculates that the possible driving force for these preferences are because:

> purchasing professionals are not privy to information about corporate strategy . . . or that they filter out aspects of strategy information that are not comparable with economic measures by which they are rewarded (Drumwright 1994: 13).

All of this compounds the argument for using whole-life costing (WLC) to differentiate between products. Using this approach, the manufacturer will be much better placed to demonstrate any down-the-line cost reductions associated with improved environmental performance and, perhaps most importantly, the risk reduction benefits that can accrue through an increase in confidence among the business-to-business's own stakeholders. This analysis confirms the views expressed, for example, by Stevels (1997).

Clearly, as suggested by Polonsky *et al.* (1998a) and Wong *et al.* (1995), price or cost may not be the key purchasing factor for business-to-business consumers. In this case, WLC can still serve a dual role by offering 'quality' assurance by identifying and quantifying related factors such as reduced repair costs.

Taking the example of inkjet cartridges, using WLC principles will help reinforce that re-usable cartridges can in fact be cheaper for the purchasing company if potential disposal costs are factored in, or if the company is struggling with its environmental image. This research has shown that, unless links such as these are made, it is unlikely that purchasing managers will be willing to sacrifice convenience for the sake of environmental performance.

Education, Training and Communications

All of this, of course, must be underpinned through a basic platform of education, training and clear communications. No matter how good the eco-improvements that designers make to products, their potential to reduce environmental impacts is usually contingent on the behaviour of others, not least of which is consumer demand that makes it possible to compete and sell into the marketplace. The simple fact is that the best products can only impact on our environmental footprint if they are actually purchased and used in preference to products with poorer environmental performance.

Future Research

This study represents a starting point in trying to quantify the importance of environmental performance as a decision criterion in purchasing. In this case, the research has concentrated on the importance of environmental performance to business-to-business consumers.

There are, of course, many other stakeholders who are interested not only in the environmental performance of the products but also of the manufacturing companies themselves. For example, as Stevels (1997) observed, company designers will benefit immensely and be better placed to develop sustainable product designs if they can integrate stakeholder priorities into the design process. So, rather than incremental product improvements, the aim must be to move towards radically rethinking the way stakeholders' needs are provided for. Trade-offs have to be made between environmental and other criteria. To increase the credibility of these choices, stakeholders must be involved in the decision process.

Future research must therefore look at ways of quantifying the priorities, values and needs of a wider set of stakeholders, and to design decision processes that will allow these factors to be integrated into the traditionally closed, internal processes by which companies reach their decisions.

Conjoint analysis, while effective with consumer-type stakeholders, is clearly neither flexible nor adaptable enough to provide the full solution. Research is therefore focusing on designing and validating a hybrid model that integrates the conjoint analysis methodology with other decision and quantification techniques (for example, Earl *et al.*'s [1998] Stakeholder Value Analysis Toolkit).

Appendix 1: Performance Attributes and Levels

Performance attribute		
Name	**Description**	**Levels**
Price	The purchase price of the printer	• £300 • £225 • £150
Printer quality	The maximum print quality of the printer	• *Laser quality* to describe a printer that can print up to 600 600 dots per inch • *Good quality* to describe a printer that prints up to 600 300 dots per inch • *Average quality* to describe a printer that prints up to 300 300 dots per inch
Printer speed	The maximum print speed of the printer when working in top-quality mode, i.e. not in draft output	• 6 pages per minute • 4 pages per minute • 2 pages per minute
Service and support	The service and support that comes as standard with the printer	• Lifetime service/support • One-year service/support • No service/support
Reliability	The printer's intrinsic reliability performance	• *High reliability* described by a 2% chance of breakdown in a year • *Medium reliability* described by 6% chance of breakdown in a year • *Low reliability* described by a 10% chance of breakdown in a year
Printer casing recyclate content	Total amount of recycled plastic material used in the manufacture of the printer's casing	• 100% recycled plastic content • 50% recycled plastic content • 0% recycled plastic content
Colour capability	The colour capability of the printer	• Black and white printing only • Colour printing capability

Table 1: PERFORMANCE ATTRIBUTES AND LEVELS: INKJET PRINTER EXPERIMENT

Performance attribute		
Name	**Description**	**Levels**
Price	The price of an inkjet cartridge for use in an average inkjet printer	• £30 • £22.50 • £15
Life	The printing lifetime of the cartridge based on best-quality print and approximately 3,000 characters per page	• 750 pages • 500 pages • 250 pages
Colour	The colour capability of the cartridge	• Black and white only • Colour
Re-usability	The ability of the cartridge to be re-used after it has been used once. There is no cost advantage from refilling or recycling a cartridge.	• *Refillable.* It is possible to refill the cartridge with ink and use it again. • *Recyclable.* The cartridge is taken back to the manufacturer for recycling. When you buy a new cartridge you will be given the option of handing in your old cartridge. • *Disposable.* These cartridges cannot be refilled and will not be taken back by the manufacturer for recycling.

Table 2: PERFORMANCE ATTRIBUTES AND LEVELS: INKJET CARTRIDGE EXPERIMENT

Appendix 2: Conjoint Analysis Results

Attribute	Level	Average utility*	Relative importance
Price	150	-0.17	20%
	225	0.89	
	300	-0.73	
Print quality	300 × 300	-0.49	16%
	600 × 300	0.81	
	600 × 600	-0.32	
Print speed			
	2 p/m	0.23	9%
	4 p/m	-0.49	
	6 p/m	0.26	
Recyclate content	0%	0.66	21%
	50%	0.38	
	100%	-1.04	
Reliability	Low	-0.17	10%
	Medium	0.50	
	High	-0.32	
Service and support	None	0.61	16%
	Limited	-0.73	
	Extended	0.12	
Colour capability	No	-0.38	9%
	Yes	0.38	

* Utilities for each attribute's performance levels are constrained to sum to 0, therefore some are negative and some positive.

Table 3: UTILITY AND RELATIVE IMPORTANCE RESULTS FOR INKJET PRINTER EXPERIMENT

Attribute	Level	Average utility*	Relative importance
Price	£30	-0.35	19%
	£22.50	0.72	
	£15	-0.37	
Lifetime	750 pages	-0.26	36%
	500 pages	1.19	
	250 pages	-0.93	
Spent cartridge option	Disposable	0.86	29%
	Refillable	-0.06	
	Recyclable	-0.80	
Colour	No	0.47	16%
	Yes	-0.47	

* Utilities for each attribute's performance levels are constrained to sum to 0, therefore some are negative and some positive.

Table 4: UTILITY AND RELATIVE IMPORTANCE RESULTS: INKJET CARTRIDGE EXPERIMENT

3 CASE STUDIES

17

Coming out of their Shell

Brent Spar

Alan Neale

> PEOPLE ARE JUST as concerned about the corporate behaviour of a company as they are about the convenient location of its sites (*Shell World*, February 1998, reporting global consumer research into brand preferences in petrol retailing).

Shell UK's 1995 attempt to dump Brent Spar, a redundant North Sea oil installation, in the Atlantic Ocean detonated a storm of protest that the company had not foreseen, despite its reputation for scenario planning. After battling at sea and in the media with Greenpeace protesters, and suffering in the marketplace, the company backed down and began to explore possibilities for recovery and re-use. This change of heart marked a turning point, not just in Shell's relationship with its stakeholders, but more generally in business attitudes to corporate greening.

The main focus in this chapter is on aspects of the Brent Spar case that are most relevant to greener marketing: communications, the international dimension, environmental innovation, and alliance building. The chapter is structured around three distinct stages in the history of the case. In Stage 1, the company's main emphasis was on compliance with UK regulatory requirements. Stage 2 was characterised by confrontation with Greenpeace, who rejected the criteria on which Shell's decision-making was based. In Stage 3, Shell placed more emphasis on communication as a two-way process, which enabled a less environmentally damaging solution to be found. General lessons from the case that may be relevant to other companies are highlighted in the appendix to this chapter.

Historical Background

Shell is a multinational grouping of companies, based in the Netherlands and the UK, and the second largest of the oil majors. In the early post-war period, the Shell

Group of companies concentrated more on downstream activities (the marketing of crude oil and oil products) than on upstream operations (exploration and production). This changed with the OPEC-induced oil price rises of the 1970s, when the group embraced new exploration on a massive scale to reduce its dependence on Middle East supplies over which it had no control. Rapid development of the Brent Field to the east of Shetland, a joint venture with Esso, was one outcome of this strategic shift upstream.

Brent Spar was installed by Shell in 1976 to bring oil ashore from the Brent Field in a way that respected the harsh climatic conditions of the North Sea. The Spar was a huge floating tank, chained to concrete blocks on the sea floor, in which crude oil was stored prior to loading onto shuttle tankers for transfer ashore. Ironically, in view of subsequent events, environmental considerations featured prominently in its design: Esso had favoured a cheaper solution involving a permanently moored tanker, but Shell UK, the operating partner, chose a giant buoy, partly because this involved less risk of marine pollution. 'Cradle-to-grave' issues were ignored, however. The main concern in the 1970s was to bring the oil ashore quickly, and, as was common at the time, no attention was given to integrating plans for end-of-life disposal into the design process, even though the understanding in international law was that redundant structures would be taken ashore for dismantling.

It was only when Brent Spar was decommissioned in 1991 that the technical problems of disposal were addressed. There were worries that, if the Spar were turned on its side to bring it onshore (reversing the sequence employed during its construction) the tank walls could rupture, risking massive inshore pollution, and the company decided to investigate alternatives.

Stage 1: Compliance (1991–95)

The process Shell UK employed to generate and evaluate alternatives followed UK government guidelines scrupulously, and the company assumed, erroneously, that this would ensure public acceptance. The narrowness of the decision-making reflected an organisational culture within the Shell Group that emphasised responsibility, while discouraging involvement with external stakeholders.

Operating companies within the Shell Group traditionally enjoyed considerable autonomy in taking their own decisions, providing these reflected Group business principles. The *Statement of General Business Principles*, first issued in 1976, recognised, uniquely for a large multinational corporation at that time, four 'inseparable' areas of responsibility—not just to shareholders, but to employees, customers and society as well. Health, safety and environmental issues were directly addressed, but the main emphasis was on Shell companies complying with the regulations of the countries in which they operated, rather than striving to achieve or exceed global best practice.

Although consensus-seeking within Shell management was encouraged, there was a split between the engineers who dominated upstream operations and the marketing personnel who concentrated on downstream sales. The prevailing organisational culture did not encourage openness with outsiders, so opportunities to find new partners to explore innovative solutions to environmental problems were missed. Aloofness often came across as insensitivity, particularly in relation to concerns expressed by environmental or human rights groups (for more on the organisational context of this narrow focus, see Neale 1997).

The decommissioning of Brent Spar provided a graphic illustration of the drawbacks of the traditional Shell approach. Shell UK spent three years studying different disposal options, the aim being to determine the 'Best Practicable Environmental Option' (BPEO), to comply with UK regulatory requirements. Under BPEO, alternative options have to be assessed, and financial considerations are supposed not to override environmental impacts, but 'practicable' is often interpreted as lowest cost, provided minimum environmental standards are adhered to. This applies especially to North Sea oil, where the same UK government department, the DTI (Department of Trade and Industry), is responsible both for promoting the industry and for regulating it, and where the Treasury has a vested interest in the lowest-cost solution (because decommissioning costs are tax-deductible).

In Shell UK's BPEO studies for Brent Spar, six viable options were identified, but only two—horizontal dismantling for onshore disposal and deep sea dumping—were considered in detail. A re-use option had been rejected on the grounds that 'no alternative users or buyers were found',[1] although this was hardly surprising given the secrecy surrounding the process. The studies suggested that, environmentally, their two main options were evenly balanced: deep-sea dumping would involve 'negligible' impact as pollutants would disperse over a wide area and, while onshore disposal might result in lower pollution if it went as planned, this technically more complex operation would risk accidental spillage into a vulnerable inshore marine environment. The deciding factor was financial, with onshore disposal costs estimated at four times those of deep-sea dumping. Shell UK convinced themselves and the UK government that they had struck the right balance between environmental, safety and financial considerations, and did not feel a need to test this against outside opinion, beyond the minimum consultation recommended by UK law and the formal notification of other European governments required in international law.

The arrogance of assuming that 'we know best' was to be Shell's downfall. The proposed dump site in the Atlantic Ocean to the west of Lewis (one of the Western Isles of Scotland) was chosen without consulting the scientists who were most familiar with its marine environment, and there was no peer review of the assumptions about dispersal of contaminants after dumping. No attempt was made to explore with engineering contractors innovative ways of reducing the risks and costs of bringing the Spar onshore, and the secrecy surrounding the process closed off possibilities for recycling or re-use. Many of Britain's neighbours had a long-standing cultural antipathy to dumping in or near the North Sea, and they had little sympathy for the primacy of financial calculations in the UK government's BPEO procedure, yet little was done, either by Shell or the UK government, to address these concerns. Nor was much attention given to Greenpeace's long-standing involvement

1 Rudall Blanchard 1996. This document was submitted to the DTI in 1994, but not made public until 1996.

in campaigns against pollution in the North Sea, and its capacity to generate unfavourable publicity for the company.

Stage 2: Confrontation (1995)

It was Greenpeace's intervention in 1995 that turned the fate of Brent Spar into an international *cause célèbre*. The organisation had been actively campaigning against ocean dumping for most of the period since 1978. Its actions had been instrumental in halting the dumping of radioactive and industrial wastes at sea, and it was particularly concerned that Shell's Brent Spar disposal would be used as a precedent to reassert the legitimacy of dumping at sea, not just for oil installations but for other industrial wastes as well (Greenpeace's position is elaborated in Rose 1998). Alongside this long-term concern, there were short-term reasons for Greenpeace's interest in Brent Spar. Early in 1995, when the UK government was announcing its intention to support dumping at sea, Greenpeace was looking for a visual focus to back its lobbying of a ministerial conference on the North Sea to be held in June 1995. Brent Spar was to provide that focus.

In April 1995, Greenpeace activists occupied Brent Spar. They mounted a massive media campaign, inviting reporters on board and providing television companies with anti-Shell video footage which regularly appeared on news programmes throughout Europe. Brent Spar, which had remained unnoticed for 19 years, suddenly appeared, day after day, on millions of television screens throughout the continent, as a symbol of business disrespect for the marine environment. Shell UK never seemed to know what had hit them, or how to respond. The company evicted the protesters and began towing the Spar to its dump site, yet protesters in inflatables and helicopters continually interfered with the operation and eventually re-boarded the Spar, providing more David-versus-Goliath images for Greenpeace to feed to European television audiences at Shell's expense.

Away from the North Sea, the Greenpeace message was clear: companies should not use the sea as a dump site, and should take responsibility for their own waste. Shell spokespeople, in contrast, seemed remote and soulless. They kept repeating to the media that they had satisfied legal requirements to determine the BPEO, without realising that this was a concept that had little meaning for most of their audience, particularly outside the UK. A related spin was that Shell managers were rational scientists, carefully weighing a range of environmental, safety and financial considerations before making a decision, unlike their opponents, whose judgement, it was suggested, was clouded by an emotional attachment to the marine environment. At times, lack of awareness of the contested nature of the scientific evidence and limited sensitivity to issues of principle turned into openly expressed contempt for public perceptions and feelings—John Wybrew, Shell UK's then Director of Public Affairs, going so far as to suggest that 'people confronted with the campaign became emotionally disturbed'.[2]

2 In a television interview on 'The Battle for Brent Spar' (BBC 2, UK, 3 September 1995).

Despite the lesson the oil industry was supposed to have learned—after the *Exxon Valdez* oil spill of 1989—that corporate complacency in the face of public concern can be fatal to a company's reputation, Shell managers continued to insist that they had made the right decision. Communication was a one-way process, and no attempt was made to understand the social values that underpinned the doubts shared by many of its stakeholders, let alone respond constructively to them.

Opinion polls showed that few people in Germany, the Netherlands or Scandinavia supported the Shell case. Many motorists in these countries boycotted Shell products, with the encouragement of Greenpeace, Green parties and church groups; Shell sales in Germany fell by 20%. At the governmental level, only Norway supported the UK in opposing calls at the June meeting of North Sea environment ministers for a halt to sea dumping of oil installations, and additional pressure was put on the UK government to change its policy at the following week's G7 summit. Some business partners such as Novo Nordisk, the Danish pharmaceuticals company and a co-signatory of the Business Charter for Sustainable Development, publicly challenged Shell to justify its decision-making over Brent Spar. Dissident shareholders added to the opposition and eventually managed to muster sufficient support to put critical resolutions, calling for more rigorous external monitoring of environmental performance, to the 1997 Annual General Meeting. Underlying all these actions was a feeling that Shell and the UK government were taking the easy option with Brent Spar, for short-term financial reasons, and riding roughshod over public opinion.

The about-turn, when it came, followed intense pressure from European partners within the Shell Group. Deutsche Shell in particular had been extremely critical of Shell UK's failure to communicate the reasoning behind its decision to dump Brent Spar, and had stressed the adverse market consequences of that failure. After intensive discussions within the Group on 20 June 1995, only hours before Brent Spar was due to be scuttled, Shell UK announced that it was abandoning the dumping operation.

Stage 3: Communication (1995–98)

In July 1995, Brent Spar was towed to Erfjord in Norway to await a new decision on its fate. At first, Shell UK managers seemed traumatised by the about-turn that had been forced upon them, but some of those who had earlier dismissed public opinion were now able to reflect on the damage that caused. Heinz Rothermund, Managing Director of Shell UK Exploration and Production, admitted a year later that 'we have too often responded grudgingly to environmental challenges, and too often seemed out of touch with popular concerns' (Rothermund 1996).

Shell UK encouraged the generation of alternative solutions by announcing an international competition. It also opened up the evaluation process by seeking greater public involvement. The company's first published environmental report stressed that 'the importance of the Brent Spar debate was to convince Shell UK that it must communicate better with those interested in its operations in order to take account of public concerns' (Shell UK 1996: 17). Instead of attempting to make the 'best' decision on narrow technical grounds, and then hoping that this would be accepted by their stakeholders, Shell managers now invited public participation in the development of a new solution. This involved significant changes in the communications function within the company. Whereas in Stages 1 and 2 communicators had not been very involved in decision-making, their role in Stage 3 was central, and communication came to involve listening to others as much as conveying the Shell position.

To support this, Shell UK produced a CD-ROM to illustrate the background data, and opened a Brent Spar website (*www.shellexpro.brentspar.com*), which made available press statements and consultants' reports, and invited comment on alternative disposal options. It collaborated, too, with the Environment Council, a leading provider of environmental mediation services, to devise a 'Dialogue Process', whose aim was to 'engage, not enrage' different stakeholder groups.

In the Dialogue Process, representative participants were invited to discuss disposal options at seminars in London, Copenhagen, Rotterdam and Hamburg. A key recommendation from these seminars was that Shell should favour solutions involving a positive energy balance (the solution saving more energy than it consumes), and a high element of re-use (to minimise waste). This was fed into the selection of seven onshore options (all having a positive energy balance, and most including elements of re-use) for detailed evaluation against the deep-sea dumping 'benchmark'. They were then used again, alongside an independent comparative assessment commissioned from DNV environmental auditors, to help decide on a new preferred solution.

In January 1998, Shell UK submitted a new decommissioning plan to the UK government, based on these deliberations. This involved lifting the Spar vertically (to minimise the risks of it breaking up), cutting it into sections, cleaning them and re-using them as a quay extension at Mekjarvik in Norway. Shell UK was confident, this time, that the environmental advantages of re-use outweighed its cost penalties, and there was widespread public acceptance that the right solution had been chosen.

Shell UK's new-found enthusiasm for dialogue spread to the Shell Group as a whole, whose public image had suffered not only from Brent Spar, but from events in Nigeria, where, in November 1995, the military regime executed Ken Saro-Wiwa and eight other activists from the Movement for the Survival of Ogoni People. Ken Saro-Wiwa had targeted Shell, as the main operating partner in the Nigerian oil industry, to highlight the human rights abuses and appalling environmental conditions in the Niger Delta. The worldwide outrage at his death raised major questions about Shell's willingness to tolerate poor environmental and human rights standards in developing countries, triggering more consumer boycotts of Shell products, particularly in the USA.

The Shell Group identified, in its annual financial report for 1995, a common thread to the problems it had faced with Brent Spar and in Nigeria: an organisational culture that was too inward-looking. 'We need,' it suggested, 'to have greater external focus if we are create a better acceptance of the Group's business among varied audiences. Group companies must consult, inform and communicate better with the public' (Shell Transport & Trading 1996: 2). The Group developed an innovative website (*www.shell.com*)[3] to actively promote debate about wider responsibilities, and the *Business Principles* were revised in 1997 to incorporate concerns about human rights and sustainable development.

Most importantly of all, Shell published its first annual social accountability report in 1998, to monitor the extent to which the *Business Principles* are adhered to in practice. 'Our traditional corporate culture has not necessarily encouraged openness,' this report admitted. 'But we are now trying hard to be more accessible and open in the way we deal with requests for information and in the style with which we communicate . . . We are also determined to listen more and get involved in debate and dialogue' (Royal Dutch/Shell Group 1998: 29). The result is a document the scope of which goes way beyond that of conventional corporate environmental reports to cover issues such as equal opportunities, human rights, corruption and political activities as well as health, safety and environmental management. Readers are encouraged to send in their views on a number of clearly identified policy dilemmas, and post them on the Shell website for wider debate. The Group also outlines plans to develop, for future reports, independently audited sustainable development accounts that integrate financial, environmental and social costs and benefits.

No More Brent Spars?

Shell's journey with Brent Spar from compliance, through confrontation, to communication, involved, as we have seen, a major change in the culture of the organisation, characterised by a greater openness to the world outside its corporate boundaries. Without this culture change, Shell UK would not have been able to understand, let alone influence, public perceptions of its operations, or to develop the alliances, political and technical, that were needed to generate an innovative re-use solution for the Spar which offered significant environmental and safety advantages over both ocean dumping and onshore scrapping.

It would be wrong, however, to exaggerate the extent of the company's greening. As Fran Morrison, Shell UK's Media Relations Manager, has acknowledged, a key objective at Stage 3 was to step back from the stand-off situation of 1995:

> You don't try to continue with this short-term over-dramatised confrontation, you just halt it, and stake out a wholly different piece of ground and start all over again. You change the agenda, you change what the debate is actually about. And you play it long.[4]

From the company's point of view, the elaborate consultation over Brent Spar was needed to take the heat out of a difficult situation and help restore the damaged reputation of the Shell brand in Europe.

---

3 The website won a Design for Business Association Award for Design Effectiveness in 1997, and was commended by the judges for its 'use of design to present complex issues and encourage debate'.

4 In a television interview on 'Sparring Partners' (BBC 2,UK, 31 January 1998).

From a green perspective, what is more significant is the extent to which Shell's recent changes will help contribute to a more sustainable future. Here there are encouraging signs, such as the Shell Group's recent strategic shift into renewable energy, its support for greenhouse gas reduction targets at the 1997 Kyoto climate summit (and the subsequent withdrawal of its US subsidiary from the Global Climate Coalition),[5] and its plans to promote greater diversity (of nationality, gender and expertise) in its senior management teams.

In the North Sea, however, Shell has not yielded any ground to Greenpeace's position that sea dumping should be rejected in principle, for all offshore installations. Heinz Rothermund has insisted that re-use of Brent Spar 'is a "one-off" solution for a "one-off" structure', and that decommissioning of other North Sea installations will continue on a 'case-by-case' basis.[6] Despite its professed enthusiasm for dialogue, Shell UK declined to take part in the 'Beyond Sparring' project, which was sponsored in 1997 by Greenpeace and the environmental consultancy SustainAbility (and later supported by the European Commission) to explore possibilities for an integrated strategy to co-ordinate disposal activities in order to reduce the costs and risks of onshore scrapping and re-use. And, as a new international treaty to prohibit further sea dumping in the North East Atlantic came up for ratification in 1998, Shell UK joined with other offshore operators to lobby for exceptions to be granted for the 50-plus large installations where substantial dismantling costs would be involved.

The broader principles articulated by Greenpeace in its Brent Spar campaign, that producers are responsible for their waste and that innovation should be directed into ways of minimising it, remain highly contested, in the offshore oil industry as elsewhere. Shell's development of greener marketing competences was central to the evolution of a process that enabled Brent Spar to become a useful ferry terminal rather than waste dumped on the ocean floor. It remains to be seen how willingly and effectively Shell can extend its application of these competences from individual projects to corporate policies, so that the organisation encourages and responds constructively to debate, not just about how environmental impacts in a particular case are assessed, but about environmental values and how they apply to its business strategies.

---

5 The Global Climate Coalition is a business lobby group, based in the USA, which expresses scepticism over scientific evidence for global warming, and is opposed to greenhouse gas reduction targets. In 1997, the Shell Group became increasingly embarrassed by Shell Oil (its US company) remaining part of this body, but argued that aggressive lobbying was integral to US political culture and that continued membership allowed Shell to present an alternative view within the Coalition. Sensitive to accusations of hypocrisy, however, Shell changed its mind, announcing in April 1998 its withdrawal from the GCC.

6 At a news conference to announce the revised decommissioning plan for Brent Spar (London, 29 January 1998).

Appendix: Some Wider Lessons from Brent Spar

- ***Design with end-of-life disposal in mind.***

Technical problems of reducing environmental damage when a facility comes to the end of its useful life are less if waste minimisation has been built into the design. In addition, where environmental standards are rising over time, life-cycle thinking may become a significant source of competitive advantage.

- ***Seek out innovative solutions.***

Innovative solutions are also called for where an organisation has to deal with environmental problems arising from poor standards in the past. Organisations that rely exclusively on their own resources are likely to be less effective than those that are open to ideas from outside.

- ***Do not rely on compliance with local regulations.***

Organisations need to anticipate how the regulatory framework may change over time, and also to recognise when regulatory requirements fail to match public expectations. Where multinational operations are involved, a company's reputation in one part of the world may be judged by its performance in another, making it imperative to take international opinion into account.

- ***Do not reduce environmental decision-making to a technique.***

Many key environmental management decisions involve value judgements, and stakeholders may distrust attempts to disguise these as technical choices. Technical competence needs to be supplemented by sensitivity to different environmental values held by different groups in society.

- ***Remember that communication is a two-way process.***

Don't see public relations as persuading people that you're doing the right thing—explore more interactive forms of communication, listen to what your stakeholders have to say, and feed this into your decision-making.

- ***Engage, not enrage, your stakeholders.***

Recognise the range of stakeholder groups who are affected by your operations and policies, and seek appropriate ways of gaining and maintaining their trust. Don't confine stakeholder involvement to individual projects that have already become contentious, but invite participation in dialogue about the environmental and social implications of the strategic choices you face.

18

The Body Shop International plc

The Marketing of Principles along with Products

Kate Kearins and Babs Klÿn

THE BODY SHOP is cited in many marketing and strategy texts as an outstanding example of an ethically and environmentally responsible business. The success of The Body Shop and the business acumen of its founder, Anita Roddick, have been recognised by the wider business community through several prestigious awards. Although The Body Shop has enjoyed largely positive attention, recent events have raised questions as to its success.

In September 1994, an award-winning American journalist, Jon Entine, published an article questioning The Body Shop's ethical and environmental practices. Even before its publication, details of the article's contents sparked worldwide media interest. A lengthy rebuttal by the company fuelled the debate, which held the media's attention for over a month. The contradictory reporting and claims and counter-claims made it difficult to decide whether The Body Shop was indeed the ethical and environmentally conscious company that the public had been led to believe.

This chapter examines whether, in the wake of the controversy, The Body Shop still presents a good model for green marketeers. The first part provides background information on The Body Shop, and the controversy sparked by Entine's article. The main issues focused on in the press are then examined. Readers are brought further up to date with some discussion on The Body Shop's highly acclaimed *Values Reports*, its other values-based initiatives, marketing strategies and restructuring plans. The broader question of whether product-focused marketing can ever be truly compatible with environmentalism is considered in the final part of this chapter, prior to some conclusions on the implications of The Body Shop case for marketing practitioners.

Background

Founded in 1976 by Anita Roddick, The Body Shop has become a major manufacturer and retailer of own-brand cosmetics and toiletries. Managed until recently by Gordon and Anita Roddick with a select group of trusted executives, it currently operates in 47 countries, with approximately 1,500 stores, many of which are franchised. It recorded pre-tax profits of £38 million on sales of £604.4 million in fiscal 1998 (Body Shop 1998a). Signs were of a loss of momentum, with underlying sales falling, although in many markets stores continued to do well. Performance in the highly competitive US market was of particular concern with declining sales figures over the past three years and some dissatisfaction among franchisees (Tannenbaum 1998). The Body Shop's image as 'one of Britain's most glamorous growth stocks' (Wallace 1996) has been tarnished as share prices fell from a peak of 370 pence in 1992 to 111 pence in March 1998.

The Body Shop's highly publicised core values, which include care for the environment, concern for human rights and opposition to the exploitation of animals, have contributed to its success. Other success factors include the inspirational leadership of Anita Roddick and her ability to communicate values with which people identify. The campaigns that the company has inspired (to save whales, to curb the spread of AIDS and to assist in the development of the third world, etc.) have contributed to the 'feel-good' factor experienced by many of its customers. Its ethical and environmental stance has enabled it to adopt a premium pricing strategy: price premiums over similar products have been as high as 40%, although prices have been cut in recent years (Cowe 1994a). Even with these premiums, the company claims that, because its products are of such a high quality, they are not expensive compared to other top brands of cosmetics. Indeed, an important part of the appeal of The Body Shop has been its comparatively low prices; however, this advantage is being eroded by newer entrants to the lower end of the market.

The Body Shop's product line comprises over 600 different products and over 600 different accessory items. Fair trading relationships are promoted in the sourcing of products and ingredients, and testing of cosmetics on animals is not supported. Product development draws from traditional wisdom, herbal remedies and modern scientific research. A product stewardship programme, incorporating product life-cycle assessment, supplier accreditation and risk assessment of downstream product impacts, has been implemented as part of the company's environmental management system (Mayers 1997).

In terms of place, The Body Shop has a global presence. On the whole, its stores conform to a tightly controlled format which offers consistency around the world. Strict franchising guidelines retain the same generic style and product range in all stores, although flagship stores for prime locations and new layouts have been trialled.

The Body Shop had originally taken an unconventional approach to marketing based on claims of no hype, no advertising, no pressurised selling and minimal packaging (Roddick 1991). However, the lack of advertising did not mean that it stayed out of the media spotlight: free editorial coverage, leaflets, shop fronts in prime locations, word of mouth and signs on trucks were some of the alternative ways in which the company was promoted.

However, many critics suggested that, if The Body Shop was to remain competitive, it would have to change its unconventional policy of not advertising its products, which Anita Roddick said would drive up product prices and would distort or make her messages look superficial. There were suggestions that the company should adopt aggressive marketing techniques such as discounting, 'gift-with-purchase' offers, and 'two-for-the-price-of-one' deals, as used by competitors such as Bath & Body Works. This company, now with more than 900 outlets in the US, and described as more 'American apple pie' than The Body Shop, sells similar products inscribed 'Made in the Heartland' to the same segment of the market (Ghazi and Tredre 1994). Bath & Body Works had also entered the UK market through a joint venture, planning to open 100 stores (*Economist* 1994)—although, so far, only five have eventuated.

The threat of the Bath & Body Works chain had, according to inside sources at The Body Shop's Littlehampton headquarters, prompted a fierce and sometimes bitter internal debate regarding the importance of advertising (Ghazi and Tredre 1994). A turning point came when Anita Roddick appeared in an American Express 'documercial'. Even though Anita was promoting social and human rights and donated her fee to charity, purists were outraged by what was essentially an advertising campaign. The negative press coverage, however, was balanced by a spectacular increase in The Body Shop's US sales (Fox 1994). With such a dramatic effect on sales, and the discovery that there may be negative effects involved with having a high media profile to gain 'free' promotion, a move towards more advertising by The Body Shop ensued. In the US, The Body Shop had taken out print and radio advertising, and initiated other in-store promotional techniques to counter the weightier marketing muscle of Bath & Body Work's major shareholder, The Limited (Wallace and Brown 1996). In Britain, a programme of direct mail and a system for phone-in purchases were instigated. It wasn't until 1997, though, that The Body Shop appointed its first above-the-line agency to run a UK advertising campaign (Crawford 1997). Evidence suggests that, until the early 1990s, the company's home-grown approach to marketing had been extremely successful but, in an increasingly competitive market where brand differentiation is vital, compromises to The Body Shop stance on marketing were inevitable.

The insistence on specialisation and a lack of diversification over time have ensured that the essential Body Shop concept is not diluted. The company has grown dramatically, largely without sacrificing its image of quality products. Basically, the Roddicks had put together a very successful formula which was being carefully replicated across the globe—the 'socially responsible McDonald's' of the cosmetics industry, you might say.

The 1994 Controversy

Despite its obvious successes, The Body Shop was not without its critics. Four years ago, it underwent quite an ordeal. On 18 August 1994, it was reported that an American ethical investment fund—The Franklin Research and Development Corporation—had sold its shares in The Body Shop, partly because it had caught wind of an investigation by American journalist Jon Entine, which raised doubts about the company's actual practice of the principles that it espoused. The investment fund later confirmed it had used financial rather than ethical criteria in making its decision, but the rumours that followed the Franklin sell-off cut The Body Shop's share price by 15% in a matter of weeks (*Economist* 1994).

Further details of the forthcoming article by Jon Entine were leaked to the media in the meantime and The Body Shop began to hit back. In a 32-page dossier, released to the media on 27 August 1994, it warned that the article would be a repetition of the same 'stale' story rejected by other magazines and television companies. The negative impact of the early release of Entine's accusations angered The Body Shop and it claimed that *Business Ethics* had launched a 'carefully orchestrated pre-publication publicity drive designed to put itself, in its own words, "on the map"—at the expense of our reputation' (Reuters 1994).

Entine's six-page exposé was published under the title of 'Shattered Image' on 1 September 1994 in the Minneapolis-based magazine, *Business Ethics,* which had a circulation of around 14,000. The editorial note accompanying the article declared that the story represented a departure from the magazine's typical editorial style, and was being published with mixed emotions. It concluded by expressing a hope that the article would provoke constructive dialogue on the issues raised.

Entine claimed that his article would reveal what sort of company The Body Shop really was. He based his allegations on more than 100 on-the-record interviews with current and former franchisees, employees, trading partners, suppliers, cosmetics experts and social researchers, and his stance was highly critical.

The Body Shop later described Entine's article as 'recycled rubbish', 'a poorly researched piece . . . riddled with errors and grossly unfair to the Body Shop and its founders', 'a mish-mash of defamatory and actionable falsehoods [that] contains distortions, shoddy reporting and the views of several unqualified or biased sources or so-called experts', and as 'rumour and scaremongering' (Body Shop 1994). This rather hostile reaction raised some eyebrows in the investment community.

When Entine's article was finally published, the general reaction was that it was rather dull and not very revealing—certainly not the 'astonishing tale' that he had claimed it would be (Gilchrist 1994), and far less interesting than the rumours preceding its publication. One reporter described the article as 'a miserable damp squib' (Fallon 1994a). Entine, however, contended that the article represented only a sampling of his findings and has continued his attack on various Internet lists.

As for The Body Shop's customers, the controversy appears to have had little negative impact. The accusations levelled at the company were not regarded as being serious enough for customers to cease their patronage. Many customers would reconsider only if it was found that it was guilty of something more serious, such as torturing animals (Ghazi and Tredre 1994). Jardine (1994) describes the (lack of) response of some of The Body Shop's customers during the crisis:

> There was no sign of disquiet among the dozen or so customers mooching around the Body Shop in London's Islington yesterday. Most were barely

> teenagers, more concerned with killing time in a pleasant-smelling, pop music-saturated shop than anything else. For them it is like going to a sweet shop. Brightly coloured bottles fill the shelves: there are things to fiddle with, open and smell.

The Body Shop's investors appeared to relax, too, in the light of the findings of the independent ethical investment monitor's investigations. After reviewing the allegations and The Body Shop's responses, the UK-based Ethical Investment Research Service concluded that the criticism of the company's environmental and ethical credentials had been unfair. This finding should have ensured that The Body Shop remained popular with environmental funds (*Observer* 1994). Brokers insisted that they remained confident in The Body Shop's earnings potential overseas, particularly in America and Asia, and highlighted that the company was one of only four British retailers capable of growing more than 5% per annum (Bernoth 1994).

British environmentalists also came to the company's support. Former Green Party leaders Jonathan Porritt and Sara Parkin wrote to *The Financial Times* stating that The Body Shop was 'one of the leading companies in the field of social, ethical, and environmental policy . . . a good example of a successful yet caring company with priorities way beyond the usual profit motive' (cited in Cowe 1994b).

The controversy, which occurred during 'a holiday news lull' (Woodward 1994), brought some interesting issues into the public arena.

Some Issues Highlighted by the Controversy

Trade not Aid

By far the strongest criticism of The Body Shop was of its 'trade not aid' programme. As explained by Anita Roddick, the company differentiates its products on the basis of the fair trade component of its operations: 'All a moisturizer is is oil and water; every one of them works . . . What makes the difference for our company is that we get our ingredients from different sources than do other companies' (cited in Schoell and Guiltinan 1992: 339). The Body Shop had created a public perception that many, if not the majority, of its ingredients come from third world countries, but its critics revealed that it had paid only around 0.3% of its worldwide sales to 'trade not aid' producers (compared with 31% by Traidcraft, a leading UK Fair Trade company) (Zagor 1994a). The Ethical Investment Research Service noted, however, that The Body Shop's fair trade-related purchases had doubled from the past year (Zagor 1994b).

Critics also claimed that indigenous cultures' patterns of living may be irrevocably changed by involving them in trade and 'introducing computers and making capitalists out of them to satisfy an artificially-created demand for faddish products' (Vidal and Brown 1994). Much of this sort of criticism could have been avoided by the company had it adopted the Fairtrade mark, administered by the independent verification body, the Fairtrade Foundation. A spokesperson from the Fairtrade Foundation stated that 'While we applaud efforts by any company to adopt fairer trading standards, such companies should not expect to gain PR benefit unless they are prepared to back claims with independent verification' (Drummond 1994).

Natural Ingredients and Animal Testing

Another of the criticisms levelled at The Body Shop was the promotion of its products as 'natural'. Entine alleged that many of the company's products contained non-renewable petrochemical ingredients, artificial colourings and flavourings and synthetic preservatives. He also revealed that, according to former employees, The Body Shop had concocted elaborate fables about the exotic origins of its products for publicity purposes. Nick Hawkins, an analyst with Kleinwort Benson, was 'a bit surprised' when he learned a few years ago that 'natural' meant less than 2%: 'If your products are no more natural than the next man's, then people might ask "why pay a big premium?" It takes away a bit of the mystique' (cited in Cowe 1994a). The Body Shop justifies its use of non-natural ingredients by saying that they are substitutes for ingredients that involve cruelty to animals, or that the costs (including environmental costs) of transporting natural ingredients are too high (Zagor 1994a).

Critics also considered The Body Shop's 'Against Animal Testing' stance inadequate. The company used a five-year rolling programme to encourage suppliers to stop testing on animals. Although the British Union for the Abolition of Vivisection, the International Fund for Animal Welfare and the European Coalition to End Animal Experiments endorsed The Body Shop's policy, both the Royal Society for the Prevention of Cruelty to Animals and Beauty Without Cruelty believed a fixed cut-off date is the most effective way of reducing demand for ingredients tested on animals (Nuttall *et al.* 1994; Cowe 1994a). A fixed cut-off date, as employed by some of the biggest supermarket and chemist chains in the UK, would have closed the loophole that enabled The Body Shop to use new ingredients five years after they have been tested on animals (Brummer 1994b). *The Guardian*, bravely, since The Body Shop had won a libel case against Fulcrum Productions and Channel 4 over a documentary that questioned the sincerity of its commitment to animal welfare, carried a front-page story saying the RSPCA would encourage consumers to switch from The Body Shop to retailers with 'higher ethical standards' (Randall 1994; Fallon 1994b).

Entine fuelled the animal testing debate by claiming that, in 1992, an internal Body Shop memo reported that the number of ingredients previously tested on animals had increased from 34% to 46.5%. The Body Shop, on the other hand, had claimed that only 17 of its 596 ingredients had been tested on animals for cosmetic purposes (Confino and Cowe 1994). Entine, however, pointed out that the company had used ingredients tested on animals for pharmaceutical purposes to circumvent its policy that restricted testing for cosmetic purposes.

Other Environmental Issues

Entine claimed that The Body Shop wasn't as environmentally friendly as it had portrayed itself to be. In support of this claim, he revealed that, in 1992, The Body Shop's New Jersey plant had had two accidental discharges. These discharges were by no means major environmental catastrophes, however. Both discharges involved harmless non-toxic finished products (30 gallons of Fuzzy Peach Shower Gel and 30 gallons of Orange Spice shampoo), rather than concentrated chemicals and merely resulted in a pleasant-smelling sewer system (Black 1994).

Perhaps more environmentally damaging, though, is the packaging of The Body Shop's products. Despite Anita Roddick's 'Reduce–Re-use–Recycle' exhortations, the company has been accused of the very practice it condemns—repackaging for the sake of repackaging—since many of its products were actually manufactured by outside suppliers. Its gift baskets, which are more environmentally damaging than its other products due to the extra packaging, are extremely popular with customers. Moreover, despite the recycling initiatives, only a small percentage of consumers actually return their bottles. This low rate of return, coupled with The Body Shop's staggering growth rate, surely outweighs many of the company's environmental initiatives by the sheer volume of waste it creates.

Disgruntled Franchisees

Franchising allowed the Roddicks to expand their empire with minimal risk, since franchisees bear the costs of opening new stores (Marks 1994). However, the franchising operations of The Body Shop also came under attack. Entine revealed that the company had been under investigation by the US Federal Trade Commission regarding its franchising practices, and he quoted Dean Sagar, an economist on the House Committee on Small Business, as saying: 'The Body Shop appears to use most of the abusive practices that are standard in the franchising industry' (Entine 1994: 27). Tran (1994) reported that several franchise operators had complained to the American Franchisee Association, but did not go public for fear of retaliation. As Cowe (1994c) points out, there will always be some disgruntled franchisees 'who discovered that it was more difficult than expected to find the seam in the gold mine'.

The Body Shop, on the other hand, claimed that it had a good relationship with franchisees and that, in the United States, more than 95% of its franchisees wrote a letter of support to the company within 24 hours of hearing the allegations (Farrand 1994).

Independent Verification

The Body Shop was further criticised for the lack of independent verification of its procedures. Although it had commissioned independent reviews of its operations and environmental reporting in the UK, such rigour was not applied consistently throughout the chain. This inconsistency, or what Amy Domini, the founder of the Domini Social Equity Fund in Boston, calls 'sloppiness and highhandedness' (Schoolman 1994), may in part be due to the company's rapid growth, but nevertheless served to intensify the controversy. Brummer (1994a) states that:

> It would not be enough to take the company's word for it, or permit the odd transgression because the company had its heart in the right place . . . because Body Shop has backed a wind farm . . . The message is the same for Body Shop as for any company professing ethical intentions; prove it, by stating clearly what those intentions are, by setting quantifiable targets, and by producing a comprehensive public report, independently verified.

The lack of independent verification contributed to The Body Shop's failure to respond adequately to Entine's article with a barrage of counter-arguments and statistics. Some people were left with the impression that the company had been economical with the truth (*Marketing Week* 1994). One reporter went so far as to state that The Body Shop was a candidate for his 1994 'appalling public relations' award (*Sunday Telegraph* 1994). It seemed, in spite of Anita Roddick's claims that she was available for interviews, such access appeared to be limited to promotional opportunities, rather than opportunities to ask The Body Shop some hard questions.

Personality and Control

The criticisms that had plagued The Body Shop, even before this controversy, had resulted in attempts by some members of the company to persuade Anita Roddick to tone down:

> It's not gagging, but she's encouraged to pack up her ideals and toddle off . . . with anthropologists, adventurers and other mavericks . . . in search of this herb or that banana skin. So Anita gives fewer interviews these days, is wheeled out only when needed and performs more at business conferences (Vidal and Brown 1994).

There was concern, however, as to what would happen to The Body Shop if it lost its figurehead. Would the company lose its enthusiasm and vibrancy to fade away and become just another run-of-the-mill cosmetics company? One London stock analyst, when asked how The Body Shop would fare if Anita Roddick retired, stated that: 'A couple of thousand years ago, you might have asked, "what's going to happen to Christianity if Jesus Christ dies?" If Anita Roddick goes, The Body Shop could potentially become even stronger. The corporate culture is very strong' (cited in Ansley 1991: 24). This assumption is a dangerous one, however, since, as explained by an employee, a large part of The Body Shop's attraction is Anita Roddick herself:

> The combination of Anita's personality and the marketing is so bloody positive and hypnotic that it sweeps people up and lets them believe for a moment that, when they buy that bottle of hair gel or navel-fluff remover made by Umba-Umba or whoever tribe, they're actually doing something for someone else (cited in Vidal and Brown 1994).

Control was also an issue. One report suggests that the company's maintenance of its small, entrepreneur-led management structure was to blame for the poor management of the crisis (Ward 1994). The Body Shop's structure in 1994 was quite hierarchical. The Roddicks, apparently reluctant to share power, maintained relatively tight control over the company and had only recently given in to pressure to adopt a 'less autocratic' structure (Jardine 1994). The Roddick's dominance is evidenced in a statement made by Anita: 'The staff put forward issues which they think we ought to tackle, but the final decision is made by me, Gordon and about half a dozen of our senior executives' (Roddick 1991: 126). The company basically worked through a top-down communication process, and there was very little participation in decision-making by staff, who were essentially isolated in their own stores and gained information primarily through promotional videos and literature. Consequently, the chain became extremely reliant on the Roddicks for its direction. This dependence was fostered by Anita Roddick, who assumed that her staff and franchisees would automatically follow her in whatever she chose to do. The top-down communication process appeared to have been relatively successful in the past and had helped to focus the efforts of many individual franchises. However, its continued success would depend on Anita Roddick retaining the loyalty, trust and admiration of her staff and franchisees—something that, according to critics, she might fail to do.

Despite her critics, Anita Roddick remained a role model and enjoyed a certain superstar status. Anita's fame and admiration is largely the result of her firmly held ethical and environmental beliefs which she has applied successfully to her business: 'I have never been able to separate Body Shop values from my own personal values' (Roddick 1991: 122-23). Many people, including consumers, identified with Anita's belief that there was a need for change, and that the business world needed to become more responsible. However, the dangers of such a high profile and such high principles have been borne out by the controversy.

The Problem of High Ideals

One report during the controversy stated that: '[The Body Shop's] performance doesn't match up to the moral high ground it has colonised . . . It is certainly true that the company's marketing hype has run ahead of reality' (*Guardian* 1994). The question is, how big is the gap between marketing hype and reality? Is the public perception of The Body Shop purer than The Body Shop itself has ever actually claimed? Unrealistic perceptions require more careful management on the part of the image-makers.

No major company can ever be environmentally benign, so The Body Shop will always be prone to criticism. Ironically, the sort of criticism levelled at the company is frightening the business world away from adopting higher ethical and environmental criteria of performance, and has some people asking, 'What hope is there for anyone else?' (Drummond 1994). The degree of 'scrutiny and potentially hostile attention from rivals, consumer groups and the Green lobby' has allowed some companies to justify their slow response to the need for more environmentally friendly and socially responsible practices (Higham, cited in Eden 1990: 14).

Supporters of The Body Shop argue that, although the company is by no means perfect, it has made genuine efforts to reach its ideals. Ronnie Morgan, a factory manager with the company, questioned the validity of its claims when he was first hired: 'I was sufficiently cynical to wonder if the whole environment thing they support was a gimmick, but I honestly don't believe it is. Body Shop is a tremendously honest company; people genuinely believe in what they are doing' (cited in Ferguson 1989: 96)

The Body Shop is certainly a lot better than the majority of companies:

> Last week several thousand acres of far eastern rainforest was offered for clear-felling on the London stock exchange and no-one batted an eyelid. Other multinationals doubtless polluted waterways or spilt their toxics,

> ram-raided poor communities or locked whole communities into debt. Yet Body Shop was hauled over the coals for spilling 30 gallons of Fuzzy Peach Shower Gel some years ago (Vidal and Brown 1994).

High ideals are well worth having but shouting about them would seem to be tantamount to asking for close and merciless scrutiny.

Quality and the Threat of Competition

In the past, The Body Shop has been able to fend off its competitors by differentiating its products from copycat products on the basis of a favourable ethical and environmental image and high quality. Gordon Roddick claimed that the company was still unique: 'We have built a market for ourselves. There is nobody else around who offers the same quality for the same prices we do' (cited in Ghazi and Tredre 1994). However, Entine attacked both The Body Shop's ethical and environmental practices and the quality of its products. He quoted a natural cosmetics distributor as saying: 'If you take The Body Shop name off the products and put "Payless Drug Store" on the label, you get an idea of the products' quality' (Entine 1994: 25). Entine also claimed that the company had a poor record of quality control and that it had sold contaminated products.

Perhaps an even greater threat to The Body Shop's survival than the above criticisms is the increasing number of competitors riding in on the green wave. This issue, largely overlooked by Entine, drew the attention of the press. Several companies, such as Boots in the UK, have mimicked The Body Shop's products (Fallon 1994b). 'Greenness' is no longer an adequate differentiating factor, since it has become fashionable in the marketplace to make references to the environment, sometimes even when those products are actually environmentally damaging.

In the UK, a number of retailers, including Boots, Superdrug, Marks & Spencer and Sainsbury, have 'natural' product ranges in competition with The Body Shop and are, in some cases, undercutting The Body Shop's prices. Even companies in the high-quality end of the cosmetics market are beginning to move into the 'green' market. Estée Lauder, for example, launched a line of natural-ingredient, no-animal-testing cosmetics in refillable bottles. In the US, The Body Shop faces more than 30 competitors, many of which are copycats, Bath & Body Works being the biggest with aggressive growth both in store numbers and sales, with the latter increasing 40% in 1997 to US$1.1 billion. Commitment to new product innovation allows Bath & Body Works to claim 30% newness in stores at any time (Intimate Brands 1998), a hard act for other chains to follow.

The controversy and the resulting stain on The Body Shop's reputation has increased the threat of competition from all sides, and there had been predictions that customers would desert the company. Analysts suggested that it would take only 10%–20% of The Body Shop's customers switching to other companies for profit margins to be under severe pressure (Ghazi and Tredre 1994). Nevertheless, the company appears to have ridden out the controversy, and still remains 'different' from most of its competitors. The colours, atmosphere and the sweet smells in its stores makes shopping at The Body Shop an attraction and an experience in itself: 'The first thing customers do when they walk in is to smell something. Nobody does that in Boots' (investment counsellor, cited in Marks 1994). Moreover, from the outset, The Body Shop had cleverly positioned itself as a company that cared about issues outside of selling product, whereas other companies' reputations for social concern were much less significant.

The Body Shop Four Years On

In the intervening four years since the Entine-inspired controversy, The Body Shop has not shied away from its founding ideals, despite the obvious challenges in meeting them in a very competitive business environment.

In 1995, it produced its first *Values Report*, comprising of three separate reports on its social, environmental and animal protection performance. Despite the obvious difficulties in measuring intangibles and reporting on the more intrinsic aspects of development, the company had made considerable efforts. Its second *Values Report*, published in 1997, again scored the highest rating out of 100 international company reports evaluated by SustainAbility for the United Nations Environmental Programme, and was noted 'as unusual in its efforts to integrate social and environmental reporting with considerable stakeholder engagement' (SustainAbility/UNEP 1997). A key feature of The Body Shop's reporting is its independent verification by external parties, an issue that had earlier been in contention. It continues to generate detailed environmental monitoring data, published in its UK site reports, as required for certification to the European Eco-Management and Auditing Scheme (EMAS), for which company registration remains voluntary. The company has also signed up to the new international standard developed by the British Union for the Abolition of Vivisection (BUAV) stating it does not test on animals. It has further increased its numbers of community trade suppliers since 1994, with the value of raw materials and accessories purchased from community trade suppliers rising to nearly £2 million in 1996/97 (Body Shop 1998b). It remains active in supply chain management. Other values-based initiatives included Anita Roddick's 1995 founding of the New Academy of Business, an independent educational body with the objective of encouraging socially and environmentally responsible business performance, and the development of a very successful campaign based around self-esteem and self-authority. Designed as a challenge to the cultural conceptions of femininity as portrayed in the beauty industry, the campaign cleverly positioned The Body Shop as a caring alternative.

Despite these successes, the company has been losing momentum with underlying sales falling. Store numbers were continuing to increase, with a net 103 new store openings during the 1997/98 financial year, 75 of which were in Asia. Profits overall remained static, with concern being voiced as to effects of the Asian economic crisis, and disappointing results in Japan and the US (Body Shop 1998a). The highly competitive US market is a source of concern, with market share being eroded and a number of franchisees either quitting or doing badly. Stuart Rose, former managing director of The Body Shop International, spoke of the need for faster product development and more attention to merchandising and advertising. As for rival competitor Bath & Body Works, he admitted, 'By American standards, they're much better merchandisers than we were' (cited in Tannenbaum 1998). Competition has remained strong from the better-run imitators across several markets, giving The Body Shop further cause to reconsider its founder's antipathy towards traditional marketing activities. Various new product launches, including the hemp and Bergamot ranges, and a broader array of marketing strategies, continue to be tried in an effort to revitalise sales. Such efforts have been particularly important to counter increased competition in major and more mature markets such as the UK, where the previous no-advertising policy has finally been revoked.

Several changes to The Body Shop's organisational structure have taken place. In 1996, after a period of restructuring, changing the top management team by bringing in other professional managers and general streamlining of procedures, Anita Roddick complained of having gone 'through a period of squashing one hell of a lot of entrepreneurial spirit' and of it taking a year or more to get new products into the stores compared with the four or five months previously (Wallace and Brown 1996). Worries concerning the company's lacklustre performance preceded the May 1998 announcement of two further major restructuring plans. First was the transfer of the management and later the ownership of the US business to a joint venture with a new CEO, with hopes of improving profitability. Second was the announcement that Anita Roddick was to step aside, with the appointment of Patrick Gournay as chief executive officer of The Body Shop on 14 July 1998. Gournay brings a record of having doubled sales and significantly improving profitability while with the Dannon company, part of the French food conglomerate Groupe Danone (Body Shop 1998a), a record The Body Shop would no doubt wish him to continue. Anita and Gordon Roddick were henceforth to share the chairperson's role, continuing their involvement in steering the company. Subsequent comment assigned much of the blame for The Body Shop's dismal financial performance with Anita Roddick, who admitted to being bored with basic retail disciplines such as distribution and more interested in spending time with the Dalai Lama (*Economist* 1998a). The Body Shop's management was described as close-knit and top heavy, with the Roddicks and a friend still retaining a 49% shareholding, and having actively explored privatising the company in recent times. Indeed, some have talked of the short life expectancy of chief executives not called Roddick (Nicolas 1998), and others complained that, according to the consensus among City and retail analysts, Anita Roddick's step-down was about three years too late (Mills 1998). Whether the Roddicks' passion for the marketing of principles will, either officially or unofficially, remain ahead of the marketing of products at The Body Shop is a question many would like to see answered.

Marketing and Environmentalism

Marketing has had an interesting role to play in The Body Shop case. Although Anita Roddick has claimed she is not a marketeer, others have a different view. At a time when conventional marketing activities were eschewed at The Body Shop, Chipperfield (1988: 54) pointed out, 'Anita Roddick is being a little coy. Either she or her husband has an innate grasp of marketing . . . [The Body Shop has a] shrewd marketing strategy which few professionals could improve on.' Anita Roddick has also been known to savage the marketing profession. For example, at a conference held in 1987, she stated:

> Marketing departments—like planning departments and advertising departments—are usually a camouflage to cover up for the lazy or worn out executive.
>
> I also know that the word marketing usually means an excuse for gathering around the table to talk. To talk on and on around every function other than the most important function which is selling the product, understanding the product and valuing the product.
>
> Whenever I've met anyone wearing the label 'marketing' they are either 25-year-old Mars or Procter & Gamble graduates who speak a language I don't understand or they are tired executives more versed in the status of profit than the art of selling (cited in Chipperfield 1988: 54).

Despite these comments and the rise and fall of traditionally conceived marketing at The Body Shop's Littlehampton headquarters, Anita Roddick has surely been aware of the potential of marketing in initiating ethical and environmental change in business. Hilton argues that the marketing profession is in 'the front line' of the type of shift in corporate thinking encouraged by Anita Roddick (cited in Eden 1990: 13). Coddington, too, believes that the marketing profession can precipitate changes in a company's non-product or product-related environmental performance by working 'backwards' (Coddington 1993: 13). He also points out that the communication skills, the ability to work through corporate bureaucracy and the sensitivity to the marketplace implications of environmental decisions possessed by marketeers are the sorts of skill needed for successful environmental management (Coddington 1993).

Critics of green consumerism, however, argue that we cannot rely solely on marketing strategies, even if they are to promote 'green products', to solve our environmental problems; green business is seen as a hoax:

> Green business tends to merely perpetuate the colonization of the mind, sapping our visions of an alternative and giving the idea that our salvation can be gained through shopping rather than through social struggle and transformation. In this respect, green business at worst is a danger and a trap (Plant and Albert 1991: 7).

The basic problem of green consumerism is, as the name implies, that it focuses on 'the selection and purchase of products on the basis of their known or perceived environmental effects and primarily by the general public for their domestic use', rather than on *how much* is consumed (Eden 1990, pp. 4-5). Plant and Albert (1991) argue that we could do without many of the so-called 'green' products, even those from 'one of the greenest businesses of them all, on the surface—The Body Shop' (1991: 3). They suggest that, while 'green' products may be welcome, these products are also a diversion from the urgent deep structural task ahead: 'We can hardly hope for societal transformation from recycled toilet paper' (1991: 7). Irvine (1991) claims that a truly green strategy would 'involve a change in thinking in terms of what is the minimum necessary to satisfy essential human needs, rather than novelty, fashion, status and all the other hooks of materialism' (1991: 28), or what Smith and Sambrook term the marketing of 'less is more' (cited in Eden 1990: 18). Plant and Albert (1991) assert that the business world is attempting to co-opt the green movement in order to prevent the implementation of radical strategies, such as reducing consumption, which threaten the continuation of business-as-usual. There is, of course, no private gain to be made in not producing. However, neither the negative nor the positive contribution that businesses such as The Body Shop can make to social and even environmental causes should be ignored.

Conclusions

The Body Shop weathered the controversy sparked by the Entine article quite well. In the intervening four years, it has continued its efforts to be environmentally and socially responsible, despite even more intense competition and the difficulties in so doing. It remains a slightly tarnished, perhaps less flagrant but still potent role model for other companies. It continues to be commended for attempting to implement procedures to deal with ethical and environmental issues that are not even considered by some.

It seems that companies that adopt more radical solutions to ethical and environmental dilemmas, rather than perhaps a safer incremental approach, can expect to come under close scrutiny. The Body Shop case highlights how scrutiny is intensified by claims of being a 'good' company, rather than one on a continuum of 'better to worse'.

One of the benefits of having companies such as The Body Shop, however, is the kinds of lesson they provide. The Body Shop has certainly raised the profile of green consumerism and boosted other environmental causes, but whether its actions will stimulate people to examine critically both their own lifestyles and society as a whole requires further evaluation. The key lesson from The Body Shop case is that marketing can, and probably should have, a far broader focus around principles and values than just changing people's perceptions around products, though, as this chapter has demonstrated, such a strategy is not without risk.

Several other implications for green marketeers, which are of a more practical nature, arise from our consideration of The Body Shop case. Of primary importance is the need to ensure that what is being marketed as green is as green as it can possibly be. Ethical and environmental claims that cannot be independently verified need to be reconsidered. Greenness is not just an add-on for marketeers. It involves total integration of an environmental ethic throughout the company's products and processes, and involves the marketeer working at a number of levels within the company, and across various disciplines. Where ethical and environmental claims are an intrinsic part of what make the company so attractive to its investors and consumers, then, provided they can be substantiated and presented meaningfully, the marketeer has an appealing proposition.

However, as in this case, the 'marketeer' may not long be alone in marketing a 'good' thing. When copycat competition ensues, brand differentiation, continued generation of new ideas and time-to-market become important. Resource-based theory (Prahalad and Hamel 1990; Barney 1991) suggests that, the more socially complex a company's resource or competency is, the more difficult it is to imitate. The Body Shop's bundling of social action, environmental protection and now self-esteem programmes with its manufacturing and retailing activities provides a potent example.

Our final comments relate to marketing based on pronouncements of high ideals or principles. If you have high ideals, if the principles become the brand, then it can be difficult and indeed risky to lower them. Moreover, public perception of the ideals you espouse may have to be managed to ensure that it does not become overblown. Here, clearly, is an important role for marketers and other image-makers. Although deflated somewhat in terms of falls in underlying sales and share price, The Body Shop's bubble has not burst—yet. Several of the issues raised both during the controversy and since, however, suggest that the future may be yet more difficult. What do you do with high ideals when the competition intensifies and traditionally conceived bottom-line imperatives impinge? Can you afford to keep blowing bubbles? But can you afford not to?

19

Greening Agroindustry in Costa Rica

A Guide to Environmental Certification

Rebecca Winthrop

ENVIRONMENTAL CERTIFICATION is a mechanism for providing marketing incentives for companies to green their production practices by harnessing consumer demand for environmentally friendly products. The Conservation Agriculture Network (CAN), an association of four conservation groups throughout the Americas, began with one initial project in Costa Rica and has now developed a group of successful environmental certification programmes in Guatemala, Brazil, Ecuador, Mexico, El Salvador, Panama, Nicaragua, Colombia and Honduras (Rainforest Alliance 1998a). Accompanying this geographical expansion has been a proliferation of programmes certifying different products. The CAN's first project, the Better Banana Project, is aimed at greening agroindustry's production of bananas.[1] Now the organisation has guidelines for the environmentally friendly production of coffee, cocoa, oranges and sugar cane (CAN 1998a).

By examining the initial development and implementation of the CAN's first Costa Rican banana project, we can learn some of the key elements for creating a successful environmental certification programme, as well as some of the potential pitfalls to avoid. This chapter will review the principles of environmental certification and explain the importance of timing, participation and flexibility in a successful environmental certification programme. Also addressed are the ways in which marketing and monitoring can pose challenges to effective programme operations. In conclu-

1 The Conservation Agriculture Network (CAN) is an association of four conservation groups throughout the Americas. The Rainforest Alliance is a member of the CAN and serves as the network's international secretariat. The CAN manages the ECO-OK Agricultural Certification Programme, of which the Better Banana Project is a part.

sion, a series of policy recommendations will be offered for those currently operating or developing environmental certification programmes. Detailed information on the farms certified by the Better Banana Project is provided in Appendix A; a summary of the project's general production standards for certification is provided in Appendix B; and a summary of the certification procedures is provided in Appendix C.

Understanding Environmental Certification

Environmental certification is a process by which industries are encouraged to change production patterns and product qualities to incorporate environmental values. This influence is often affected when independent, third-party organisations, either private or public, inspect a company's operations and grant 'green labels' if, among other things, the production and product meet established, environmentally friendly standards. The company can then use the green label in product marketing, a practice that gives the company several advantages: access to the growing group of consumers who will only buy environmentally friendly products, a potential price premium for the product as a speciality item, and an increase in product quality along with less quality variation of the product.

Environmental certification programmes can also encourage social benefits: in theory, many such programmes attempt to contribute to the growth of green consumerism (Schmidt 1995), the idea being that consumer awareness of environmental issues is stimulated through exposure to products with environmental labels. For example, when shoppers are faced with two brands of a product but only one has a green label, they are forced to think, at least in passing, about the environmental impacts associated with that product. Many environmental certification programmes also engage in consumer education campaigns. One of the first, the Blue Angel programme, founded in Germany in 1978, is accredited with, among other things, 'stimulating consumer awareness' through direct action as well as through public exposure from label use (Salzhauer 1995). Many, including the Good Environmental Choice programme in Sweden and the Green Label programme in Singapore, also attempt to educate consumers as a policy goal (Schmidt 1995).

There is also a variety of programmes often working with environmentally friendly products but the main goal of which is to contribute to social justice. Often called Alternative Trading Organisations (ATOs), these usually belong to the Fair Trade movement. Fair Trade is the support of an 'equitable and fair partnership' predominantly between marketers in North America and Europe and 'producers in Asia, Africa, Latin America, and other parts of the world' (Fair Trade Federation 1997). ATOs and other members of the Fair Trade movement seek to eliminate the middleman and provide financial support directly to small-scale agricultural producers and craftsmen in less developed countries. Such producers may sell crafts, textiles or 'food items such as tea, coffee, chocolate, and honey' (Oxfam 1996). While most agricultural products bought by Fair Trade programmes are organic, the labels they attach to such products are not environmental certification labels: predominantly, they guarantee only that the products were fairly traded.

While there are many types of label that can be attached to products, environmental labels fall into three main categories. First there are the 'single-issue voluntary labels' which only reflect information about one aspect of the product or production process: for example 'CFC-free' (Schmidt 1995). Companies often make such information available to consumers voluntarily. Government regulations may set standards, allowing manufacturers to claim compliance without using certification programmes.

A second type is the 'single-issue mandatory label' which does not, in most cases, involve certification programmes. Such labels are put on products by manufacturers because of legal requirements—for example, most governments have regulations mandating that fire hazardous products be labelled 'flammable' (Schmidt 1995).

The 'eco-label', the third type, may be used on products or in advertising and is always granted through environmental certification programmes: the Better Banana Project is an eco-label programme. An eco-label provides consumers with an idea of the overall environmental quality of a product. Certification programmes use criteria intended to judge the total environmental impact of the product from production to disposal. These labels help consumers determine which among competing products is environmentally 'best' or less environmentally harmful (Schmidt 1995). Programmes that award eco-labels are voluntary and typically use an identifying symbol or logo.

The increase in environmentally aware consumers is the key driver for eco-labelling programmes. Particularly in the US and Europe, a growing number of people are seeking to strengthen environmental protection: not by lobbying but by purchasing choices. A 1990 Gallup poll showed that more than 90% of US consumers not only wanted to buy green products but would also be willing to pay more for them (Schmidt 1995). However, not everyone who supports environmental products in theory translates conviction into action by buying green goods once at the checkout counter. Nevertheless, a recent survey shows that between 40 and 50 million Americans, approximately one-quarter of the adult population, are buying socially responsible products when price and quality are comparable (Hartman Group 1996). In the words of Harvey Hartman, 'the green consumer is now mainstream. There is significant market potential for earth-sustainable products' (Hartman Group 1996).

In the 1990s, however, consumers became concerned that manufacturers' environmental claims were merely marketing ploys rather than objective verification of products' environmental quality (Haddon 1993). Thus, an eco-labelling programme meets consumers' demands in two ways. First, it identifies which products are environmentally friendly and, second, it ensures that the labelling claims are true.

The growth of environmental consumerism leads directly to an expansion of the green market by providing increased demand. The mere existence of environmental certification programmes also contributes to the expansion of the green market because such programmes offer a visible mechanism for products to enter that market. Companies are more likely to pay attention to consumers' concerns for the environment when consumers' preferences show up as a label on a competitor's product. With certification programmes awarding companies or products for an element that was previously not counted in market success, the race is, or theoretically should be, for all companies to strive for certification in order to maintain competitiveness (International Ecolabeling Forum 1994). Together, these elements of eco-labelling work toward creating a 'green trade strategy' in which 'consumer choice, rather than government intervention, is creating pressure for environmental protection' (Schmidt 1995). Environmental certification provides a mechanism that is able to link environmental protection with consumer concern and business practice.

The Importance of Timing

It is precisely this link between environmental protection, consumer concern and business practice that dictates the success of an environmental certification programme. With the exception of government regulation, if consumers are not concerned about the environmental quality of a specific product, then the company producing the product has very little incentive to participate in an environmental certification programme.

The founders of the CAN, the Rainforest Alliance and its Costa Rican partner organisation, Fundación Ambio, understood the importance of this link. They introduced their banana environmental certification programme at a time when steady consumer outrage, in the form of lawsuits, protests and denouncements, made the industry ripe for change. Before this period of forceful consumer discontent, most of the national and multinational corporations involved in the banana agroindustry in Costa Rica were not actively involved in greening their production practices 'nor did they see any reason to be'.[2]

Bananas were introduced into Costa Rica in the 1850s. Today Costa Rica is the world's second-largest banana producer. Yet, since its introduction, cultivation of the fruit has been at the expense of Costa Rica's once-plentiful natural resources (Wille 1992). Large-scale commercial banana farming is characterised by endless rows of banana plants, and deforestation is one of the primary impacts of banana farming as rainforests are being cut down to plant monocultures (Wille 1992). High levels of pesticides and fertilisers are used, killing almost everything but the banana plants themselves. Deforestation and pesticide use results in soil erosion and contamination which, during the daily rains in the tropics, washes topsoil and chemicals into rivers. Fish are often killed from sedimentation and pollution in nearby waterways (Wille 1992). Improper waste disposal leads to rivers clogged with discarded bananas and the plastic bags that are used in production, which later leads to the contamination of marine habitats. Soil quality is also destroyed by layers of plastic cord that are used to tie up banana plants. When workers harvest bananas, they cut the cord and leave it lying on the ground. In one year, the Costa Rican banana industry uses enough plastic cord to wrap around the earth 45 times (*Nius Ecos del Atlántico* 1995). Workers often use hazardous chemicals without proper safety techniques and families

---

2 Personal communication: e-mail interview with Chris Wille, Regional Director of the Better Banana Project, 21 January 1995.

usually live in substandard housing with poor water quality and poor sanitary facilities (CAN 1998d).

It is only in the last 20 years that consumer discontent with these environmental practices has been strong enough to be noticed by the banana agroindustry. In 1985, a lawsuit on behalf of 101 sterile Costa Rican banana workers focused international attention on the production practices of banana companies. Shell Oil, Dow Chemical and the Standard Fruit Company were all sued in the US—where environmental laws are in general stricter than in Latin America—for manufacturing and using the nematicide DBCP (dibromochloropropane) (*Mesoamerica* 1992).

This judicial precedent paved the way for national organisations within Costa Rica to exert pressure on the banana industry. In 1994, SITRAP, a Costa Rican banana workers' union, brought legal action against the Standard Fruit Company on behalf of an alleged 6,000 workers rendered sterile from DBCP (Castillo 1994). The union also lashed out at a British multinational, Geest, for illegally using minors and children in pesticide handling on the company's banana plantations, highlighting the case of a 16-year-old boy who died from pesticide poisoning on one of Geest's farms (Castillo 1994).

Responding to local complaints of pesticide contamination, the Netherlands-based Second International Tribunal of Water denounced Standard Fruit for its abuse of the environment in Costa Rica and recommended that the company change its production practices (*Mesoamerica* 1992).

Standard Fruit retaliated to the denouncement by saying that the environmental groups were trying to start a boycott of the company's bananas. While consumer discontent continued to be articulated through local and international lawsuits, distribution of press stories, environmental and social denouncements, and pressure group activity, the threat of a boycott became very real. A global network composed of over 70 environmental groups was indeed considering a boycott. 'At this moment we could create a boycott,' said ADCH spokesman Javior Bogantes in 1992, whose group is part of the network, 'We have the strength to do it' (*Mesoamerica* 1992). A boycott could have done much damage to the Costa Rican banana industry: the US imports 45% of all Costa Rican bananas produced and the remainder is exported to Europe and a boycott even in a single country could be devastating for the industry (*Mesoamerica* 1992). While a boycott would have been a powerful tool, many consumers were against the strategy because those hardest hit by a boycott would undoubtedly be the plantation workers whose livelihoods depend on the industry—Costa Rica's banana industry employs approximately 45,000 workers directly and an estimated 100,000 workers indirectly (Rossi Umaña 1994).

It was in this political context that the CAN introduced the idea of an eco-labelling programme as a proactive marketing mechanism that was beneficial to both the banana companies and the environment while allowing workers to retain their jobs. National and multinational companies could participate in the Better Banana Project's eco-labelling programme by reforming production practices and in return they would be able to market their bananas with the programme's green seal of approval. This new environmentally certified banana would simultaneously appease consumer discontent and gain a competitive edge by appealing to green consumers. Compared to a boycott, this alternative was by far more beneficial for banana producers and banana workers.

The rapid expansion of banana farms participating in the Better Banana Project shows that the CAN's introduction of a banana eco-label came at an excellent time. Existing and new farms that reform and develop operation procedures to meet the project's criteria are able to pass inspection tests for certification and, in August 1993, the first farm was certified with a total of 110 hectares of environmentally friendly bananas (Rainforest Alliance 1995a). Just two years and four months later, by November 1995, one in ten Costa Rican banana farms had been certified (Rainforest Alliance 1995b). In August 1996, 8,841 hectares were certified (Rainforest Alliance 1995a), increasing dramatically to 20,000 certified hectares in 1997 (CAN 1998a).[3]

> At present, nearly 20 per cent of banana production in Costa Rica and 50 per cent in Panama has been awarded certification; farm evaluations are under way in Ecuador and Colombia; and banana growers in Honduras and Guatemala are making necessary improvements to achieve certification (Rainforest Alliance 1998b).

Judging by the rapid increase in level of participation, it would seem that the Better Banana Project has been successful in capitalising on the consumer discontent with the banana industry by presenting environmental certification as a mechanism beneficial to both business and the environment.

The Importance of Participation

While the introduction of the Better Banana Project was auspiciously timed, the inclusive methodology used to develop the programme also contributed greatly to its success. Research for the programme started in 1991 (Rainforest Alliance 1995a) and, from the outset, the CAN sought the input of the banana industry. To develop the programme guidelines, the CAN convened a series of meetings involving representatives from the banana industry, the Costa Rican government and other environmentalists and scientists (Rainforest Alliance 1995c). These guidelines form the heart of the programme and provide 'clear and objective criteria against which a farm can be evaluated' (Rainforest Alliance 1998c). The goal was to design guidelines that would be strict yet manageable, and, most importantly, acceptable to everyone involved in programme operations.

The variety of input used to create the guidelines facilitated agroindustry's acceptance of the Better Banana Project. In 1995, both the Central American coordinator and the technical director of the programme believed that this focus on participation effectively illustrated the programme's willingness to work *with* rather than *against* the banana industry.[4] It is this attitude especially that helped to win

3 For more detailed information on the certified farms participating in the Better Banana Project, see Appendix A, a summary taken directly from Rainforest Alliance literature.

4 Personal communication: e-mail interview with Chris Wille, Regional Director of the Better Banana Project, January 1995.

over the support of the multinational banana producers. Both the co-ordinator and the director recall that, when the Better Banana Project began, it was met with indifference and even hostility from the large multinational producers in Costa Rica who rejected environmentalists as radical and impractical.[5] However, by 1995, Chiquita Brands International, formerly United Fruit and currently the world's largest banana producer, had joined the Better Banana Project and Geest had employed the programme's technical director to review its agrochemical practices. Also around 1995, Dole Food Corporation, Del Monte Fruit Company and Geest all hired environmental staff and the companies began to talk 'glowingly of their environmental activities and accomplishments . . . [the companies] are willing to meet with environmentalists and they are busy sponsoring environmental research, education and conservation projects'.[6] The Better Banana Project has been successful in influencing corporate attitudes, claims the co-ordinator, 'because it does not lob criticism from outside the walls, but earns credibility with the industry and then proves to them that change is necessary, possible, practical, good business, and can even enhance profits' (Rainforest Alliance 1995d).[7] In this respect, involving industry in the development of a certification programme can be an important key to unlocking corporate scepticism and gaining company support.

The Importance of Flexibility

The Better Banana Project differs from many other agricultural environmental certification programmes in that it allows a 'circumspect use of agrochemicals' (Rainforest Alliance 1995d). According to Eric Holst, manager of the Better Banana Project, the philosophy behind the project's pesticide policy is to 'minimise the use of agrochemicals and maximise biological control of pests'.[8] Practically speaking, this means that producers are required to 'minimise exposure of humans and ecosystems' to chemicals, to train workers properly, and to continually reduce agrochemical applications in order to ensure the active pursuit of biological controls.[9]

The agrochemical policy of the Better Banana Project stems from the extreme difficulty of cultivating commercially successful, large-scale monocultures without pesticides. While criticised by some organic certification programmes, it is its flexibility on the pesticide issue that enables agroindustry to participate in the project without losing production quantity.[10] Indeed, it is doubtful whether many companies would choose a mechanism for greening production practices that resulted in decreased output and revenue loss. While financial success is important to the banana companies themselves, it also is a crucial part of the Central American economies.

5 E-mail interview with Chris Wille, January 1995.
6 E-mail interview with Chris Wille, January 1995.
7 E-mail interview with Chris Wille, January 1995.
8 E-mail correspondence, Eric Holst, 8 July 1998.
9 E-mail correspondence, Eric Holst, 8 July 1998.
10 Personal communication with Elizabeth Skinner, Manager of the ECO-OK Programme, January 1996.

For example, for Costa Rica in 1993, export earnings from bananas reached US$500 million making it the primary export product and the second-greatest source of government revenue (Rossi Umaña 1994). Hence, by allowing some agrochemical use, the programme aims to alter production practices while maintaining economic conditions.

The Better Banana Project justifies its permission of limited pesticide use on two grounds. First, it claims that 'if only organic operations were certified, the positive effects of [the project] would benefit only a few, small farms. Instead, the project aims to make entire [national] and multinational industries partners in conserving tropical ecosystems for the benefit of people and wildlife who deserve a wholesome, productive environment' (Rainforest Alliance 1995d). Second, the project claims that consumers are already protected by international laws from excessive chemical contamination of produce. The programme seeks to shift the emphasis of the agrochemical policy to 'the protection of workers, who have a right to training programs, proper safety equipment, medical check-ups and healthy living and working conditions' (Rainforest Alliance 1995d).

While the guidelines are flexible in allowing agrochemical use, they also encompass rigorous environmental and ecosystem standards. The nine main principles of the guidelines are:

- Protect natural ecosystems
- Conserve wildlife
- Conserve water resources on and around farms
- Conserve the productivity of soils
- Minimise and strictly manage use of agrochemicals
- Employ complete, integrated management of wastes
- Ensure fair treatment and good conditions for workers
- Maintain good relations with neighbouring communities
- Develop environmental planning and monitoring protocols (Rainforest Alliance 1998d).[11]

The Better Banana Project is also flexible regarding the cost of participation in the programme. In an effort to encourage all interested farms to take part, the programme maintains a sliding-scale fee structure. While most farms 'pay the direct costs of the inspection and an annual certification fee between $5 to $10 per hectare, [the project] never refuses any farm simply for financial reasons'.[12] However, all farms in the programme do pay some fee, even those receiving discounted rates.[13]

---

11 For more detailed information on the Better Banana Project's certification standards, see Appendix B, a summary taken directly from Rainforest Alliance literature.

12 E-mail correspondence, Eric Holst, 8 July 1998.

13 E-mail correspondence, Eric Holst, 8 July 1998.

Ultimately, participation in the Better Banana Project has led to significant environmental changes. Certified farms have modified banana monocultures by, among other things, planting trees along roads and waterways, recycling plastic ropes and bags, altering banana row lengths to allow streams and rivers to flow naturally, creating buffer zones along waterways so chemical run-off will be minimised, developing waste disposal and worker safety systems, and monitoring water quality.[14] Since the outset of the programme, the rate of this ecosystem restoration has been quite rapid. In December 1993, 500 hectares per month were restored, rapidly increasing to 5,000 in December in 1994. Also in December 1994, 140,000 kilos of plastic from bags and cords had been recycled. By December 1995, the programme led to the protection of more than 500 hectares of forests and marshes (Rainforest Alliance 1995a). Today, these benefits have increased with the expansion of farms participating in the programme and, by 1998, the CAN claims that 'enrollment in the program has dramatically decreased pollution of rivers and beaches from the effects of agriculture production' (Rainforest Alliance 1998b).

While some may criticise the Better Banana Project for not requiring organic production methods, the programme's willingness to devise guidelines that allow a wide variety of producers to participate is commendable. If agroindustry was effectively excluded from the certification programme due to organic production requirements, many of the companies that have made great environmental strides would not have had the marketing incentives to do so and thus probably would not have greened production of their own accord.

The Challenges of Marketing

While examining the development and implementation of the Better Banana Project has shown us key elements, such as timing, participation, and flexibility, for a successful environmental certification programme, further examination can also illuminate potential challenges to successful programme operation. In many ways, much can be learned from the programme's mistakes.

In its initial years of operation, the project experienced some difficulties in marketing. When first founded, the Better Banana Project was entitled the 'ECO-OK Banana Project'. When a farm was certified, it was awarded the ECO-OK seal of approval and bananas were marketed with an 'ECO-OK' label (Rainforest Alliance 1995d). Unfortunately, in the summer of 1995, the programme was accused of unfair trade practices by Arbeitsgemeinschaft Ökologischer Landbau (AGOL), a German union of rural organic growers, developers and farmers.[15] The issue revolved around a certified banana producer who was exporting fruit to Darmstadt, Germany, an area where AGOL worked. There were two main points in the accusation. First, the prefix 'eco', as well as the words 'ecological' and 'biological', are designated by European

14 Personal communication while the author was a visiting member of the Costa Rican Better Banana Project inspection team, August 1995.

15 Rainforest Alliance, fax to Fundación Ambio, re: Germany, 25 July 1995.

Union legislation to mean 'organic' and thus the 'ECO-OK' label was misleading to consumers. Second, all organic or perceived organic products must, according to German law in 1995, comply with set guidelines for organic production or take part in a government 'control system' that enables German officials to inspect the certified products. If the producer does not comply with these regulations, it is in breach of the Food and Necessities Act and subject to stiff fines and imprisonment.[16]

The CAN resolved the issue by changing the programme's name from the 'ECO-OK Banana Project' to the 'Better Banana Project'. Now, when farms are certified, they are able to place a label on their fruit that says 'Better Bananas, Rainforest Alliance Certified' (Rainforest Alliance 1998e). While the CAN has retained the 'ECO-OK' label, it is no longer used in Europe (CAN 1998b). The programme also created a certification contract as a step in the certification process (CAN 1998c):[17] currently, the farm wishing to be certified must 'enter into a contract with the Rainforest Alliance, which governs the use of the certification mark, the handling of certified products and marketplace promotion' (CAN 1998c). As the secretariat of the CAN, the Rainforest Alliance manages the network's administration, owns the trademarks on behalf of the network, and holds the licensing contract with the producer.[18] The certification contract allows the certifying organisation to make sure that the use of the certification label and promotional materials do not violate any marketing regulations.

The Better Banana Project also seeks to educate others about its certification process. These promotional activities take several forms, such as trade shows, letter-writing campaigns, publications, an extensive website, and developing personal contacts.[19] The programme's marketing has targeted industries and retailers because direct consumer advertising is extremely expensive and, claims Eric Holst, 'better done by the vendors'.[20] Of all the promotional tactics, the most effective has been the 'one-on-one contact with people within the industry' which has led to close relationships with various producers.[21]

The marketing problem faced by the Better Banana Project should be noted by those wishing to develop environmental certification programmes. It is important to remember that the certification programme does not decide where bananas will be shipped. Each certified company needs to have the flexibility to adjust to market conditions, which may include changing the countries to which it exports. In this respect, it is important that the certification programme monitors 'packaging and other promotional materials using the certification mark' (CAN 1998c).

---

16 Rainforest Alliance, fax to Fundación Ambio, re: Germany, 25 July 1995.

17 For a summary of the Better Banana Project's certification procedures, see Appendix C, a summary taken directly from Rainforest Alliance literature.

18 E-mail correspondence, Eric Holst, 8 July 1998.

19 E-mail correspondence, Eric Holst, 8 July 1998.

20 E-mail correspondence, Eric Holst, 8 July 1998.

21 E-mail correspondence, Eric Holst, 8 July 1998.

The Challenges of Monitoring

All environmental certification programmes are faced with the burden of making sure that certified producers actually adhere to the set standards and procedures. For example, in 1995, one of the Better Banana Project's inspection teams conducted a surprise site visit of a certified farm. While walking through the rows of banana plants, looking for morphological changes, recycling of plastic ropes and bags, cleaning of debris between plants, etc., the team came upon a row of workers' safety equipment hanging on the transportation cable. The equipment was all there—shirts, goggles, gloves—and right beside the equipment were workers spraying chemicals bare-chested and bare-faced. Members of the team asked the workers why they were not wearing their protective clothing. The workers answered that it was too hot. It was 2 pm in a hot and humid section of the tropics, and the workers were afraid of heatstroke if they put on full-length plastic clothes with thick elbow-length rubber gloves and face masks.[22]

Recently, several certified farms owned by subsidiaries of Chiquita Brands International have been accused of using dangerous levels of pesticides (Frantz 1998). The accusation, made in a controversial article in the *Cincinnati Enquirer*, cited a coroner's report 'on the death of one worker from exposure to toxic chemicals' along with reports of 'hundreds of people in Costa Rica who were exposed to toxic chemicals from the factory of a Chiquita subsidiary' (Frantz 1998).[23] Chiquita officials refuted the accusation claiming that the company is 'a model in its attempts to reduce the use of pesticides' and the Rainforest Alliance supported the company, noting that almost half of the company's farms have been environmentally certified compared to 'other big American producers', namely Dole and Del Monte, who do not participate in the certification programme (Frantz 1998).

Greening agroindustry via environmental certification programmes is only effective if monitoring systems ensure that producers follow programme guidelines. The Better Banana Project has a lengthy evaluation and certification process that includes a monitoring mandate. Producers undergo a year-long inspection during which technical teams visit the farms and 'inspect all aspects of production and interview workers and their families' (Rainforest Alliance 1998f). Based on the team's findings, farmers are issued a report detailing the changes that need to be made to receive certification. Farms can be certified when 'they meet a majority of the standards and the farm manager has an approved work plan for continued improvements' (Rainforest Alliance 1998f). In order to retain certification, farms must comply with monitoring inspections and producers must provide access to their farms and records on an annual basis. Besides the annual evaluation of the farm's compliance with the

22 Personal communication while the author was a visiting member of the Costa Rican Better Banana Project inspection team, August 1995.

23 On 3 May 1998, the *Cincinnati Enquirer* published an 18-page section of articles criticising Chiquita Brands International. On 28 June 1998, the *Cincinnati Enquirer* published an apology to Chiquita, repudiating the series of articles because of the author's questionable reporting methods.

certification guidelines, the project also reserves the right to conduct surprise inspections (Rainforest Alliance 1998f).

The Better Banana Project, however, cites education of workers and building close relationships with farm managers as the key to its monitoring process.[24] After all, the project claims it is built on 'trust, transparency, involvement, and consensus' between certification experts and producers (Rainforest Alliance 1998g). However, on the farm visited by the aforementioned inspection team, the education and trust elements of the monitoring system were obviously not fully developed. If the farm manager had been attentive to all the programme requirements, he would have consulted the project about the workers' problem or he would have directly solved the situation by providing cooler protective gear, or scheduling chemical spraying at cooler times of the day, such as the early morning.

It is true that many of the elements in the programme guidelines—for example, water quality or vegetative growth—can be assured to a certain extent by testing. Water does not run clean the day before the inspector comes nor do trees spontaneously appear. Surprise inspections also guarantee guideline compliance but only to the extent that farm owners are worried about being caught violating programme requirements. Perhaps more frequent and strict surprise inspections would encourage more faithful adherence to programme regulations. Ultimately, however, full compliance with programme standards lies in the hands of the farm's manager and workers. In this respect, compliance with certification guidelines will be much more successful if programme criteria are feasible, understandable and easily learned by the people responsible for the day-to-day operations associated with certification standards. This somewhat obvious point is important for policy-makers, practitioners and technicians to remember when designing, developing and implementing an environmental certification programme. No matter how sophisticated the programme is on paper, it will only be successful if it can effectively translate to production operations.

Recommendations

Through this examination of the development and implementation of the Better Banana Project, several lessons can be learned for those wishing to develop or reform similar programmes.

First, **timing** turned out to be an important element contributing to the Better Banana Project's success. Pressure group activity and consumer opinion made the banana agroindustry ripe for change during the CAN's introduction of its environmental certification programme. It is possible that corporate producers might not have paid attention to the Better Banana Project without the widespread demand for environmental reforms. Introducing certification programmes can be a difficult process and, in many ways, may benefit from careful timing. Industries that receive

24 Personal communication, Elizabeth Skinner, January 1996.

pressure to change may be more open to the idea of certification. Similarly, if an industry resists the idea of certification, stimulating public pressure may be one way of getting companies to listen.

Second, **participation** was a key to winning corporate support for the Better Banana Project. By involving the banana industry in the development of the certification guidelines, companies felt a sense of ownership in the project and important relationships were formed between producers and programme members. The process of certification is complicated and relies heavily on good communication between certifier and producer. Building trust is an important aspect of a successful certification programme. One of the best ways of building this trust is by involving industry in programme development from the very beginning.

Third, **flexibility** on the part of the Better Banana Project allowed it to meet programme goals. By using a lenient fee structure and by allowing some agrochemical use, largely unheard of for environmental certification programmes, the project was able to meet its goal of reforming the production practices of large- and small-scale producers without altering economic conditions. The ability to listen to industry input and compromise unrealistic criteria is no doubt essential for the development of a feasible certification programme. That is not to say that certification programmes should compromise their standards, but rather find ways of reaching common ground with producers. For example, although the Better Banana Project allows the use of some agrochemicals, a majority of the certification criteria address chemical safety measures for the environment, the workers and their families. The ability to be flexible and find creative solutions is an asset for any certification programme.

Fourth, **marketing** mistakes caused difficulties for the Better Banana Project. Marketing misrepresentation can cause serious problems for certification programmes and producers. It is important for certification programmes to remember that each certified producer needs to have the flexibility to adjust to market conditions, which often entails changing the countries to which it exports. Certification programmes must ensure, either through direct control of trademarks or by other means, that the marketing that certified producers engage in does not misrepresent the certified product.

Fifth, **monitoring** breakdowns have sparked some criticism of the Better Banana Project. Poor compliance with certification criteria could cause a programme's label to become meaningless. In order to safeguard integrity, certification programmes must carefully monitor producers' adherence to programme guidelines. Monitoring techniques, such as frequent and strict surprise inspections, should be integrated into certification criteria. Also, involving workers in the development of certification guidelines and the implementation of certification reforms can help ensure that a programme's requirements are feasible, understandable and easily learned by those responsible for day-to-day operations. After all, no matter how sophisticated a programme is on paper, it will only be successful if it can effectively translate into production operations.

Sixth, additional **research** should be undertaken. The above lessons are useful in the development of environmental certification programmes, but they are the result of a review of only one such programme and, undoubtedly, many issues involved in eco-labelling have been left untouched. For example, further research needs to be undertaken on supply chain issues involved in eco-labelling programmes. Specifically, more information is needed on the activities of all stakeholders in the process of environmental certification—from producer, to retailer, to the end-consumer. Additionally, there is a need for further research on existing eco-labelling schemes in order to provide precise and comprehensive information for those wishing to develop, implement and monitor environmental certification programmes.

Appendix A: Certified Farms Participating in the Better Banana Project

Name/Company	No. of farms	Area (hectares)	Country
Banadex*	13	1,732	Colombia
Chiriqui Land Co.*	32	8, 427	Panama
COBAL*	29	7,757	Costa Rica
EARTH University	2	315	Costa Rica
ESSA	3	660	Costa Rica
Platanera Rio Sixaola	1	110	Costa Rica
Probana, SA	1	349	Costa Rica
Reybancorp, SA	4	1,092	Ecuador

* These companies are subsidiaries of Chiquita Brands International.

Source: Rainforest Alliance 1998h

Appendix B: Summary of the Better Banana Project's General Production Standards

1. Legislation

Comply with national legislation corresponding to natural resource management, agrochemical use, solid and liquid waste management, labour conditions and human rights in all activities related to agricultural production systems.

2. Natural resource management

2.1. *Forests.* Conserve and recuperate forested areas in a manner that ensures the socio-environmental benefits they offer.

2.2. *Water resources*. Protect water resources by adopting measures of control in agricultural, industrial and domestic activities.
2.3. *Soils*. Promote a system of soil conservation that ensures that resources functions of support and nutrition over the short, medium and long term.
2.4. *Air*. Avoid the production of solid particles, dust, smoke, gases, odours, noise and other atmospheric pollution.

3. Crop management

3.1. *Planning and establishing crops*. Establish plantations at those sites most appropriate for agriculture, where the desired yields can be obtained while minimising socio-environmental impacts.
3.2. *Fertilisation programmes*. Base fertilisation programmes on the conservation and increased productivity of the land, while protecting human health and the environment.
3.3. *Controlling pest populations*. Base control of pest populations on the principles of integrated pest management in a way that reduces the environmental impact caused by pesticides, improves the biodiversity of the plantation and increases the farm's productivity.
3.4. *Handling agrochemicals*. Decrease agrochemical use through the use of less toxic products in order to reduce damage to human health and the environment.
  3.4.1. General considerations
  3.4.2. Transport of agrochemicals
  3.4.3. Storage of agrochemicals
  3.4.4. Application of agrochemicals
  3.4.5. Crop dusting
  3.4.6. Showers and changing areas
  3.4.7. Uniform cleaning zones
  3.4.8. Maintenance of application and protective equipment
3.5. *Agricultural machinery*. Maintain safety measures in the storage area for fuel and lubricants and in the workshops that reduce the risk of accidents and problems with environmental contamination.
  3.5.1. Fuel and lubricant storage
  3.5.2. Equipment maintenance and storage area
3.6. *Managing the cardboard storage area*. Design the cardboard storage area according to safety standards that decrease the probability of accidents and damage to the health of workers.

4. Solid and liquid waste management

Establish an integral plan for solid and liquid waste management based on reduction, re-use, recycling and ecologically adequate disposal.

5. Environmental education programme

Implement a permanent education process for workers and their families with the goal of helping them accept values, clarify concepts and develop the abilities and attitudes necessary for establishing a harmonious co-existence between human beings, their culture and the environment.

6. Prevailing social and work conditions

Improve the quality of life for workers and their families.
6.1. Neighbouring populations
6.2. Living quarters of farm employees
6.3. Occupational health

Source: Rainforest Alliance 1997

Appendix C: Summary of the Better Banana Project's Main Certification Procedures

1. Preliminary site visit
Farmers may request a preliminary site visit by project staff to determine what changes must be made to achieve certification. A detailed report is prepared and sent to the producer within six weeks.

2. Evaluation
An evaluation is an official visit by two or three auditors who conduct a comprehensive review of farm operations; this includes travelling throughout the farm and meeting with workers and farm managers. The auditors prepare a report analysing the farm on all certification criteria.

3. Certification committee
A committee of representatives of the Conservation Agriculture Network makes a determination, based on the evaluation report, of whether the farm achieves certification. Farms awarded certification will receive written notification of their approval and a certificate.

4. Certification contract
In order to complete the certification process, the producer must enter into a contract with the Rainforest Alliance, which governs the use of the certification mark, the handling of certified products and marketplace promotion.

5. Annual audits
Evaluation occurs once a year. The programme also reserves the right to conduct random audits.

6. Monitoring of promotions
All packaging and other promotional materials using the certification mark or describing the programme must be pre-approved by ECO-OK staff. This includes brochures, packaging and other materials used by the certified company to promote their involvement in the programme.

Source: CAN 1998c

20 ja!Natürlich

A Success Story

Sonja Grabner-Kräuter and Alexander Schwarz-Musch

THIS SHORT CASE STUDY deals with the performance of the store-brand line *ja!Natürlich* ('yes!Naturally'). Introduced by the Billa/Merkur Group in Austria in October 1994, *ja!Natürlich* was the first store-brand label of organic food products to be marketed in that country. Up to that time, the market for organic food in Austria was considered to be very small and was available only in health food stores. There was no market research that indicated a large potential for organically grown and processed food products but, nevertheless, the sales figures of most *ja!Natürlich* products have exceeded all expectations.

The purpose of this case study is to describe the launch of the *ja!Natürlich* line and to analyse potential factors in its success.

Some Background Information on the Company

The history of the Billa/Merkur Group begins in 1953. Karl Wlaschek, a former pianist, opened the first discount perfume shop in Vienna. This revolutionary idea—to sell perfumes and other cosmetics at discount prices—proved very successful and the number of outlets grew. In 1960 Wlaschek decided to transfer his successful idea into another market sector and opened the first discount food and grocery outlet. The chain was named 'Billa' (an abbreviation of the German for 'cheap store') in 1961. It was the first self-service store in Austria.

In 1969 he opened the first large-scale outlet and founded a new chain of superstores—Merkur—to offer customers a larger product range in over 1,000 m² of retail space. In 1977 the group became a limited company in order to co-ordinate the distinct areas and to benefit from potential synergies. With the famous founder's

70th birthday in 1987, the group had every reason to celebrate: at that time Billa/Merkur were already running more than 330 outlets with 3,300 employees and generated annual sales revenue of ATS 8.6 billion (about US$700 million). In 1994, the group came under a new type of business ownership: its assets now belong to a private foundation. Two years later the sales of the group topped ATS 51 billion, generated in over 1,400 outlets.

In July 1996, Wlaschek sold his life's work Billa and Merkur to the German group Rewe for about ATS 16 billion. This unexpected transaction made the 80-year-old the richest senior citizen in Austria. Rewe is the largest food retailer in Europe (11,000 outlets, 200,000 employees and annual sales of ATS 350 billion), and an important reason for its expansion into the Austrian market was the growing importance of discount stores in Germany. An escalating price war has resulted in lower margins, so the only way of maintaining the company's profitability was to increase its sales—which seemed easier abroad than in the German domestic market.

The Austrian Food Retail Market: Some Market Trends

The trend in almost all industrialised countries toward larger units and a dwindling number of retail outlets can also be observed in Austria. This trend means a stronger retailing sector with greater influence over manufacturers (see e.g. Terpstra 1997: 572). Because of their scale, Austrian as well as other European retailers wield considerable power in price negotiations and demand more store-brand products. Store-brand lines already command a larger share of the market in Europe than in the US—the UK being ahead of other countries. Chain stores such Marks & Spencer, Tesco and Sainsbury's have demonstrated how to earn high profits with store-brand lines: in 1996 their profitability was the highest among all European retailers (Lang and Trippolt 1997: 44).

During the last few years, the Austrian food retailing sector has also seen a strong trend in the concentration and growth of large-scale retailing. The number of food and grocery stores has decreased by 15% in the last five years; in 1996, 64.5% of all food and grocery retail sales in Austria were conducted in stores with over 400 m^2 of retail space.[1] In 1997, two groups—Billa/Merkur and Spar—held almost 60% of the market. With a market share of more than 32%, Billa/Merkur is the leading food and grocery supermarket chain in Austria, 5% ahead of Spar, its nearest competitor.[2]

There are no data available on the sales of organic food products in health food stores in Austria, so it is difficult to characterise the organic food market *before* the launch of *ja!Natürlich*. Up to that time, health food stores were the only places where Austrian consumers could actually buy organic foods—not in supermarkets. The majority of consumers associated organic food with health, but also with expense

1 ACNielsen Press Release (Vienna: ACNielsen, June 1997).
2 ACNielsen Sales Barometer (Vienna: ACNielsen, 1997).

and consumption without pleasure, explaining its marginal position in the Austrian market.

Launch of the ja!Natürlich *Line*

In October 1994, the *ja!Natürlich* line was launched as the first store-brand organic food line in Austria. It initially comprised 30 products: yoghurts, fresh milk and cream, butter, cheese, eggs, some vegetables, fruit and bread. The basic concept was to offer delicious organic food products to a majority of consumers at a reasonable price.

Organic refers not to the food itself, but how it is produced. Organic food production is based on a system of farming that maintains and replenishes the fertility of the soil. Organic food is produced without the use of synthetic pesticides and fertilisers. 'Certified' means that the food has been grown according to strict uniform standards that are verified by independent state or private organisations. It includes inspections of farm fields and processing facilities, detailed record-keeping and periodic testing of soil and water to ensure that growers are meeting the standards. For Billa/Merkur, it has been very important to find farmers in ecologically intact rural regions who could meet the high standards required for growing certified organic food. Contracts for a five-year period with fixed prices have been offered to selected farmers.

To ensure the new product line's success , it was essential to guarantee that only organically grown food was processed. Each farmer wishing to sign a delivery contract with Billa/Merkur has to prove that he has been working according to the EU rules for organic food (EU Directive 2092/91) *and* according to the regulations of the Austrian Food Codex (chapter A8) for at least two years. Additionally, each farmer has to be a member of a recognised association of organic farmers. 'Harvest for Life' (*Ernte für das Leben*) or 'Demeter' are the most two important organic farming associations in Austria, with tight restrictions on membership. These organisations also provide training and continuing education for their members. Apart from legally required inspections, Billa/Merkur carries out quality audits of farm fields, pastures and processing facilities three to five times a year. The founder, Karl Wlaschek, adds a personalised guarantee of the organic provenance of all *ja!Natürlich* products by having his signature printed on all the packaging.

The traditional target group for organic foods—people who already bought their food at health food stores—was not intended to be the main target for *ja!Natürlich* products. This sector was estimated at only 3% of the total population and, with such a narrow focus, the market would have been too small to justify the launch of a store-brand line. The new brand was projected to achieve at least a 5% share of the total sales of Billa/Merkur within the relevant product group during its first three years, otherwise it would be taken off the market.

To appeal to potential customers, the concept that 'organic food tastes better' was proposed as the fundamental message for product development and advertising strategies.

Product group	Proportion of total sales (%)
Yoghurt	35%
Fresh milk	27%
Cheese	10%
Frozen vegetables	26%
Fruit and Vegetables	15%
Pizza	18%
Bread	80%*

* This high proportion of total sales was achieved only in 'Merkur Bakeries'

Table 1: PROPORTION OF JA!NATÜRLICH PRODUCTS IN TOTAL BILLA/MERKUR SALES IN 1997

The First Three Years

Within its first five months, *ja!Natürlich* exceeded all expectations. For example, up to February 1995, 20%–25% of the total sales of yoghurt by Billa/Merkur were for *ja!Natürlich*. Table 1 shows the actual proportion of *ja!Natürlich* sales in 1997 relative to the relevant product group.

With a 35% share of total Billa/Merkur yoghurt sales, *ja!Natürlich* yoghurt was the most successful product in terms of percentage share in 1997, followed by fresh milk and frozen vegetables. Remarkably, the proportion of *ja!Natürlich* frozen vegetables of total Billa/Merkur frozen vegetables sales (26%) was higher than the share of *ja!Natürlich* fresh fruit and vegetables (15%).

Initially, consumers had their doubts whether *ja!Natürlich* products were really organically grown. To dispel these suspicions, Billa/Merkur made use of one of the product line's disadvantages. Billa/Merkur decided to introduce the new store-brand label into all its Austrian stores, but could not guarantee that all types of product would always be available all over the country. They pointed out that, because the produce is grown in harmony with nature, it is to be expected that certain products will be seasonal. Consumers took this as proof of the organic provenance of *ja!Natürlich* products. Supply shortages actually served to increase consumers' faith in the new line.

Gradually, more delivery contracts were signed with organic farmers, thus reducing supply problems. The capacity to deliver the products on time is now considered an important factor in the increasing sales of *ja!Natürlich* products. Today, about one-fifth of all organic farmers in Austria (4,000 out of 19,433) have signed delivery

Period	Sales (ATS)
October 1994–February 1995	200,000,000
March 1995–February 1996	560,000,000
March 1996–February 1997	1,100,000,000
March 1997–February 1998	1,500,000,000
1998 (budget)	2,000,000,000

Table 2: SALES OF JA!NATÜRLICH PRODUCTS

contracts with Billa/Merkur. However, because most *ja!Natürlich* products are of Austrian origin, the availability of fruit and vegetables is still seasonal: for example, organically grown tomatoes are usually available only in July and August. Additionally, some special organic fruits and vegetables are imported from other countries (e.g. avocados, grapefruits, lemons and papayas).

Line extension is believed to be another reason for *ja!Natürlich*'s rising sales figures—because of the success of the basic products, new ones were developed. For example, in 1996 Billa/Merkur introduced organically grown frozen vegetables and in 1997 *ja!Natürlich* frozen pizza. As was the case in 1994, when the new store-brand product line was launched, market research to support the decision was not sought: people at Billa/Merkur simply believed that there was a market for these products. Despite the evidently successful expansion into other lines, the *ja!Natürlich* brand is still associated primarily with dairy products—an association that was made by 54% of people who were aware of the *ja!Natürlich* brand.[3]

In 1998, Billa/Merkur intends to expand into meat products. One of the potential effects of a number of scandals connected with the meat industry (e.g. the side-effects of intensive farming; routine hormonal treatment; BSE) is a large demand for organically processed meat products. However, at the time of writing (August 1998), actual sales of *ja!Natürlich* meat products have failed to meet expectations. One reason could be that consumers who prefer organic food eat (and therefore buy) less or even no meat products. According to Billa general manager Wolfgang Wimmer, the situation regarding organically processed meat products is considered 'difficult' (Loser 1998).

Table 2 charts sales from 1994 to 1998. It is estimated that sales of *ja!Natürlich* products in 1998 will be ten times higher than they were only four years ago.[4] In fact, 75% of *ja!Natürlich* sales are derived from three product groups, each of them similar in significance: dairy products, bread, fruit and vegetables.

3 ACNielsen Multi-Client Study (Vienna: ACNielsen, 1998).

4 Interview with Werner Lampert, *ja!Natürlich* product manager.

Period	Budget (ATS)
October 1994–February 1995	20,000,000
March 1995–February 1996	30,000,000
March 1996–February 1997	35,000,000
March 1997–February 1998	40,000,000

Table 3: STAFF AND PROMOTIONAL BUDGET OF JA!NATÜRLICH

Surveys carried out by Billa/Merkur showed that, at the end of the year of the *ja!Natürlich* launch, the new product line was the main reason for 2% of all Billa/Merkur customers to visit its stores. This number increased to 5% in the second year. Now about 7% of all customers indicate that *ja!Natürlich* products are the main incentive for them to shop at Billa/Merkur. Another study indicated that 53% of customers who have tried *ja!Natürlich* continue to buy these products regularly or quite often. Only 2% of 'trial buyers' said that they would not buy *ja!Natürlich* products again.[5]

Werner Lampert, brand manager at Billa/Merkur, with full responsibility for the whole product line, cites the high quality of *ja!Natürlich* products as the main reason behind their success. The products are actually much better than their own advertising would suggest. The slogan 'organic food tastes better' was the core message behind product development and advertising strategies, and the budget for *ja!Natürlich*'s market introduction was fixed at a relatively low level (see Table 3). Within this budget, all expenditures for promoting the brand as well as for staff (three people when it was launched) had to be met.

For the first three years, promotional activities were concentrated on outdoor advertising. Every two months, new posters were designed focusing on the new product. At the end of 1996, public interest in *ja!Natürlich* posters declined, and Billa/Merkur decided to concentrate more on television commercials. In March 1997, the first commercial was aired and in summer 1997 Billa/Merkur decided to use television advertising as its principal promotional medium.

Billa/Merkur was the first chain store in Austria to co-operate with organic farmers and to launch a store-brand line for organic food products. This meant public interest in *ja!Natürlich* was extremely high and generating favourable publicity was comparatively easy. For example, in January 1995, *ja!Natürlich* was introduced and discussed on a popular Austrian talk show, at no cost to the company. Such opportunities provided invaluable exposure for the new brand, as well as for the Billa/Merkur group in general. It is likely that the positive contribution that the new brand made to Billa/Merkur's image could not have been achieved with an expensive advertising campaign.

5 ACNielsen Multi-Client Study (Vienna: ACNielsen, 1998).

Decisions about the price structure of *ja!Natürlich* were not easy. First of all, there were no comparable products and it was open to question what prices consumers would accept for organic food sold in a supermarket. Of course, in addition, Billa/Merkur had to consider higher processing and production costs. Before *ja!Natürlich*'s launch, the product manager identified that consumers would accept price differentials of up to 20% compared to non-organic foods. However, a 20% price increase would not have covered costs *and* turned in acceptable profits. It was therefore decided to reduce the packaging size of several *ja!Natürlich* products and thus disguise price differences. For example, yoghurts, cream and sour cream are sold in 200 g packs instead of the usual 250 g packs for non-organic products. In this way, the majority of consumers do not immediately notice that the average price difference is actually higher than 20%.

Identification of Potential Success Factors

It is quite remarkable that, when the new product line was launched, no analyses of market potential and consumer purchasing intentions were made to support the decision. According to Werner Lampert, the product manager, there was absolutely no market research undertaken to indicate a promising market potential. Nevertheless, the introduction of *ja!Natürlich* was given top priority. Karl Wlaschek gave full support to Lampert's concept of an organic store-brand line, for example by allowing his signature to be used on *ja!Natürlich* packaging. Both shared a vision of a product that can provide a symbiosis of freshness, health, delight and ecologically responsible consumption.

A pivotal question now is: What are the reasons for *ja!Natürlich*'s success, despite the lack of market research or a clear definition of the target group?

All product categories were highly competitive. As already mentioned above, Billa/Merkur's advertising and promotional strategy could not have been the key success factor. In 1995, a total of ATS 15.9 billion was spent on advertising in Austria; in 1996 it increased to ATS 16.6 billion (*Bestseller* 1997). In the midst of this incessant bombardment of advertising, Billa/Merkur successfully launched a new product line on a relatively small budget. Only about one-tenth of Billa/Merkur's total advertising budget goes on *ja!Natürlich* products. The proportion of advertising to sales in fact decreased from 10% in the launch period to 5.4% in the second year, 3.2% in the third year and only 2.7% in the fourth. Nevertheless, today over two-thirds of Austrian consumers are aware of this store-brand line, compared to less than one-third only two years ago.[6]

At Billa/Merkur, the high quality of *ja!Natürlich* products is believed to play a crucial part in their success. But are consumers aware of the difference in quality compared to other products? Is it possible to communicate effectively the better taste and higher quality in product categories where many competitors spend heavily on advertising strategies with very similar messages, in which 'high quality', 'freshness' and 'taste' are emphasised?

6 ACNielsen Multi-Client Study (Vienna: ACNielsen, 1998).

It is dubious whether a successful product launch can be achieved by adopting the same positioning as the majority of competitors.[7] The only difference in Billa/Merkur's concept that 'organic food tastes better' is the organic dimension. Consuming organic food means a 'double responsibility'—both for the environment and for one's own health. In fact, Billa/Merkur has been successful in differentiating its store-brand line not only via the responsibility dimension, but also via the enjoyment dimension.

In discovering the new competitive dimension 'organically grown', Billa/Merkur was able to acquire distinctiveness for *ja!Natürlich* in a society overloaded with communication. Preconditions for the success of *ja!Natürlich* products were their high quality, their good taste and their freshness. However, when these attributes were complemented by the new 'organic' dimension, a unique selling proposition of 'pleasurable as well as responsible consumption' resulted. In this way, Billa/Merkur has been able to communicate its message to its target group and to attract customers with a relatively small advertising budget.

Obviously many consumers do believe that organic food tastes better and is more nutritious. Consumers are willing to pay more for a good, healthy product, when they are aware of its higher quality. With *ja!Natürlich*, Billa/Merkur managed to find a new unoccupied position that was valued by a large enough number of consumers. The company was able to acquire distinctiveness for its store-brand line and thus improve customer loyalty *and* profitability at the same time.

Management decided to follow a premium pricing strategy for the *ja!Natürlich* line. As already mentioned above, the prices for *ja!Natürlich* products are about 20% higher than the prices for other products in the same category. However, this is true only for supermarkets. In health food stores, the prices for organically grown food are much higher, making organic food virtually unaffordable for the average family. Thus another reason for *ja!Natürlich*'s success is the fact that Billa/Merkur made certified organic food available to the general shopper.

Differences from the store-brands of other chain stores can be found in *ja!Natürlich*'s price as well as in its image. Store-brand lines are generally positioned in the lower price segment, chain stores often adopting a good-value strategy: for example, Spar Austria promotes 'Spar American Cola' with the slogan 'Why should you pay the difference if you can't taste it?' Unlike Spar and other chain stores, Billa/Merkur is able actually to charge higher prices for its *ja!Natürlich* store-brand because these products are of a higher quality than most manufacturers' brands.

Billa/Merkur is going from strength to strength in Austria, having adapted though not forsaken its original strategy: its success formula still is based on high and consistent quality of all products. Meanwhile, the *ja!Natürlich* brand comprises 200 products in 12 product lines. In the dairy product category, as much as 30% of Billa/Merkur's total sales are *ja!Natürlich* products. The Rewe Group that acquired Billa/Merkur two years ago now wants to benefit from the Austrian experience and plans to remix its successful recipe for cheese and yoghurt in Germany. However, at

7 A product's position is the way in which a product is defined by consumers on important product attributes, benefits or usage. It is the place the product occupies in the minds of consumers relative to competing products (Kotler and Armstrong 1996: 409).

the moment the German market for organic foods is believed to be very small. Rewe manager Otto Kalmbach assumes that German consumers are not yet accustomed to organic food and not willing to spend more money on it (Loser 1998) and therefore decided not to transfer the *ja!Natürlich* concept to Germany. In a first step, the product strategy for cheese and yoghurt in Germany will not focus on the organic origin but on the freshness of the Austrian dairy products.

These authors are reluctant to accept that there is no promising market potential in Germany and that German consumers really are that different from Austrian consumers. Besides, the company could find themselves pleasantly surprised because, as we have seen, there was no market research that would have led one to expect a promising market potential in Austria before the very successful launch of *ja!Natürlich*.

Conclusion

In this case study, the launch of the store-brand line *ja!Natürlich* was analysed and potential reasons behind its huge success were identified. The background to the launch indicates that sometimes management's vision and a basic concept about realising this vision are still more important for a product line's success than carrying out careful marketing analysis to support each decision. Apparently, more Austrian consumers are willing to pay a premium price for organic food than was generally believed before the launch of *ja!Natürlich*. One reason might be that organic food invokes a feeling of 'double responsibility'—both for the environment and for one's own health.

With the concept of 'pleasurable as well as responsible consumption', Billa/Merkur has been able to identify an important benefit that could be captured convincingly by *ja!Natürlich*. In this way 'green' products have competed successfully with non-'green' products in the same product category. Billa/Merkur was the first company offering organic food in Austrian chain stores, and this first-to-market strategy has been the reason for the enormous public interest in *ja!Natürlich* products, which no subsequent product was able to achieve. Meanwhile, other Austrian chain stores are now also offering organically grown food, e.g. Spar with 'Natur pur' ('Pure nature'), but have not been nearly as successful.

21

Green Marketing of Green Places

The Tasmania Experience

Dallas G. Hanson, Rhett H. Walker and John Steen

THE PURPOSE OF THIS CHAPTER is to explain and discuss how three different service providers in Tasmania are facing the challenge of environmental sustainability within the particular context of eco-tourism. Each case serves to reveal a number of environmental management challenges faced by each service provider, and offers insight into how these diverse challenges have been satisfactorily resolved for all interest groups or stakeholders. Consequently, these insights may beneficially inform comparable practice elsewhere.

The chapter begins by defining what is meant by 'green marketing' and environmental sustainability in a nature-based tourism context, and then explore implications of this for service providers. We then show how these implications have been accommodated by each of the three service providers selected, and discuss how this experience may prove useful to others.

Green Marketing: Definitions and Practical Implications

Green marketing can be defined in many ways and is presented in different ways by different firms; it also varies in practice between regions and between industries. A definition based on the available academic literature that encompasses these differences and carries the broad meaning of the term is:

> **Green marketing** is a holistic process that anticipates, identifies and satisfies the requirements of customers and society in an ecologically

sustainable manner (see also Peattie 1995; Button 1989; Charter 1992a; Davis 1991a, 1991b, 1993; Klafter 1992).

The ideas of holism and sustainability are keys to the definition. Holism means taking into account all the factors that affect the service or product manufacture and delivery as well as the marketing process. This means that a firm's source of supply, its distribution policies, product disposability or recyclability, pollution management and marketing communications must all fit the sustainability requirement. Moreover, holism suggests taking into account the interests and welfare of all stakeholders, that is all those identifiable individuals and groups with a significant interest in the marketing effort and its consequences. These include, for example, a manufacturer's suppliers and distributors, its customers, consumers and relevant community interest groups.

Ecological sustainability means that the impact on the environment of all marketing- or product/service-related activities must be carefully managed so that the activities can continue into the future without decline in the natural resource. This is not meant to imply no impact on nature. It does, however, require careful scrutiny and management of all aspects of the system so that impacts can be minimised over time. Green marketing practice is therefore characterised by a genuine concern for the impact on nature of inputs (the raw materials used in production, design features, production processes) as well as outputs (products, disposability, durability). It is also concerned about the impact of products and marketing on consumer behaviour: for example, whether consumers are encouraged, by marketing and design, to use products and to dispose of waste from product use in an ecologically sensitive fashion.

The requirement that customers and society be borne in mind suggests that, while profit-oriented organisations need to survive in a competitive world, they must also bear in mind the ecological interests of customers and the interests of future as well as current generations. This is a challenging tension requiring a form of careful thinking and management that often goes beyond regulatory compliance, particularly in regions where regulations do not reflect ecological best practice.

The application of these ideas to nature-based tourism and destination marketing presents different challenges to those associated with product marketing or standard services marketing. Nature-based tourism means tourism based on the natural environment and, consequently, with both direct and indirect impact on nature. This can be called 'eco-tourism' in cases where nature-based tourism operations are demonstrably environmentally sensitive and ecologically sustainable. There is a large academic literature and debate on this distinction (see, for example, Ziffer 1989; Boo 1990; Ryel and Grosse 1991; Joy and Motzney 1992; Valentine 1993; Western 1993; Poon 1993). We avoid this debate here but, as the definition we have offered makes clear, our arguments emphasise the eco-tourism end of any continuum between simple nature-based tourism and ecologically sensitive tourism, and are based on a philosophical position that requires that the natural world be given very high value. This means that the integrity of the natural environment be given the highest possible importance in decision-making. In terms of the literature in this area, we write from an ecocentric perspective (see Shrivastava 1995b), that is, one that values harmony with nature, quality of life and the value of nature for its own sake.

Green Marketing of Green Places: Implications and Challenges

All nature-based tourism involves the use of nature. It is better termed 'use' than 'consumption' because, while somewhat changed by every individual's experience, the basic resource remains. In a green marketing programme, a primary challenge is to maintain the natural environment, acting as a steward rather than an exploitative user. Protecting and sustaining the natural environment are central imperatives.

The green destination marketer has no real control over what is ultimately experienced by the tourist. The marketing effort sets the context and raises expectations, but the experience is individual and, in any case, is often under the general control of those directly involved in service delivery—tour guides, bus drivers and so on—rather than the destination marketer. In cases where marketing is regional or national in scope, this contrast is most marked (for example, with marketing of New Zealand as a whole, or of Tasmania) but the pattern of separation of marketing and service delivery is consistent across almost all of the nature-based tourism industry.

The notion of ownership is foreign to the destination marketer and to the nature-based or eco-tourism service provider. It is also foreign to the visitors whose experiences come within this same context. Tourists must physically (or electronically) visit a destination in order to experience it and learn first-hand something about it, and to avail themselves of the goods and services on offer there. They may then depart with the ownership of nothing more than a few mementoes or souvenirs of their experience and the memories that these evoke. Certainly, unless they have invested in property there, they do not acquire 'ownership' of the destination *per se*. They do, however, own whatever tangible and intangible reminders of their visit that they have acquired. There is also another kind of 'ownership' peculiar to nature-based tourism, and that is the sense of personal responsibility and accountability that might be associated with visiting particular places such as sacred sites, natural reserves and world heritage wilderness areas. Correspondingly, the nature-based tourism service provider is cast much in the role of steward faced with accommodating multiple interests, including those of visitors, residents, the local cultural heritage and the natural environment, its flora and fauna. These are key issues to which we will return later in our discussion.

In order to meet these challenges successfully, the green destination marketer and service provider require wide-ranging knowledge and a particular mind-set, and must also be prepared to assume the role of educator. The mind-set has already been implied. A green marketer of a green place will need to have a philosophical stance opposed to human-centredness or anthropocentrism. The role of environmental steward requires a genuine concern for the natural world and the capacity to think of nature as having an innate value, rather than as a commodity for human use. The natural world is thereby given the status of a legitimate stakeholder in the tourism operation. (For more on anthropocentrism, see Fox 1990; Hanson 1996; for more on stewardship, see Hanson and Tapp 1994.)

The knowledge requirement is implied by the education role played by those involved in destination and nature-based tourism marketing, and by the requirement for holistic thinking. The educator role implies a usable level of ecological understanding and specific knowledge about the natural environment. The natural world is fragile, and many of its delights are hidden from those without a basic understanding of how it operates. Accordingly, provision of information about that part of the natural world that provides the focus of eco-tourism activity may be necessary. For example, information about wildlife, about impact of walking on fragile vegetation, about basic ecology, serves both to protect the environment and to add to the client's enjoyment.

Holistic thinking requires an ability to think in terms of links between the many elements that make up the total operation and its relationship with the natural environment. This requires a sophisticated understanding that is still unusual in most organisations but is becoming more common because the allied notion of systems thinking has been popularised as part of the learning organisation movement (Senge 1990). A key implication of this is the shared responsibility of those involved in nature-based tourism to provide a full analysis of environmental impacts that result from their efforts, and to disseminate this to other stakeholders. In this way, a long-term state-of-the-environment record is created and maintained which allows assessment of the extent of cumulative impacts of activities over time.

In the next section of this study, we discuss the relevance of these implications for eco-tourism in Tasmania and indicate how they are being addressed.

The Tasmania Experience

The Nature of the Place

Tasmania is generally considered, by residents and visitors alike, to be a clean, green, pure and largely unspoiled environment that is further distinguished by its colonial history and rich architectural heritage. The tourism industry is well developed, with a great diversity of accommodation, attractions, historical sites, scenic cruises and flights, and throughout the year there are numerous special events staged specifically to attract visitors to particular parts of the state.

The significance of the natural environment cannot be overstated. The special character and diversity of Tasmania's natural environment combine in such a way as to afford visitors a unique experience. The state's natural attributes include an abundance of inland waterways, forests, mountain ranges and beaches, and an environment conducive to numerous sporting and recreational activities, including all water sports, skiing, fishing, hunting, bushwalking, caving and camping. The state also has a wealth of parks and reserved areas of wilderness. About 20% (almost 1.4 million hectares) of the state's total landmass is made up of the Tasmanian Wilderness World Heritage Area, one of the last temperate wildernesses in the world. This includes most of the mountainous central region of the island-state and is highlighted in numerous bushwalking tracks. At least one of these, the 80 km Overland Track, is widely considered by experienced travellers and bushwalkers to be one of the world's best walks. Another 10% of the state is made up of other reserves, conservation areas and Crown (i.e. government) land. In addition, the Forestry Commission, a statutory authority, controls around 24% (approximately 1.8 million hectares) of state forests which have multiple uses, including eco-tourism.

Tourism forms a major part of Tasmania's economy and is a significant contributor to economic development in the state, being fourth in line behind forestry and timber processing, agriculture and food processing, and mining and mineral processing. It is estimated by the Centre for Regional and Economic Analysis (CREA) that tourism revenue presently exceeds A$400 million per annum and that this represents a contribution of approximately 8% to Gross State Product (GSP). This has been steadily increasing over the past ten years (CREA 1993). Furthermore, direct and indirect employment generated by tourism has also been steadily increasing, and is currently estimated to be approximately 10% of aggregate employment in the state. This means that tourism plays a central role in Tasmania's economic growth and development. The place, significance and value of tourism in and for Tasmania, however, are grounded in more than economic considerations alone.

The innate value of Tasmania as a tourism destination, and therefore the value of tourism in and for Tasmania, is directly attributable to its natural environment. This, in effect, is the marketable 'product'. Consequently, the marketing and management of Tasmania as a tourism destination is faced with a dual challenge. This involves, on the one hand, attracting visitors and making the environment accessible to them; and, on the other, due regard for the unavoidable impact of this on the natural environment and its residents. This, in turn, raises a number of quality of life issues. These include the quality of what is experienced by visitors, the quality of life desired by residents, and the quality of life of the natural environment itself, its flora and fauna. All of these issues and concerns must be accommodated and reconciled if the interests of these respective stakeholders are to be mutually satisfied.

The Respective Role and Responsibilities of Visitors, Residents and Independent Service Providers

Visitors are, of course, essential to the economic welfare and growth of any tourism destination. Furthermore, tourism has the potential to contribute positively to management and preservation of the environment by generating the financial means that enable suitable maintenance and improvement of a destination's infrastructure, facilities and services (Coccossis 1996). At the same time, however, tourism can have the effect of degenerating this same environment. For example, bushwalkers, rock climbers and campers can easily, albeit unintentionally, impair the natural environment as a consequence of ignorance or careless behaviour; water-based sports and recreational activities have the potential to pollute waterways, disturb or injure native wildlife; and the place and manner of use of recreational vehicles also has the potential to damage flora and fauna. This creates implications for residents and tourism service providers as well as visitors, their respective attitudes, values and behaviour.

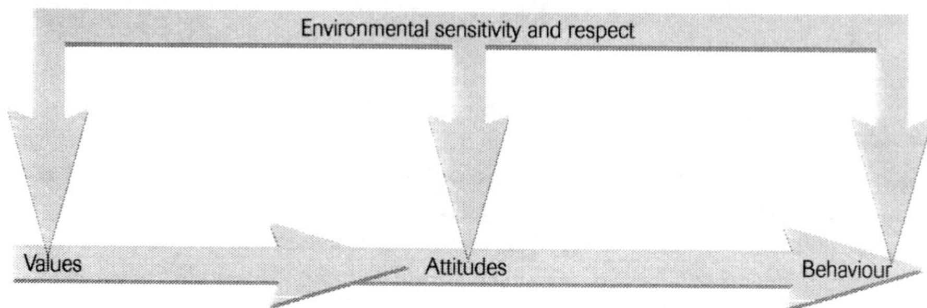

Figure 1: A MODEL OF ENVIRONMENTAL ORIENTATION

As we have already stated, at the heart of 'green behaviour' must lie a fundamental environmental philosophy or ethos that leads to respect for the natural environment, its flora and fauna, and its innate and inherent value (Berry 1993; Naess 1993; Taylor 1993). This means regarding the natural environment as something to be valued in its own right, to be experienced, shared and protected for its own good as well as the good of others, now and for the future. Put another way, this shift in attitude, which is manifest in individual attitudes and behaviour, is the result of a reorientation in how one values, regards and reacts to one's natural environment. This reorientation is illustrated in Figure 1.

The value system that is grounded in, or predicated on, a genuine concern for one's natural environment and its sustainability results in attitudes and behaviour that then beneficially serve the interests of the environment and all of its stakeholders. These attitudes include, for example, recognition of the fact that each being is supported by every other being (Berry 1993) and of the need to protect and preserve that which nurtures and sustains all. The resulting behaviour is thus characterised by prudent and sustainable consumption or use of what is regarded as a finite natural resource. In this way, service provider, visitor and resident alike share a sense of ownership of environmental responsibility and of the consequences of personal actions.

In Tasmania, this sense of ownership and personal responsibility is evident in a variety of forms. Residents who value the natural environment and its unique character maintain an active public presence and voice, also manifested in strong political representation, with the express aim of preserving as much of the status quo as possible and not succumbing to the pressures of industrial development at the expense of the environment. Visitors, attracted to the state because of the uniqueness of its natural environment and the richness of what it has to offer recreationally, willingly observe anti-litter requirements and regulations in parks, forests and other natural reserves; participate enthusiastically in helping to clean beaches, tracks and caves that they have visited; and readily comply with environmental protection guidelines provided by tour guides, rangers and others. Moreover, because of this behaviour, the environmental sensitivity of residents and visitors alike is heightened and made more acute. Education and personal experience combine to enhance one's appreciation of the innate value of the natural environment and commitment to preserving it.

Correspondingly, service providers whose enterprises are most closely aligned with, indeed inextricably linked to and dependent on, the natural environment customarily share an ethos that casts them in the role of environmental steward and protector. Examples include providers of guided walks and tours, camping, climbing and canoeing expeditions, wilderness flights and cruises. For instance, one major operator of nature-based backpacking treks through the central highlands of Tasmania has made a particular point of mapping a wide variety of walks in order to spread the environmental impact of their tours; an operator of wilderness scenic flights has recently invested heavily in new seaplanes which, because of their improved design, are more environmentally friendly in terms of their quieter engines, landing impact, and the nature of fuel used; an operator of wilderness cruises up the Franklin and Gordon rivers on the west coast now travels at reduced speeds in order to minimise the wash created and its impact on the river banks; and virtually all providers of guided walks and camping expeditions now limit tour sizes, use natural products (soaps and cleansers, etc.) and collect and carry out with them all rubbish and waste. Moreover, there is a commonly shared commitment to developing means of positive rather than negative environmental impact (for example, through the prudent use of duckboarding in marshy areas, and the design of cabins that are aesthetically as well as functionally compatible with the environment in which they are located). In these ways, care is taken to ensure that the natural environment is suitably protected, doesn't become degraded or over-used, retains as much of its integrity as possible, and remains appealing. In turn, this means that, although the natural environment is being used profitably by a wide range of stakeholders, it is being done so with minimum cost to the environment, and it is being done in such a way that is characterised more by stewardship and sharing than by domination and possession (Berry 1993; Taylor 1993; Canadian Environmental Advisory Council 1991).

In the following section, we illustrate some of these practices with reference to three particular and quite different eco-tourism service providers, with the aim of extending understanding of the theory just discussed.

Cradle Mountain Huts

Cradle Mountain Huts (CMH) is a private commercial organisation that provides guided walks through part of a national park located in Tasmania's north-west. These walks typically take six to seven days to complete. Overnight accommodation and meals are provided in huts established and maintained by CMH under the auspices of Parks and Wildlife Tasmania (PWT). Separate public huts are also provided and maintained by PWT.

The notion of a guided walk along the overland track based on privately owned huts was a government initiative that was subject to a tendering process. One of the main reasons for the need for a commercial operator was the steadily growing numbers of people walking the length of the track. At first glance, it would appear counter-intuitive to seek a commercial operator as a response to growing visitation pressures. Indeed, initial opposition to the idea from some conservation groups and members of the public was considerable. However, CMH's tender was actually designed to minimise impact on the park to the extent that one of CMH's walks results in less environmental impact than does a private walker who stays in tents and government-owned huts. It must be remembered that the annual traffic on the overland track is currently around 5,000 people and the impact of this without regulation or infrastructure could be considerable.

To achieve a viable commercial operation, with minimal impact on the national park, CMH has invested substantially in a number of initiatives. For example, their huts were designed in such a way as to be as unobtrusively complementary to the natural environment as possible. Most of these huts are no larger than the public huts and are certainly less conspicuous, with only the last few being visible from the track. Most of them are a long distance from the public camping areas with the desirable effect of diffusing human presence in the park. This assists in maintaining the walk as a true wilderness experience. The CMH huts are fully self-contained and maintained in pristine condition at all times by guides. The fuel, food and maintenance requirements for the huts are flown or carried into the park under PWT guidelines. The largest proportion of restocking is carried out by helicopter twice every year in early spring and mid-summer, a restriction imposed by PWT to prevent the intrusion of helicopters on the park. After supplies are delivered, all disposable and waste material is carried out of the park on the same day.

One of the biggest problems faced by PWT on the overland track is sanitation. The best solution to date has been to install composting toilets at camp sites. These toilets are fitted with solar-powered fans that dry waste and allow it to compost to a topsoil-like material. The alternatives to composting toilets are wet pits or individually dug holes which, apart from being unsightly, have been a primary contributor to gastric illness. The CMH composting toilets are maintained and raked on a daily basis by guides to ensure that the composting is efficient. Any residual material at the end of the season is flown out by helicopter.

Selection and training of guides is a key factor in the sustainable operation of CMH. Job advertisements are placed in the local newspapers, and young Tasmanians always respond well. These include university students who frequently have formal training in botany, zoology, geology or ecology and who are therefore able to bring an extra degree of environmental knowledge into CMH. Staff are selected on the basis of walking experience, service industry experience, knowledge of the SW Tasmania World Heritage area and social skills. During pre-season guide training, CMH employ PWT personnel to discuss issues such as World Heritage listing, flora and fauna, threats that face the area and the principle of minimum impact bushwalking (MIB).

MIB is a code of behaviour that has been established by PWT and covers issues of track erosion, garbage removal, sanitation and consideration for other walkers. CMH guides routinely stress the importance of MIB to each group of guests at the beginning of each walk. Frequently, the guides will explain to the group that practising MIB will ensure that the park will be able to be enjoyed by many other people for many years to come. Enforcing MIB is not always easy and often requires diplomacy from the guides as well as co-operation from the guests. For example, PWT track-monitoring studies have indicated that tracks can erode into very wide quagmires when visitors walk around muddy sections of track. The MIB response is, therefore, to walk directly through the mud. This, however, means that walkers get wet, muddy, and often cold. Most of the time, the desired response can be achieved if the guide sets the example and perhaps makes an offer to clean the boots of the muddiest guest that evening. CMH guides also collaborate with park rangers in removing any rubbish from tracks and monitoring track erosion.

Another feature of CMH is the restriction of guest numbers to ten per group, each of which is escorted by two guides. The temptation to achieve economies of scale with a much larger group has been resisted in order to create an experience for the guest rather than a simple tour. The small group size allows the guides to cater for individual interests such as peak climbing, photography, botany or simply relaxation and a little solitude at the end of the day's walk.

Consequently, through these and other similar initiatives, an eco-sensitive paradigm has been created by CMH which has a direct bearing on everything the organisation does. In this way, use of the natural environment is managed sustainably, visitors are educated about the natural environment and what is acceptable behaviour, environmental problems are identified and rectified early, and the CMH infrastructure and operation is managed with minimal visual and physical impact. All of this is underpinned and driven by an ethos and commitment to environmental sustainability that is shared by a variety of stakeholders, including CMH and its personnel, PWT and other regulatory bodies. In the most recent seasons, CMH walkers have represented approximately 20% of the total walking traffic on the overland track. However, this 20% would be by far the lowest-impact user group due to the appropriate infrastructure and supervision from guides. At the same time, the CMH operation makes consistent high-level profits and maintains a good reputation with government and regulatory authorities, clients and fellow eco-tourism operators.

Freycinet Lodge

Situated on the eastern coast of Tasmania, the Freycinet Peninsula and National Park offers dramatic coastal scenery, pristine beaches and year-round opportunities for walking, rock climbing and a range of water-based recreational activities. Freycinet Lodge (FL) is a privately owned business that has been specially licensed to operate within the boundaries of the Freycinet National Park. The lodge offers quality accommodation in the form of self-contained cabins, as well as a host of recreational amenities.

Prior to the Freycinet Lodge development, the site was occupied by an older-style lodge known as The Chateau, which had been operating for many years. A comparison between The Chateau and FL provides an insight into the changes that have occurred in the nature-based tourism industry, and the increased demand by customers for sophisticated and environmentally sensitive services. A subtle but instructive example can be found in their names: 'The Chateau' conjures up an image of a European country retreat; there is absolutely nothing in the name that identifies it with an environmentally unique national park. FL has addressed a number of pollution problems left behind from The Chateau. For example, the old waste-water system that discharged waste into the bay has been replaced by a water treatment process, and electric power is provided on-site by dedicated energy-efficient generators. Decorative but foreign plants and flowers that surrounded The Chateau have been removed and replaced by native species. The original buildings were white and trees were removed to open up the area. In contrast, the lodge buildings are a green colour and tree-planting programmes have reversed much of the damage done by the former operation. Indeed, during the initial building stage, plans were altered to preserve a number of large casuarina trees to further conceal the development.

FL's research suggests that their clientele visit the lodge for three main reasons. First, the lodge is situated in a unique location. Well-known places, including bays for swimming, scenic lookouts and walking tracks, are only a short walk from the resort. Second, FL has earned a reputation for being a quality destination, particularly with affluent well-travelled visitors. Third, FL is able to offer experiential activity programmes that assist guests in obtaining an appreciation of the environment in which the lodge is situated. For example, the activities officer/guide has a degree in environmental science, and lodge guests who are taken on guided walks are inducted with an appreciation of not only the natural environment *per se*, but also its ecosystems, their fragility and their sustainability. During a recent international conference for professional chefs, FL was able to arrange a talk from a University of Tasmania botanist on traditional bush foods. Similarly, other special programmes are offered throughout the year to offer visitors an opportunity to find out more about the special nature of Freycinet Peninsula, its origins, history and ecosystems, and its current management.

FL's relationship with PWT rangers appears to take the form of an informal alliance. While the rangers have little direct influence over the daily business of the lodge, they frequently consult FL guides on the management of the National Park. For example, a very popular rock climbing site known as White Water Wall was facing dramatic degradation from camp fires, poor sanitation and inappropriate vehicular access. In this case, PWT and FL guides are working together to devise a management strategy for the area, even though White Water Wall is very rarely visited by FL guests.

FL recognises that guests are frequently interested in the way in which the lodge operates on a sustainable basis. Staff members, all of whom share a common environmental interest, frequently receive comments from guests about the absence of plastic packaging and chemically based detergents, the use of natural soaps in the cabins and the design of the lodge which is, to a large degree, integrated with the immediate surrounds. It is very difficult to get an idea of the size of the complex (60 cabins) from the road that passes the lodge, or from across the bay, because of the low-profile cabins and the effective use of native vegetation.

FL recognises that the guests need to become partners in the sustainable operation of the lodge. While this relationship cannot be forced, it is encouraged by the distribution of literature designed to outline responsible use of the park, such as not feeding animals and keeping to marked tracks, as well as slide shows designed to stimulate awareness of the environment. FL management is keen to emphasise that these initiatives are not simply a reactive façade, but instead reflect a deeper commitment to environmental sustainability. The strategic mission of the lodge management is directed towards preservation rather than large profit margins. The following quote from an FL manager articulates this succinctly:

> I have known this area for many, many, years and it is very precious to us. A lot of people believe think that you are building just money-making concerns. If you wanted to make money, you would be far better investing money in Central CBD in Sydney or Melbourne where you would get a good return on it. You actually have to love the place to put up this type of lodge.

Wild Cave Tours

Wild Cave Tours (WCT) is a small guided tour operation which takes clients through undeveloped or natural caves in the Mole Creek region in the north of Tasmania. Caves in this area are among the most famous in Australia and are widely regarded by cave photographers and speleologists as some of the best-decorated caves in the world. This is to say that these caves include a variety of attractive stalagmites, stalactites and mineral formations. Their potential as a tourism venture was recognised by the Tasmanian government earlier in the century when the land over two large caves, Kubla Khan and King Solomon's Cave, was purchased and the caves were subsequently developed with concrete walkways, steel staircases and, most recently, electric lighting.

WCT is a very interesting case in several respects: first because WCT is essentially providing a purely experiential service product; the tangible dimensions of the service product offered are not owned by the company. Also, unlike the previous case studies, this operation does not offer traditional services such as accommodation or significant dining facilities. Second, none of the caves that are used by WCT for tours is developed in any way, but all are purposefully maintained as near as possible in their natural state. This aspect takes on additional significance given that all of the caves used by WCT are accessible by the public and are frequently visited by caving club groups and amateur speleologists.

The size of WCT tour groups is necessarily small because of the need for close supervision of the guests for safety and cave preservation reasons. As there is no protective infrastructure such as stairs and walkways in these caves, visitors must be carefully educated in where to walk and what can be touched without leaving marks. According to the owner/manager, this allows for a more immediate and affecting caving experience, one far more appealing than that offered by larger tours of developed caves. By international standards (such as cave tour companies in Europe and New Zealand), the annual clientele is extremely small. For example, one of the wild cave tours in New Zealand processes 27,000 people every year compared with 200 for WCT. New Zealand caves such as this, characterised by a very high client throughput, are active river caves that contain few decorations along the main thoroughfare streamways, are frequently (and beneficially) washed by natural flooding, and are very tolerant of large numbers of visitors. In contrast, the decorated caves of the Mole Creek area are extremely sensitive to human damage. Oil from a human hand can permanently stunt growth of a stalactite, and a smear of mud on a decoration can remain forever if left unattended.

WCT's manager asserts that, for an operation to be environmentally sustainable, the resource must dictate the style of business rather than trying to impose a successful business formula from elsewhere on the environment. In accordance with this resource preservation mind-set, WCT operates a cave-cleaning programme, despite the fact that WCT is the smallest user group of the resource and is therefore one of the lowest contributors to wear and tear on the caves. Cave cleaning is a practice that is growing in caving regions such as those in the United States where large numbers of visitors are having a detrimental impact on caves. It involves the removal of mud and algae from formations by a process of gentle scrubbing, which is very time-consuming but widely regarded as an excellent way of preserving caves in their natural state. Mud that covers cave formations comes mainly from visitors' clothing and boots, as well as from water run-off from cultivated land. Algae can be a particular problem in frequently visited caves and is caused by unnatural levels of light introduced into them. Interestingly, the manager of WCT is eager to point out that guests are frequently willing participants in the cave-cleaning programme.

Concluding Remarks

For nature-based destination marketing to be truly green, i.e. sustainable, all stakeholders must share a fundamental regard for the intrinsic value of the natural environment—and all that this implies—on which all are dependent. All stakeholders are, therefore, also cast inescapably in the role of environmental stewards and protectors, each with their own set of respective responsibilities and each charged with accommodating the interests and welfare of all others as well as of the environment itself. These stakeholders include service providers, visitors and residents, as well as those charged with policy-making, management and protection of the natural environment. The nature and extent of this regard, and the manner in which it is manifest will, of course, vary across different interest groups and individuals. It is also to be expected that one's personal ethos and set of environmental values will be enriched over time through education and experience. Nonetheless, sustainable nature-based destination marketing requires of its stakeholders some form of subscription to the ideal that the environment is for all to enjoy but for none to destroy.

With green product marketing, where sustainability is a key issue in terms of supply, design, production and consumption, the green manufacturer aims to produce and market something in an environmentally compatible way. The green destination marketer and service provider, however, produces nothing but must ensure that exploitation of the natural environment is sustainable and acceptable to all stakeholders. Correspondingly, consumers of goods do not necessarily interact with, or impact upon, the natural environment and its stakeholders in the same way as visitors to a destination do. For example, the environmental impact of someone using environmentally friendly shampoos or detergents packaged in recyclable containers is quite different to that of someone trekking unsupervised through sensitive wilderness areas. Nature-based recreational activities necessarily involve the consumer in the natural environment; therefore, some form of impact on this resource and other stakeholders is unavoidable. Furthermore, this impact is potentially not only in terms of the careless disposal of packaging, product residue and waste materials, but is born directly of human presence and activity in a natural environment that might otherwise remain untouched. This line of thought can be extended. Although it is possible to consume green products without necessarily a personal green ethos or philosophy, and with no adverse environmental effect, exploitation and use of a destination without a green philosophy and value system runs the risk of environmental ill-effect with ramifications for all stakeholders. Therefore, if this natural resource is valued and its integrity is to be maintained, great care needs to be exercised in how it is used, by whom, how frequently and under what conditions, and with due consideration for the interests and welfare of all stakeholders. Philosophical commitment must lead business practice for green marketing to succeed.

The experience provided by the three cases discussed above illustrates the special sensitivities and difficulties associated with answering the challenge of environmental sustainability within the eco-tourism industry. They also indicate that there is no set of rules that can be applied to all situations. Instead, unique answers must be developed to resolve the problems thrown up by the particular ecosystems that make up the focus of the eco-tourism operation. In all three cases, however, the commitment to ecological sustainability is clear. This is fundamental to the long-term success of eco-tourism not only in these cases but in all eco-tourism operations. Furthermore, these three cases also provide useful insights for comparable practice elsewhere. These insights include:

- An internalised orientation that combines environmental and commercial interests synergistically and equivalently
- The mutual value to be derived from partnering with local environmental management and regulatory authorities
- Ways in which a complementary 'fit' with the natural environment can be achieved
- Examples of how the natural resource may be used prudently and in such a manner that the impact on it is minimised
- The importance of controlling visitor numbers to a level that is environmentally sustainable as well as manageable
- The mutual value to be gained from educating stakeholders in the natural environment of their respective yet shared responsibilities

Green Strategies in Developing Economies

A South-East Asian Perspective

John E. Butler and Suthisak Kraisornsuthasinee

ENVIRONMENTAL CONCERNS emerged as important governmental policy issues in economically developed countries during the early 1960s (e.g. Baumol and Oates 1971; Carson 1962). Businesses initially viewed environmental legislation with suspicion because they associated it with increased costs, impinging on their operations and reducing profitability (Duerksen 1983; Gladwin and Walters 1976). The profit side of the equation was not given sufficient consideration until the late 1980s, when academic and applied research began to highlight ways firms could and did enhance their performance by stressing the environmental qualities of their products and services. The term 'green marketing' became associated with firms' efforts to formulate and implement marketing strategies that highlighted what they were doing. This included the development of more environmentally sensitive products and their promotion based on these characteristics (e.g. Chase 1991; Kleiner 1991; Polonsky and Mintu-Wimsatt 1995; Wasik 1996).

The size of the market segment of consumers who incorporate environmental factors into their purchasing decisions has become quite large. There has not been any surveys published on the size of the green market segment in South-East Asian countries, but organic food is much more available and commercial advertisements often stress environmental issues, which was quite rare before 1990. In Europe, Charter's (1992b) research reports a growing environmental awareness among younger people in the late 1980s, and the Roper Survey (Ottman 1998) showed that about 15% of the people in the United States were in the active green consumer group. Active green consumers were defined as those willing to seek out and pay more for environmentally friendly products. Ottman (1998) also reported that sales of organic foods has gone from US$178 million in 1980 to US$3.6 billion in 1996. The size of these market segments and the rates of sales growth for green products provided an incentive for firms to develop green products and formulate green marketing strategies (e.g. Miles and Munilla 1995).

Marketing remains an essential element in the successful implementation of green strategies, because a product's environmental qualities are not always obvious and one's support for environmental programmes must be communicated to the relevant green consumer segments if it is to have an effect. In South-East Asian and in other less developed countries, marketing is even more important because avenues for disseminating product information are often not independent. Newspapers, television and radio stations, and even organisations designed to protect consumers, are often owned or controlled by the government. Environmental regulations are also less comprehensive than in developed countries, and the general population has a lower level of knowledge with respect to environmental issues (e.g. Serirat 1996).

Initially, some firms made product changes because of governmental regulations (Polonsky 1991). Their original motivation is no longer relevant if they have been able to develop an organisational culture and strategy that supports their green marketing effort. They must effectively communicate their product's characteristics to consumers who use environmental criteria when making product purchases (Fergus 1991).

Targeting and reaching consumers in developing economies such as Thailand, Malaysia, the Philippines and Indonesia is extremely difficult. The populations are still largely rural, poor, and have less access to the various media than is true in more developed countries. This results in them having less information about new products, which reduces their ability to pressure retailers to carry new products. For instance, Green Cott markets natural cotton products in Thailand and minimises external packaging. However, its products are directly marketed to hotels and hospitals because it was unable to identify or develop a sufficient number of retail outlets to carry their product. 'Green and Clean' had a related problem with its non-bleached toilet tissue. It was unable to obtain shelf space in traditional retail outlets. Retail stores still tend to be small in South-East Asia, usually less than 30 m^2. Owners are reluctant to allocate shelf space to new products, but they are usually willing to order products requested by their customers. The retail scene is changing in larger cities and it should be easier to obtain some shelf space as more large stores are opened. However, the key to sales of green products involves more than shelf space, and firms have to be proactive about educating consumers about their products' environmental benefits.

How firms develop, advertise, manage and promote environmentally friendly products is related to both internal firm characteristics and external factors (Polonsky 1995a). Internally, the firm needs sufficient technical expertise and a supporting culture to support this type of product development. External factors, especially those related to the size of the market segment willing to purchase these products, are also important. External factors that vary by nation, especially those related to stage of economic development, need to be incorporated into products' design and the strategies formulated for these products. What is optimal in a developed country may prove to be too expensive in a developing economy because developing-country consumers have lower salaries. A green marketing strategy can be successful in both developed and developed countries, but this requires that firms adapt to cultural, economic, legal and distribution channel differences that affect consumer behaviour.

Strategic and Tactical Considerations

Several issues related to multinational firms' green marketing strategies are relevant to green marketing efforts in the South-East Asian region discussed in this chapter. These include:

- The formulation of a strategy that includes adoption of a global environmental standard, such as ISO 14001, and a recognition of the degree to which unified manufacturing processes and product designs can be used when firms are operating in countries with both developed and developing economies.
- How to compete effectively with both locally and regionally based firms and other multinational corporations with facilities in the region who have not incurred development costs associated with making environmental product because they only market in developing-economy countries that have weaker environmental regulations.
- Identifying ways of effectively communicating one's environmental efforts or products using public relations and advertising, given the varying degree of awareness and interest in developing and developed economies.
- Identifying and targeting customers who will be receptive to green products, or who will respond to the firm's environmental support programmes, in order to expand one's sales in South-East Asia.

Product Design and Manufacturing Dilemmas

Some developed economies' governments have begun to focus on comprehensive product policies rather than just on emissions and other waste by-products (e.g. Oosterhuis *et al.* 1996). This involves requiring firms to produce products that last longer, and that are manufactured with materials that do not harm the environment either at the manufacturing or at the disposal stage. In parallel, they are concentrating more on what goes into the pipeline, rather than just coping with emissions and other waste by-products after the fact. However, this shift has not yet occurred in many developed countries or in any South-East Asian countries.

Government policy in Thailand, the Philippines and Malaysia is more likely to be focused on stopping the disposal of dangerous waste by-products into waterways, and the reduction of factory emissions, especially those related to power production. There has been some discussion about the cradle-to-grave approach to environmental issues by Bangkok's 'Green Theme' governor, Dr Pichit Rathakul. His recent election success may suggest that a political constituency is developing in Bangkok that incorporates candidates' environmental programmes into their voting decisions.

However, his Green Theme message stressed problems related to air and water pollution in Bangkok, and only briefly touched on broader issues such as sustainable development. Organic food has also received a considerable amount of press attention (e.g. Kraisornsuthasinee 1997), but this interest in organically produced food has yet to filter down to other types of green product.

South-East Asian countries are moving towards more comprehensive environmental regulations, although it will probably be another decade before these regulations are in place and enforced. Green political constituencies are beginning to emerge. However, for the present, multinational firms must cope with multiple environmental standards. In many respects, the multinational firms may not be competing on a level playing field in developing countries. Multinational firms can absorb the higher costs associated with making their products more environmentally friendly and try to generate a higher absolute level of sales. Alternatively, they can pass on their higher costs to consumers and sell less at a higher price. The degree to which the playing field is not level is important, but the more important strategic question is how long this difference will continue. Multinational firms have to consider this time-frame issue when formulating strategies for product introductions in developing-economy countries.

Developing products that can be adapted to suit conditions in countries with different levels of economic development is an effective way for multinational firms to compete against lower-costs firms. This requires a more comprehensive approach to product development that considers variations in product configurations. If product variation cannot be built into the product, marketing has to be adjusted to reflect this element of added value. Consumers in developing economies may be willing to purchase higher-cost green items, but they will have to be informed and educated about the reasons for the higher costs and the broader societal benefits associated with their purchase.

Singapore has the most developed economy in the South-East Asian region, and it serves as a good lead indicator of where the region is heading. This is especially important in calculating the time-frame over which other countries in the region move towards tougher environmental regulations. Singapore's government is focused on visible pollution and on educating the public about the causes of ecological damage (Holland 1998). However, recycling has begun. Although recycling does not automatically increase the size of the consumer segment willing to purchase green products, the government uses advertising to stress the importance of recycling to the environment and schoolchildren receive lessons on its importance. This should lead to an increased awareness of the benefits of green products, which could have some 'spillover' effects that increase demand. Cars are also required to have state-of-the-art emission systems and air and water quality is monitored. Noise and dust standards for the construction industry are in place, which is also true in Malaysia and Thailand.

The pace at which environmental regulations have emerged in the region has also been affected by well-publicised and visible environmental hazards and pollution (Kleeman 1994). For instance, the recent haze from Sumatra's forest fires heightened pollution awareness in the region, especially in neighbouring Malaysia and Singapore, which were overcast by haze for several months. Political leaders in the region were forced to stress an increased commitment to health and safety after schools were closed and people were forced to remain at home because of hazardous air quality. The decline in tourism highlighted the fact that there are additional economic costs associated with this pollution. For example, in Malaysia and Singapore, the government considered additional environmental legislation.

South-East Asian governments' attempts to deal with environmental issues are not always successful. For instance, Thailand opened disposal sites for hazardous waste, but these official disposal sites have been under-used. This is because it is easy to dump hazardous waste illegally and many firms choose to do this and avoid the cost of change involved in using the legal disposal sites (Inchukul 1998). Collection of toxic waste at the government's hazardous disposal sites has been 75% lower than expected, and numerous cases of illegal dumping have been reported (Inchukul 1998). Firms that dump illegally have lower costs, which give them a competitive advantage over firms that pay to use the government's dumps. Illegal dumpers also lack any cost reduction incentives to use more environmentally friendly material, although this may change if the government is successful in its recently announced attempts to identify and prosecute illegal dumping (Inchukul 1998).

The trend towards stricter environmental regulations in the region is not uniform, nor is enforcement. Some countries are progressing at faster rates than others, and enforcement is often a function of the dedication and honesty of local enforcement officials. This variability in legislation and enforcement makes it difficult for multinational firms to implement a single strategy in the region, and yet it is impractical to have separate approaches for each country in the region.

Local and Regional Competition

Local and regional firms that are not involved in direct or indirect exporting have fewer incentives to develop products that meet the environment regulations of economically developed countries. Local green strategies can provide positive financial returns, but local firms have fewer incentives to develop products or employ manufacturing processes that exceed local regulations until the green consumer segment grows larger.

Green multinational firms with product name recognition and loyalty have some competitive advantages in South-East Asian markets. Local competitors of non-environmentally sensitive products may have lower production costs, but South-East Asian consumers have shown a strong preference for and loyalty to established brands. This consumer preference could help cushion multinational firms to a certain degree because they can pass some of their higher costs on to consumers in the form of higher prices. Multinationals have to be realistic about the degree to which they can pass on costs, and should attempt to relate the premium to the product's green characteristics.

As mentioned previously, South-East Asian consumers have a strong preference for some international products, and it has been extremely difficult to get them to purchase local or new brands. Goods purchased in other countries are often perceived as superior, even when these items have been manufactured in the South-East Asian region and are available locally. Local consumers will often say they can see or feel the differences in these foreign products, when the language used on the labels is in fact the only difference. This suggests that consumers in these countries may see green products as superior simply because they were manufactured in economically developed countries. This perception can help support the green marketing efforts of multinational firms, while discouraging local firms from developing such products. For instance, Oriental Princess has modelled its product line and shops after The Body Shop. They have stressed their commitment to the environment and located their shops in upscale shopping centres. However, in South-East Asia they have not done well in competing against The Body Shop, because many consumers believe the quality of the domestic producer must be lower.

Analysing environmental trends is also important, because the time-frame over which environmental regulations converge affects the degree to which the green marketing efforts in one country are likely to be effective in another. Multinational firms have a distinct advantage over local and regional competitors in forecasting regulatory trends because they have large planning staffs and access to global information sources. South-East Asian governments are issuing more environmental regulations and politicians are stressing environmental issues, which suggests that the environmental regulation gap between the developed and developing countries is closing.

Firms are also beginning to attempt to exploit this enhanced level of environmental concern through advertising. Increased levels of green advertising are also likely to lead to stiffer environmental regulations because this will heighten awareness and lead to public pressure for stricter regulations.

The product-based regulations being developed in Europe and North America will spread to the South-East Asian region, although this will probably take another ten years. A significant number of South-East Asian firms have already achieved ISO 14001 environmental certification, and there are over 200 environmental consultancy firms in Hong Kong alone. This suggests that firms have a high level of environmental awareness and are actively working to enlarge their knowledge bases. It also indicates that they expect product-based regulations to begin to appear in the region. In fact, the majority of business executives in South-East Asia felt that environmental regulations were too weak (*Far Eastern Economic Review* 1996). Firms that engage in their own environmental efforts have a vested interest in seeing competitors absorb these same costs, and will attempt to get governments in the region to pass tougher environmental regulations.

Coping with Governmental Regulations

Governmental regulations result in compliance and reporting problems for most firms, and the reporting requirement costs are very high in some countries. In other cases, variations in the application of standards makes it difficult for firms to make design and manufacturing decisions. This is also true in South-East Asia. For instance, in 1992, Thailand passed an environmental act that required firms to undertake an environmental impact assessment for certain types of project. However, a recent pipeline project by the Petroleum Authority of Thailand showed that the law was vague and subject to multiple interpretations. Firms that operate in multiple countries have to deal with a variety of enforcement and reporting regulations, which adds to their administrative costs. The vagueness issue is especially important in South-East Asian countries because local government legislators are still learning how to formulate product legislation and government bureaucrats are learning how to enforce technical environmental regulations. For instance, the new Thai constitution states that citizens have a right to a clean environment, although the government has yet to determine what this means in terms of legislation. Once the legislation is passed, the bureaucrats have to write the regulations required to ensure enforcement is comprehensive and uniform.

It may be possible to enjoy some consistency in environmental regulations within the European Union, where the members have reasonably similar economies. Clearly, this is what ISO 14001 and the Eco-Management and Auditing Scheme (EMAS) are attempting to achieve. Although EMAS has only made headway in Germany, some people are beginning to be certified as EMAS auditors and this is probably based on their belief that there will be a growing demand for their services (Tibor and Feldman 1995). However, other political/economic unions such as the Asia-Pacific Economic Co-operation (APEC) group have far too diverse a membership to expect consistency in environmental regulations in the near term. This means that firms with international strategies must deal with a wide variation in regulations when developing their product and manufacturing strategies. They must also simultaneously develop green marketing strategies that effectively communicate to different sets of consumers, or industrial purchasers, in ways that educate and convince these consumer groups that purchasing environmentally friendly products makes sense, even if not legally mandated.

There are some issues to be considered on the manufacturing side, but Thailand, China, Malaysia, Indonesia, the Philippines and Vietnam have legislation related to air and water pollution and restrictions on dumping manufacturing waste. Currently, multinational firms may be able to install manufacturing equipment in their South-East Asian plants that they could not legally install elsewhere. However, environmental regulations related to manufacturing seem to be converging quickly because these countries are adopting legislation similar to that of developed countries. A related reason for having a global manufacturing standard is that it can be used to justify the firm's research and development expenses. If the returns from research and development expenditure come from all operations, rather than a selected set in developed countries, then it will be easier to recover the associated expenses.

Since local competitors are probably already facing or will soon face strict regulations in the manufacturing area, multinational firms will have difficulty gaining an advantage with green consumers based on their manufacturing processes alone. It makes sense for local and regional firms to ensure that their manufacturing processes incorporate the highest level of environmental standards when they purchase new equipment because regulations are likely to quickly diffuse from countries with strict regulations to those with weaker regulations. When the Electric Generating Authority of Thailand began a campaign for energy-saving air conditioners, domestic manufacturers moved quickly to meet the new standard (Thongrung 1998). However, ozone air cleaners—which use special lamps and passive air flow to create ozone that prevents bacteria from forming in food and ice storage areas—were ignored by local manufacturers. A foreign manufacturer, the Bright Group, sold over 100 million baht-worth of product in its first year (Bunyamanee 1997).

Product standardisation presents multinational firms with a much more difficult set of choices. For instance, in Europe, attention is being directed towards the disposal problem and towards manufacturing-based pollution. This is especially true for products such as batteries, which have a heavy metal content that causes disposal problems. Clocks that operate with electric current or with rechargeable batteries are a preferred alternative. However, it is likely that developing economies have fewer product restrictions, which presents a manufacturing dilemma for many firms as they attempt to configure their products based on the local environmental regulations of different countries. One approach to this problem involves using technology to develop superior products for developing countries, which are also environmentally sensitive. This is what British inventor Trevor Baylis did when he invented the Baylis generator, which uses a wind-up mechanism capable of generating power sufficient for about 30 minutes of radio listening. His company BayGen Power Group has an extended line of radios and lanterns which are selling well in developing countries.

The issue of environmental certification is more complex in South-East Asia because public awareness of these labels is low and there is no comprehensive attempt being made to increase awareness. Programmes such as Germany's Blue Angel certification are accepted as a signal that the product has valid environmental claims in developed economies. South-East Asian consumers are more sceptical about this type of certification because, even though the government is not performing the certification, governments seldom prosecute firms for making false claims. In Thailand, the Foundation for the Environment was founded and funded by the Central Department Store chain in 1995. This company foundation then certified that Central's plastic bags were environmentally friendly. Since the foundation is funded by Central, its labels are more like the firm's own eco-label, rather than an independent certification. The independent Thai Environment Institute also awards certification labels to products that promote environmental conservation, although it is easy to understand how consumers could get the two groups confused.

There is also some indication that consumers are willing to pay more for green products. The Thailand Business Council for Sustainable Development conducted a study on green-label products, in which 712 consumers and 78 businesses participated (Chaimusik and Nivatpumin 1996). Forty-eight per cent of the consumers said they would purchase a green-label product at the same price, and 25% of the consumers said they would purchase green-label products even if they were a little more expensive. Sixty-four per cent of the business respondents said they believed that complying with green-label criteria would result in higher prices, and 55% of producers and 44% of the distributors felt the label would lead to higher sales.

However, neither of the existing independent or Central Department Store labels are well known among the general consumer population (Chaimusik and Nivatpumin 1996). This means that the public has to be educated, but the independent Thai

Environment Institute lacks the funds to do this. Central Department Store has little incentive to make its label an independent national one, because they have to provide the funding, and their direct competitors would also benefit. Firms that earn the independent green label may have to invest in educating the public by participating in publicity efforts so that public awareness and acceptance of the green label increases, which will ultimately benefit their products.

When it makes sense to transfer products that are more environmentally sensitive in their design and material configuration to countries where they could legally market and produce products that are less environmentally sensitive, is a key green marketing issue in the South-East Asian region. Modern forms of information transmission and the high level of international travel means that South-East Asian consumers will quickly become aware of any obvious product variations. This could cause a firm to suffer a loss of reputation if they market products that are environmentally inferior in developing countries. Green consumers in developed countries may react negatively to firms that attempt to exploit their proactive environmental stance in one country while ignoring it in another.

The trend toward global products can also help support multinationals' green marketing strategies. Consumers in developing economies increasingly expect the same product specifications as are offered in economically developed countries, and are willing to pay more for these products. This means that green products may be able to recoup their higher costs in these markets more easily than they could in their domestic markets where they usually face stiffer competition.

Advertising and Public Relations

Firms that invest in production processes or products that are environmentally sensitive can benefit directly if they can effectively communicate to consumers. In some cases, their investment may be objectively demonstrated, such as when restaurants changed from Styrofoam to paper plates, or when grocery stores stopped using plastic bags. In these cases, consumers can see the differences, but South-East Asian consumers must be told the reasons underlying the change. When firms use components that can be repaired rather than discarded, or make minor product modifications that reduce environmental harm, it is necessary to communicate these efforts aggressively if they are part of a more comprehensive green effort. For instance, MAKRO, a Thai discount store, requires customers to purchase large re-usable plastic bags at their stores. This concept has been a big success, but few users, and perhaps not even MAKRO, have attached any environmental significance to this practice. Some firms, such as Lever Bros, promote their products' environmental characteristics more in developed economies than in developing ones but, as the green segment grows in South-East Asia, it will make more sense to advertise one's green products.

Firms must also decide if public relations or paid advertising is the best medium of communication. Firms that want to build a general reputation for being sensitive to green issues may find that public relations is the better approach in this region. South-East Asian newspapers, television and radio are appreciative of and receptive to using well-written press releases, often exactly as submitted. This enhances the returns from a public relations effort. In some South-East Asian countries, these forms of communication are not viewed as being completely objective, which means each message has a lower level of credibility, so care must be taken to ensure that public relations is used in a credible and consistent fashion.

Green advertisements, in developed markets, must pay a great deal of attention to their claims and the firm's ability to substantiate these claims (Polonsky 1995a). There is also the assumption that there is an existing segment of green consumers that is fully aware of what they want in a product (Carlson *et al.* 1995). However, this is less true in developing economies: in many South-East Asian countries, consumers are still relatively uninformed about what makes a product environmentally friendly. Thus, advertisements must both broadly educate consumers and also sell products. The time-frame in which to establish validity with consumers may be longer than it is in a developed country, because advertisements are viewed as being less credible.

Identifying and Targeting Customers

Identifying market segments in developed countries is expensive, but often presents few technical problems for experienced market researchers. Broad segments of the general population have been sensitised to environmental issues and these segments can be identified using marketing research surveys. In addition, published research provides information on likely green-segment characteristics and buying behaviour. This can benefit all firms, but smaller and medium-sized firms, who often cannot afford to conduct their own research, benefit especially from this published research.

However, this is not true in South-East Asia, nor in most developing countries. Consumer segments are smaller and more fragmented. This makes them more difficult to reach, unless one targets the entire population. Consumer surveys are seldom undertaken with poorer or rural people and results for published surveys are best interpreted as applying only to the higher- and middle-income groups in the larger cities. Firms that need information on buying attitudes of all income and geographic groups—because their product introduction would only be warranted if it had broad consumer support—have to make sure that those responsible for market research include rural households and lower-income groups.

Targeted communication avenues are beginning to emerge, which will make it easier for firms to target green consumers. For instance, one of Thailand's easy-listening music stations, Green Wave (106.5 FM), provides a good communication channel for producers with green products or messages, and several newspapers have 'green pages', which are devoted to environmentally friendly products and firms.

Higher prices remain a problem for some green products in the South-East Asia region. It has proved difficult to get consumers to pay more for generic products, although there are some signs this is changing. For instance, initial efforts to market organic fruits and vegetables in Thailand ran into stiff price resistance. Consumers were not willing in large numbers to buy these higher-priced products, although the credibility of the organic certification was also an issue in their market acceptance. However, recently, an organic produce store called 'Lemon Farm' opened in Bangkok and it has been doing well, even though it opened during the current economic crisis. Sales at the initial outlet were sufficiently high to encourage the owners to develop an expansion plan to open multiple retail sites, although they indicated that their ability to expand is constrained by the limited supply of organic produce.

In the short term, higher-income and middle-income groups in the major metropolitan areas can be most easily targeted. However, this requires placing advertisements in a combination of upscale magazines, business newspapers and in store displays. Rural consumers in most villages also have access to televisions and radios, although stations do not specifically target this audience. In addition, rural residents may have difficulty buying products that are advertised. Rural retail outlets tend to carry very restricted sets of product lines, usually consisting of lower-priced but high-turnover products. This is because of their small store size and the lower incomes of rural residents. Normally, green products must have broad appeal and be competitively priced to gain access to these rural retail outlets, unless the firm can make their product sound sufficiently attractive to villagers so that they request their local retailer to carry the product.

Economic growth will continue in the South-East Asian region, despite its recent financial setback. This means that the potential green market segment will also grow each year. Sensitising these consumers to environmental issues before they join the segment makes sense. Attempting to educate consumers about the links between a clean environment and the minimisation and re-use of waste and recycling and remanufacturing can do this. In this way, they will be convinced about the longer-term benefits associated with green products before they have the income to buy these products. However, once they reach the critical income level, they will not have to be convinced that green products are the optimal choice. Osterhus (1997: 27) suggested that 'subtle repetition of modest claims', especially by objective third parties, may be the most effective way of achieving this goal.

Market Development and Green Policies

Obviously, keeping one's market growing is an important issue in market development. Formulating a green strategy that can achieve these aims is possible and the South-East Asian countries seem to present a good opportunity for market growth, with a regional population of approximately 500 million people. While the level of personal disposable income recently decreased, this is a temporary effect and it will continue to increase once the current economic crisis ends. South-East Asian markets are increasingly tied to those of South and East Asian countries, which have large populations and thus tremendous market potential. Access to a college education has also increased, and is continuing to increase, which is contributing to the building of a market segment that is more likely to be receptive to green products.

This means that the financial returns from an effective green marketing strategy may eventually be larger in this region than in some economically developed regions. Firms that have to absorb the costs of product modification in developed-economy countries may be able to extract returns from a larger market. Being a market leader is just as important in a green marketing effort as in any new product introduction, and South-East Asian consumers will discriminate in the same fashion as those in developed countries.

Conclusions

South-East Asian green consumers will increase in number and governments will increasingly respond to their environmental concerns. This will eventually also be true in Myanmar (Burma), Vietnam, Cambodia and Laos when these economies begin to grow. Firms that respond quickly and effectively will be rewarded with enhanced financial performance. However, green marketing is a new field and there is no set of well-established approaches that marketing managers can access. This means that firms must not only respond to these changes, but they also have to find new ways of benefiting from these trends. Several are discussed in this volume, but they still require some variations because of national or cultural differences. As with other business skills, competence will affect the level of one's success in implementing a green marketing strategy.

The issue of social responsibility will also enter into the green marketing equation. Nike, Reebok, several designer clothes manufacturers and Pepsi have discovered that issues related to human rights, child labour and exploitation of workers can affect their sales in more industrialised countries. This interaction between social responsibility and product sales can also negatively impact sales of products that are environmentally friendly. This suggests that green marketing will be part of a much broader social responsibility posture for firms in the future.

One of these broader issues relates to sustainable development. The World Commission on Environment and Development (1987: 43) define sustainable development as actions 'that meet the needs of the present generation without compromising the ability of future generations to meet their needs'. At the firm level, this may also be an issue because the longer-term survival of a firm also rests on its ability to produce and deliver goods in the future.

Green learning is also an issue that firms need to consider. Most firms have thought of learning in a manufacturing context, where learning is measured by the rate of reduction in labour-hours per unit. A 20% learning curve implies that labour-hours or material inputs have been reduced by 20% over a period in which the cumulative number of units manufactured doubled. Green marketing also needs a steep learning curve if strategies are to be implemented successfully. Measuring inputs and outputs will not be as precise as in manufacturing, but firms need to find ways of gauging how well they are doing. Successful green marketing firms need to be able to educate consumers faster, communicate product information more efficiently and make the same types of manufacturing improvement for their green products as for other types of product.

Early entry and learning are not a guarantee of permanent competitive advantage. This is because later entrants can benefit from their experience by avoiding their mistakes. In some cases, later entrants may be able to benefit from more advanced or cheaper technology, which may allow them to have a lower cost structure than the original entrant.

Designing, producing and selling green products requires a certain amount of learning-by-doing. Firms that acquire knowledge early have a learning advantage over those who join the market later, in many cases. This is because firms get better at doing things as they move along the green marketing learning curve. This advantage is not automatic, but if an early-entry firm keeps improving at the same rate as new entrants, they should be able to stay ahead of competitors. These firms are more capable of developing products that will appeal to green purchasers and in developing advertisements that will reach these customers. They are also more likely to have appropriate channels of distribution and retail outlets in place for the new products that they develop.

Another aspect of competition that is occasionally forgotten is that implementation is often the key to success. Technology and product modifications will not be sufficient in the green consumer market because implementation is equally important to a green marketing effort. For example, Xerox developed much of the early technology for personal computers, but they were never able to bring their products to market successfully. The same thing can happen to a green product, because even the most committed green consumer is not going to be willing to spend large amounts of time trying to stay aware of new green product introductions.

In the final analysis, success in the South-East Asian market will require many of the same things that are required in other markets. There is a growing green segment of consumers and firms can enhance their performance by appealing to this segment based on their environmental proactiveness, clean production processes and products that are environmentally friendly. However, their message will have to be adapted to the local market, and they will have to discover new ways of reaching this segment. Different environmental issues may have to be stressed, and managers responsible for green marketing efforts will have to remain in an adaptive mode as they learn how to direct green marketing efforts in these developing economies.

Ultimately, they will have to develop green products that meet the needs of the region and educate potential consumers about the benefits of these products. While it is likely that some of what has been learned in other markets can be transferred, different channels of distribution, retail practices and government legislation will require that firms develop marketing programmes that incorporate these differences.

23

The Tainting of a Green Titan

The Petroleum Authority of Thailand

Suthisak Kraisornsuthasinee and John E. Butler

MANY THAIS were surprised to learn, in late 1997, that the Petroleum Authority of Thailand (PTT)'s natural gas pipeline from Myanmar (Burma) would pass through and harm a sensitive wilderness area. They were even more surprised that PTT had not been fully forthcoming in revealing the details of the project, because PTT was one of the few firms with a strong positive image among Thailand's green constituencies. The firm had taken a strong stance towards the need to protect the environment, supported environmental causes and used environmental themes in their advertising. While the public was aware that a pipeline was being constructed, Thais with environmental concerns generally felt there was little need to monitor the actions of PTT closely.

The pipeline construction project of PTT presents an interesting case because it depicts a firm with an outstanding record of environmental concern that had been using its record to build support with a segment of the population that was most likely to be receptive to a 'green marketing' message.

Corporate Background

PTT was established as the nation's fully integrated petroleum company in 1978, through a merger of the Natural Gas Organisation of Thailand, the Oil Fuel Organisation and the Bangchak Refinery. It was restructured in 1992, in response to the government's Seventh National Economic and Social Development Plan, which loosened government controls, increased private investment and made the PTT more subject to market mechanisms. PTT was decentralised into four major sector groups, with PTT Holding serving as the corporate arm. In 1996, the corporation went through a re-engineering effort and transformed business units into subsidiaries, which could be listed on the Stock Exchange of Thailand

During this period, PTT also achieved rapid financial growth. It was 170th in *Asia Week*'s rankings of firms in 1997, up from 269th the previous year. Within the oil and gas industry alone, PTT moved from 14th to 3rd among the ASEAN firms.

Image Building and PTT's Green Portfolio

The visible 'greening' of PTT began in 1985, when it formulated a top-down commitment to make sure that its operations minimised adverse environmental impacts, and it was active in implementing its commitment in terms of focusing innovation on improved products and supporting the efforts of others that were focused on improving the environment.

PTT played a leading role in reducing pollution from cars and buses, which began when it helped develop 'Gasohol' in 1985. In 1988, it launched the first low-leaded regular gasoline and introduced the first unleaded super gasoline in 1991, followed by the first unleaded regular gasoline in 1993. During this period, it conducted several promotional campaigns that warned consumers of the danger of air pollution caused by lead in petrol. In 1995, PTT reached the older car market with the debut of their unleaded regular gasoline, which was extensively promoted with an award-winning 'talk-of-the-town' advertising campaign.

PTT developed a low-sulphur diesel fuel in 1991, specifically for the Bangkok Mass Transit Authority's fleet of buses, but to support all diesel-fuel vehicles. Its latest formula, brought to the market in 1997, contains 0.05% sulphur, which is 0.20% lower than the legal requirement. Pollution from buses has also been reduced through its pilot project for natural gas vehicles. These efforts have earned the company praise because Bangkok has a high level of vehicle emissions.

PTT's most prominent environmental support activity was joining in the reforestation campaign in commemoration of the Royal Golden Jubilee of the King of Thailand, who enjoys loyal support from the Thai people. This three-year project targeted the reforestation of 5 million rai of land (1 rai = 1,600 m^2). PTT injected 3 billion baht to support this project. During the same period, it launched a public relations campaign to encourage local villagers to place a higher value on forest resources. As a result of these and other activities, Thai Prime Minister Chuan Leekpai praised PTT for being 'the best state agency in carrying out the reforestation programme to mark HM the King's Golden Jubilee' (Unnarat and Techawongtham 1998).

PTT also has an aggressive environmental programme on the production side. They have instituted strict control on toxic air and water emissions at their oil refineries and storage sites, and have procedures in place to prevent accidental oil spills. PTT plants attempt to meet the ISO 14001 environmental certification standards, which was awarded to their petroleum storage site in Songkla Province in 1997.

PTT also has the reputation of being philosophically committed to being a world-class green company. This was articulated as part of their mission, and was perceived as a genuine commitment by those in the various green constituencies. Their internal publications promoted environmental behaviour among all their stakeholders and their desire to be a green company. For instance, Viset Jupibal, Director of PTT's Oil Sector and Acting President stated:

> ... we want the entire business sector to recognise us as a 'Green Company' ... However, we cannot just call ourselves a 'Green Company', we must be perceived [as such] by others (PTT 1997).

Natural Gas: A Greener Option?

In 1977, Thai government officials were delighted to learn of Unocal Thailand's discovery of natural gas in its concession area in the Gulf of Thailand. This was because they felt this gas reserve would reduce the country's dependence on imported fuel oil. In terms of the environment, natural gas was also perceived as cleaner-burning than most fossil fuels, and air pollution was emerging as a serious concern. When the Natural Gas Authority of Thailand was merged into PTT, the initial natural gas was distributed to a series of interlinked power plants operated by the Electricity Generating Authority of Thailand (EGAT). Although this preceded the proactive 'green' building phase of PTT, the cleaner natural gas was consistent with their effort to be an environmentally responsible firm.

Thailand's rapid economic growth during the 1987–97 period resulted in the country's commercial energy use increasing from the equivalent of 330,000 barrels per day of crude oil in 1986 to 1 million by 1997. This meant that the reserves of natural gas in the Gulf of Thailand were not adequate to sustain future growth and new sources of natural gas would have to be found.

Myanmar: A New Source of Natural Gas

On 30 October 1990, Thai Prime Minister Chartichai Chunhawan learned that there was an estimated 5.7 trillion ft^3 of natural gas in Myanmar (Burma)'s Gulf of Martabun, and internal discussions began on ways of accessing this natural gas. On 5 October 1993, the cabinet of Prime Minister Chuan Leekpai approved PTT's request to procure natural gas from Myanmar's two offshore natural gas fields, Yadana and Yetagun, to supply EGAT's new power plants in Ratchaburi Province. These plants would run on two systems, with natural gas as the main fuel and diesel fuel as backup source. A trade ambassador for PTT Exploration and Production was sent to work out an agreement with Myanmar's State Law and Order Restoration Council (SLORC), which has been widely criticised for human rights violations to the democratic movement since 1988.

On 9 September 1994, a memorandum of understanding was signed in Yangoon to purchase gas from Yadana gas field, to the extent of 525 million ft^3 per day for 30 years. This agreement also had some political implications, because of opposition to business agreements that support the SLORC regime in Myanmar. These political opponents of SLORC eventually became allies with the environmental activists. At the time, financier George Soros announced, 'The [Yadana] pipeline, when completed, will be the SLORC's single largest source of foreign currency' (Haq 1997). The annual income from Yadana was projected as US$400 million, or up to 65% of the national income in total (Suamapuddhi 1998). This injection of foreign investment was vital for Myanmar to run a project of this size, with an estimated cost of US$1.2 billion. The concession to explore and develop the submerged energy was won by a group of companies, led by the French oil firm Total (31.24%) and the US firm Unocal Corporation (28.26%). Two other regional partners, Thailand's PTTEP (25.5%) and the Myanmar Oil and Gas Enterprise (15%) also participated.

Transporting the Invisible Energy

The route for the pipelines, from the gas wells to the power plants, was separated into two construction projects that were to meet at the border of the two countries. The consortium of oil firms was responsible for laying the 413 km pipeline via sea bed and over land from the Yadana gas field to the point of consignment at Baan I-Thong village in Thailand's Kanchanaburi Province (see Fig. 1), where it would connect with PTT's pipeline from the EGAT power plant in Ratchaburi Province (see Fig. 2).

The cabinet of Thai Prime Minister Banharn Silapaarcha agreed in principle to allow PTT to begin constructing the 260 km pipeline on 7 May 1996. The estimate of the Thai investment needed was 16.5 billion baht: PTT invested 3.6 billion baht and the rest of the US$100 million funding came from an Asian development bank: Japan's EXIM Bank loaned ¥42.8 billion to EGAT to build the gas power plant in Ratchaburi, which was scheduled to be completed by 1998.

At this point, the project seemed to be going well, and several independent power producers indicated that they were interested in PTT's natural gas. Industrial users along the path of the pipeline also expressed an interest in accessing the natural gas. PTT also seemed capable of handling the project. It had a 19-year clean record of no pipeline accidents, and at a technical level that met or exceeded world-class safety standards for pipelines. Its 1.5 m underground pipes were designed to bear the intensity of an earthquake of up to eight points on the Richter scale and they were to be protected from corrosion by a fusion bond epoxy coating—claimed by PTT to be the world's best technology—and the world standard electrochemical cathodic protection.

In addition to these precautions, PTT had X-ray instruments, called pipeline inspection gauges, which would run inside the pipeline and monitor it continuously. Block valves were planned at 20 km intervals along the line, which would be automatically and continuously controlled via a satellite system called 'SCADA'.

PTT announced that its construction project would operate only during summer to prevent soil erosion, and that the width of a ditch for laying a pipeline would be

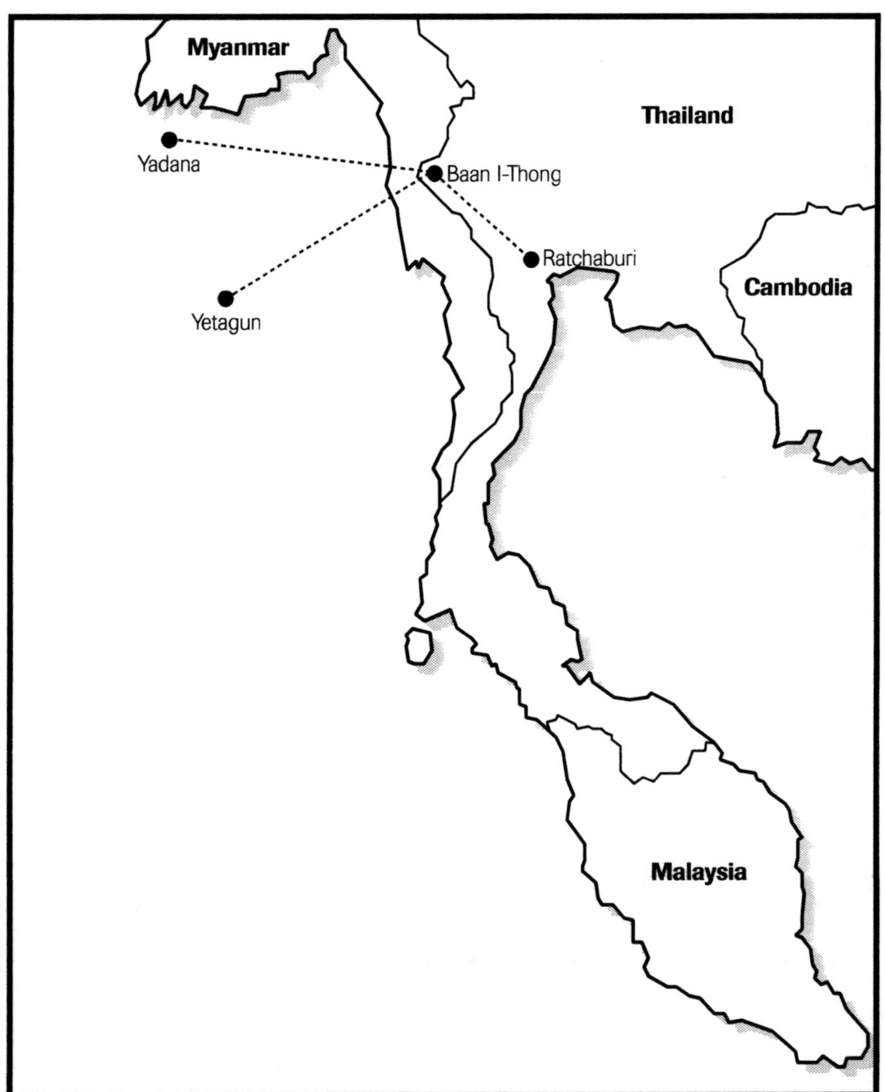

Figure 1: MAP OF THE PIPELINE SYSTEM

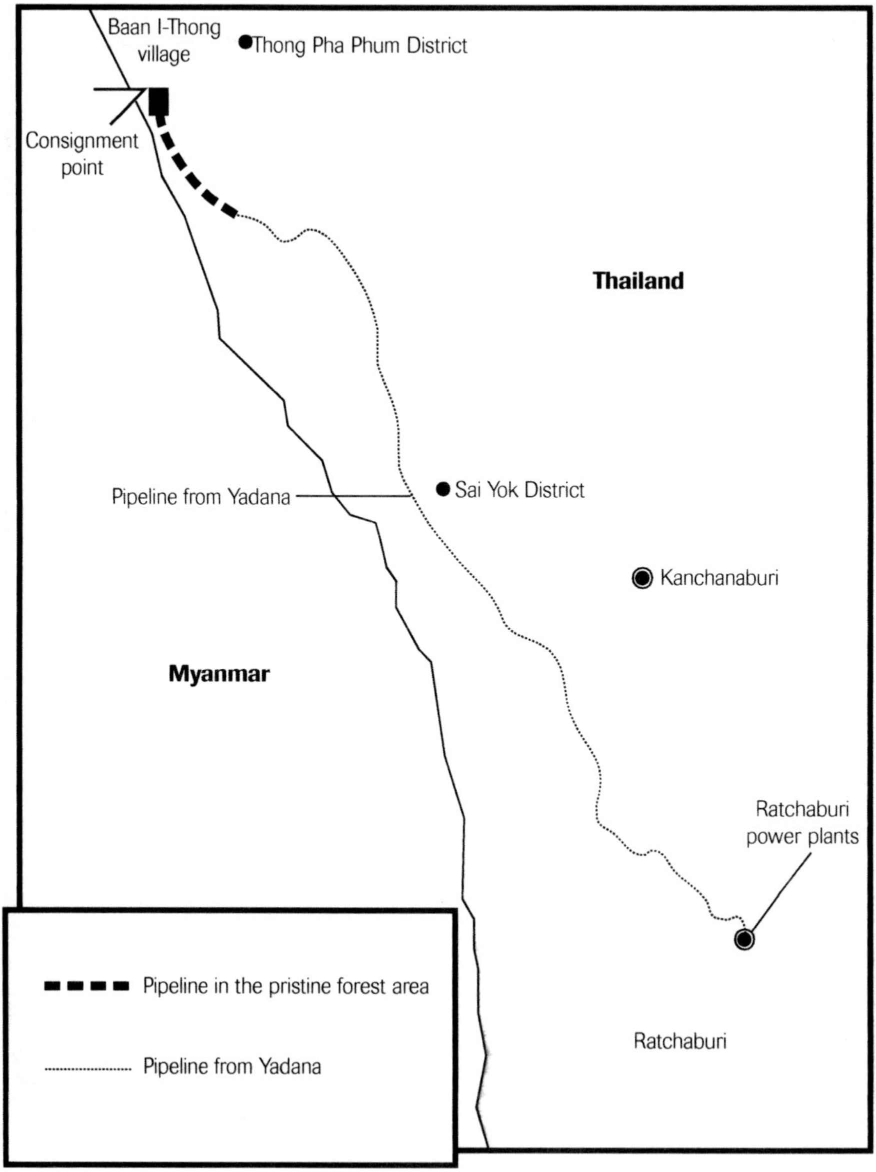

Figure 2: MAP OF PIPELINE FROM THE CONSIGNMENT POINT TO RATCHABURI POWER PLANTS

narrower in forest area thanks to new construction technology. No inspection road would be needed, which would make it more difficult for those engaged in illegal logging to operate in the area. Wherever a stream was crossed, the pipeline would be installed underneath using a directional drilling technique. Tall trees would be felled as little as possible, and reforestation would be implemented at the end of the operation. At night, the labourers would camp out in nearby villages, not in the forest. These actions supported PTT's image as an environmentally friendly firm, and minimised the pipeline's impact on the environment.

The Green Image Problem Begins

PTT selection of the routing for the pipeline provided the first indication that there might eventually be some problems with the project. PTT said it had to lay the first section of the line through a dry evergreen forest area of Huay Kayeng Forest Reserve, which is part of the Thong Pha Phum National Park. This meant that the pipeline would cut through the Western Forest Complex, starting from Huay Kayeng Forest Reserve to the Sai Yok National Park, which is further connected to the Thung Yai Naresuan Wildlife Sanctuary, one of the world's Natural Heritage sites.

PTT had explored three possible routes, and hired TEAM Consulting Engineers, an external environmental consulting firm, to conduct an environmental impact assessment (EIA). The option chosen was based on the criteria that it would cause the least damage to the environment. It would run through 26 km of conservation reserve and another 18 km of first-grade watershed area, while other options were longer. In addition, the proposed route would pass through only 6 km of pristine forest (see Fig. 2). To facilitate the operation, the cabinet approved the revocation of Kanchanaburi's Sai Yok National Park status, so it could be crossed by the pipeline.

The EIA included a social impact assessment (SIA). This involved the survey of 10% of the 1,350 households in the 40 villages affected by the project, which were defined as those within a 100 m of the pipeline. However, the sample size turned out later to be an underestimate of the number of households that would be affected by construction. Public participation, especially in terms of risk perception, had not been included at the beginning of the assessment and, also, communication between PTT and villagers, which actually served to highlight the potential danger of natural gas, did little to improve relations.

PTT had problems in gaining approval for their preferred route. The chronological sequence of procedures that were outlined in the EIA report seemed to be as complicated as its content. It was revealed that the consulting firm, TEAM, started the EIA study on 31 May 1995, which was just after PTT had learned that the 1992 Environmental Act required an EIA before pipeline construction approval could be given. After a four-month study, TEAM submitted their report to the Office of Environmental Policy and Planning (OEPP) on 22 January 1996. The Cabinet of Prime Minister Banharn Silapaarcha agreed in principle to let the pipeline construction commence on 7 May 1996 on the condition that the EIA was approved. On 21 May 1996, the EIA report was rejected by the OEPP because the project contained insufficient information on adverse ecological effects, particularly on key wildlife species. The OEPP required TEAM to collect more information for the technical hearing session to be held during 10–12 February 1997, and stipulated that the information presented at this time would be used to make a final decision on approval of the EIA. At this point, PTT was still optimistic and declared that the operation was still viable.

PTT then asked the Wildlife Fund of Thailand and the Royal Forestry Department to revise the impact of the construction on endangered animals. The EIA report was resubmitted to the OEPP on 26 February 1997. However, the effort was again rejected because of its failure to evaluate completely the impact of the project on wildlife. A week later, on 5 March 1997, PTT handed over a final report to the National Environment Board chaired by Prime Minister Chavalit Yongchaiyudth. This EIA was approved by the NEB on 24 March 1997, the same day that construction had started—on the grounds that any further delay would cost PTT a fine of up to a 100 million baht per day to the consortium of oil firms. To help improve the situation, a monitoring committee was set up, chaired by the Governor of Kanchanaburi. Later, on 21 October 1997, the cabinet of Prime Minister Chavalit approved the revocation of Sai Yok National Park to help unclog the operation. PTT decided to leave the area of greatest environmental sensitivity as the final stage of the pipeline construction project.

While these procedural issues did not appear to be a major factor at the time, they eventually had an adverse effect on the green constituency that PTT had so carefully cultivated over the previous decade. This sector, to a large extent, viewed the actions and omissions of the environmental consultant firm as deliberate PTT policy rather than attribute it to inadvertent behaviour.

The Fight for Change

During the second administration of Thai Prime Minister Chuan Leekpai, which began on 9 November 1997, scrutiny of the approval process of PTT's EIA increased. Opposition groups formed an alliance, which included 88 environmental and social organisations, the Law Society of Thailand, academics and students. These allies teamed up with villagers in the affected region in an attempt to halt the pipeline project, or at least delay it until an in-depth revision of the EIA has been completed. They pointed out that the existing study did not even cover the whole year, and a seasonal approach, they felt, was essential for an ecological study. As an alternative, they called for a re-routing of the pipeline.

Legal challenges were raised in the courts while protesters called for a boycott of PTT's, Unocal's and Total's products and for a closure of the forest to pipeline construction. A human shield of protesters was set up to block entry to the forest by the workers of PTT's contractor Tasco-Manessmann. All of this resulted in increased press coverage, which began to tarnish PTT's image as a green company.

PTT extended a nationwide integrated marketing campaign to convey their long-term corporate sponsorship of the reforestation project. Public relations, newspaper advertisements and infomercials were used to outline the positive points of the pipeline project. These efforts may have been effective with the general public, but they tended to be placed in response to criticism. At the technical level, PTT continued to be a good environmental citizen and hired university experts to manage the removal and transplantation of trees, but this did little to help its national-level credibility with those opposing the pipeline. PTT did, however, manage its position more effectively at the local level in Kanchanaburi Province. Their local support was broader and included the governor, business leaders and some local villagers.

Negotiation between the opposition groups and PTT reached an impasse. However, the fact that 80% of the pipeline was completed and that PTT faced a large fine for breaching their 'take-or-pay' gas contract if the pipeline was not completed by 1 July 1998 were very important factors. The debate also became somewhat less substantive as parties on both sides began to move away from issues directly related to the pipeline's impact on the forest area. For instance, the press reported a kidnap threat against an opposition leader. A week later, the head of the association of the village chiefs lead a rally of 30,000 pipeline supporters. Other advocates of PTT's position submitted thousands of signatures to the Kanchanaburi governor supporting the pipeline. Both sides lost some credibility with these press antics but, for PTT, the loss of credibility was, crucially, with political and environmental activists whom they had cultivated in the past.

Army commander-in-chief General Chettha Thanajaro pressed for public hearings and blamed the dispute on the failure of PTT's public relations efforts. The Prime Minister stated that opponents had voiced their concerns too late in the process, after the bilateral contract has been signed. He attempted to mediate the crisis by setting up a public information panel to sort out the facts, and former Prime Minister, Anand Panyarachun, who has a reputation in Thailand as a man of integrity, was appointed chair. The information-gathering sessions were broadcast live on radio. The leaders of environmental groups promised to withdraw from the site after the findings were publicised and justified by the Prime Minister, but this resolution did little to restore PTT's lost green credibility with the very sector that would most appreciate its past efforts on green product development.

Panel Findings

Anand Panyarachun's panel felt that PTT should have been more open with respect to the pipeline project findings and expressed their disappointment in PTT's lack of transparency. On the second day of the information-gathering sessions, PTT admitted that they might have misled the public as to their deadline for natural gas delivery, as well as the extent of any potential fine. PTT also pointed out that they would face a severe liquidity problem if the project had to be delayed, which would be reflected in final fuel costs. PTT also revealed the reason that they agreed on the point of consignment being Baan I-Thong was that it was the only location that the Burmese military junta would guarantee security, rather than because of any environmental issues.

The public information panel also pointed out that time constraints prevented them from undertaking a detailed analysis of all the issues. They did decide, however, that both the EIA and PTT's public relations efforts needed to be more accurate. Their report to Prime Minister Chuan Leekpai found that:

1. There were many weaknesses and shortcomings in PTT's decision-making process, due to lack of transparency and failure to consider fully issues relevant to people directly affected by the project.
2. Local people have the right to access information related to the pipeline and to hold peaceful protests.
3. PTT ignored problems and the impacts related to the environment, local communities and human rights.
4. The failure to resolve some of the disputes was related to poor communication between the parties.
5. Thai society lacks effective mechanisms for dispute resolution and consensus building.

The panel's report also made a number of proposals related to the EIA and its approval. However, the impression the report left was that PTT was in the best position to know the facts and that they could, and should, have done more. The panel did not specifically urge the government to halt, re-route or scrap the project but, more broadly, urged it to resolve the pipeline problems as quickly as possible.

The PTT Gas President responded to the question about transparency in a magazine interview. He stated:

> I admit that we [PTT] were less than open with our information to the public because it was not required by law. The project started four years ago, but the new constitution was just ratified in October 1997 . . . Don't call it a lesson. I am amazed with the term 'lesson' because PTT did everything in compliance with the law. The lesson is, if we stop this project, Thailand has to pay out an awful lot. And if we can't complete it on time, there wouldn't be any money going back to the state. As you know, the state earns 2–3 billion a year from PTT (Pracamthong 1998).

The Prime Minister decided to conduct an aerial inspection of the pipeline site on 28 February 1998. Prime Minister Chuan Leekpai stated, after completing his inspection, 'I have decided that the project must continue. I don't have the right to stop it' (*Bangkok Post* 1998).

Aftershocks

Sulak Sivaraksa, a prominent social critic, visited the protest site to replace the protesters when they left. He claimed that he acted in the name of justice for the environmental opponents, in conscience against the Burmese dictatorship, and in the name of truth. He vowed to stage a lone protest and insisted the project must be scrapped. The Prime Minster responded that, 'The pipeline must continue. Whoever wants to go and stay in the forest is free to do so' (*Bangkok Post* 1998).

Later, Sulak persuaded a group of some 40 allies to join him. PTT lodged a complaint, and the confrontation ended with the arrest of Sulak and other protesters for obstructing the construction. Most of the protesters were quickly released, except for Sulak, who was detained for further interrogation. The main impact of this was to keep the pipeline in the news, which further hurt PTT's green image.

PTT then appeared to be in the process of rebuilding its green reputation. In a rare public appearance, PTT's governor Pala Sukawech thanked the conservationists for moving out of the construction site and claimed that PTT had followed the panel's recommendations. Compensation for 1,300 of the 1,350 families affected by the project had been settled and PTT expected to reach compromises with a further 23 of the remaining 50 families. He then promised that PTT would work with environmental organisations to declare the forest along the pipeline route a National Park in order to protect wildlife.

In the public information panel's official report, PTT President, Piti Yimprasert, indicated that he wished he could turn back time, that he could have done things better, and that he deeply regretted that incomplete information was given to the people affected by the project. In addition, PTT Public Relations Director, Songkiert Tansamrit, stated that 'PTT will use this process of public hearings as an integral part of future projects' (*Matichon*, 26 February 1998).

Lessons Learned

Many firms are attempting to build a green reputation in Thailand and South-East Asia, but the experience of PTT shows that inadvertent action can damage reputations with the emerging and active green constituency. The green product market segment, environmental faction, social issues faction, legal faction, and various political faction, are interconnected. PTT's gas pipeline impacted their green product reputation because the faction interested in lead-free gasoline and lower levels of pollution is also interested in ecological preservation. PTT's experience suggests that firms have to broaden their decision analysis process greatly if they are to maintain a green reputation with the environmental constituency effectively.

They also have to consider marketing and not just legal issues when deciding how much information to make public. Besides financial support, providing green groups with information about the pipeline and including them in the discussion about alternatives from the outset might have been a better way of keeping them as allies in other environmental projects and as customers.

Finally, firms need to ensure that they learn from the crisis situations they face. PTT and other firms have to think about designing their organisations so that they can deal with crises. The next crisis will involve a different project and the managers who responded to this crisis will probably have retired. Organisations need to find ways of ensuring that what they learn from coping with one green crisis makes them better equipped to handle the next one.

24

Green Power

Designing a Green Electricity Marketing Strategy

Norbert Wohlgemuth, Michael Getzner and Jacob Park

> MOST UTILITIES offer as much choice in how your electricity is created as Henry Ford offered to those buying his Model T: you can have any color you want, as long as it is black (*Scientific American* 1997).

Consumer choice and consumer sovereignty in general is—at least in theory—a guarantor of efficient and competitive markets for private goods (e.g. consumer goods). This observation is also true for the electricity industry which, to date, has been characterised by a monopolistic structure. Before deregulation, the concept of electricity marketing barely existed, since consumers and companies had little choice except to purchase their power from their local electric utilities. Worldwide, the electricity industry[1] is in the process of a crucial industrial reorganisation which is

1 The electricity industry includes four primary components. Following the general direction of electricity flows, it begins with **generators**, which utilise a host of technologies to drive turbines and produce electricity. This electricity is shipped over long distances via high-voltage **transmission** lines. Large industrial users of electricity may take delivery directly, but most electricity is then transformed down in voltage and made available to customers through a **local distribution** network. The local distribution network provides services related to the **marketing** of electricity.

The cumulative effect of technological progress and public concerns began to erode the myth of natural monopoly and public good rationales for government intervention (Jaccard 1995). Technological advances have lowered the costs of smaller-scale technologies (small natural gas combined-cycle units; alternative renewable generation technologies). These developments imply that the assumptions about economies of scale (resulting in a horizontal integration of the industry) in power generation no longer hold. Also, the assumption about the existence of economies of scope (vertical integration) could no longer be sustained (Hunt and Shuttleworth 1996). At the same time, the ecological costs of electricity supply—which are normally not reflected in electricity prices—have increasingly achieved much economic and public attention (cf. most recently for external costs of electricity the voluminous collection by Hohmeyer [1997]).

driven primarily by a competitive industrial environment (cf. O'Reilly 1997; Farhar 1996) and by technological developments (e.g. energy-saving technologies, efficient use of biofuels; cf. EIA 1997c; Goldstein 1995). This move towards liberalisation of electricity markets is part of a greater global process (Newbery 1997). At the same time, environmental dimensions of electricity generation have received a lot of public attention in recent years due to greater international awareness of climate change as an international business and public policy dilemma. Electricity generation and global climate change has become intricately linked in recent years because the production and distribution of power is responsible for a large share of greenhouse gas emissions, accounting, for example, for 35% of all US emissions of carbon dioxide (CO_2), 75% of sulphur dioxide (SO_2) and 38% of nitrogen oxides (NO_x). No global environmental issue comes close in terms of complexity, the scope of potential impact, and environmental significance (WBCSD 1997; EPA 1998).

Energy market liberalisation is not something new, although at least for Europe the trend towards more competitive electricity and gas markets is relatively recent. In countries such as the USA, the UK, Chile, Argentina, Norway, Australia and New Zealand, processes have already taken place aiming at liberalising electricity and/or natural gas markets and exposing them to competition (Midttun 1997).

The scope and pace of market transformation can arguably be best seen in the American electricity sector. The US$230 billion electricity market, which is twice the size of the long-distance telephone market, is the single largest US industry to be deregulated and the success or failure of this market transformation is likely to influence electricity liberalisation efforts that are currently under way in a number of industrialised countries.

Consumers[2] will be able to choose among a range of electricity service providers, services, pricing options and payment terms.[3] Competition is also expected to provide electricity consumers with a greater choice of how their electricity is produced, and this opens a new market for renewable sources of energy.[4]

2 Throughout this chapter the term 'consumer' refers to 'consumer of electricity'. Therefore, there is—in principle—no differentiation between households and other groups of electricity consumers. Non-household consumers are also eligible for subscription to green pricing programmes. However, experience shows that their willingness to voluntarily pay more for green electricity is lower than that of households (see Table 1 on page 375).

3 There should, however, be a clear distinction between the (voluntary) participation in green pricing programmes and the 'new rules of the game' of a liberalised electricity market organisation which allow electricity consumers to shop around freely for their ('conventional') electricity. In some cases, consumers are free to choose their electricity supplier only if they consume a certain amount of electricity annually; in other cases, e.g. in California, households too are 'liberalised'.

4 Compared to conventional fuels, renewable sources of energy are advantageous not only from an environmental perspective (e.g. they are CO_2-neutral), but also in terms of their economic consequences: e.g. a substitution of fossil fuels by biofuels generally reduces a country's import dependency, and the use of domestic renewable sources of energy also has significant employment effects. Effective use of these resources can help lower consumers' electricity bills, create more choices for consumers, reduce the environmental impact of providing energy service, contribute to overall system reliability, increase resource diversity, and help achieve important local, provincial and national energy goals (IEA 1997a, 1997b; EC 1997b; EIA 1997b).

In the face of competition, utilities have a growing impetus to strengthen their image with their customers and build customer loyalty in order not to lose them. Two important factors in customer loyalty are: many customers want their utilities to pursue environmentally benign options for generating electricity; and some (mostly private) customers are willing to pay an additional 'ecological fee' to receive (or even just to fund) electricity generated from renewable sources of energy.[5] From a marketing point of view, the instrument has for a long time been known as market segmentation and price differentiation (cf. Kotler and Armstrong 1996; Simon 1989). **Market segmentation** consists of the splitting of the market (in our case of an economically homogenous product, i.e. electricity) into different groups of customers with a different willingness to pay (e.g. customers with and without a willingness to pay for green electricity). **Price differentiation** is charging different prices for these different customer groups (e.g. higher electricity prices for customers with a higher willingness to pay for green electricity). Liberalisation of electricity markets has thus upgraded marketing as a key business strategy issue for many private electric utilities, independent power producers and renewable energy firms.

As fossil fuel prices are at historic low levels (EIA 1997a), green pricing can be considered as a way of raising capital for renewable, i.e. environmentally preferable, projects. With self-selecting customers subscribing to a renewable premium, electricity utilities once averse to wind, biomass or solar energy supply can now diversify their electricity generation fuel mix at no risk to their basic rate structure.

Surveys have consistently revealed a strong environmental consciousness and public preference for renewable energy—and a willingness to pay a premium (i.e. a higher price) for electricity derived from these sources.[6] This idea has been getting significant attention since its conception in 1992 (Moskovitz 1992). In this context, 'green pricing' is a relatively new scheme to find new markets for renewable energy sources and to penetrate existing markets.

While green marketing programmes[7] have been introduced for example in more than 30 states in the US (NREL 1998), there is a number of issues that need to be resolved before green marketing is able to realise its business potential. Consumers participating in such institutionalised green pricing programmes may have questions (Holt 1997a; cf. Holt 1997b; Ottman 1997): Where is the 'juice' coming from? Is it really adding renewable energy? Is it making an environmental difference? Is it fairly priced? Will the utility use my premiums wisely? Independent authorities can ensure that these questions are answered to their satisfaction, but consumers may continue

5 It must be noted, though, that some forms of renewable energy have already achieved full competitiveness. Some others are close to it (*Economist* 1998b).

6 Renewable sources of energy also play a key issue in the context of global warming (Kyoto agreement). Apart from measures to increase technical efficiency of energy use, they are another option in the set of instruments to achieve CO_2 emissions reduction.

7 Although green pricing and green marketing is used interchangeably in this chapter, it should be noted that they are conceptually different. While green marketing touts the existing environmental qualities of a given product or service, green pricing expects customers to pay an additional amount for the environmental benefits of a product or service. For instance, an electric utility in a green pricing programme may ask its customers to pay 10¢ a kilowatt more for a 10 megawatt solar/photovoltaic plant.

to have doubts. Overcoming these credibility concerns is, on the one hand, an issue of recognised eco-labelling; on the other hand, especially, a marketing issue.

Market Segmentation and Product Differentiation

Market research often includes demographic questions in order to discover whether some groups of consumers are more interested in a product than others. For example, age, income, education level, family size and geographic location are often asked in customer surveys. In addition to demographic information, some electric utilities have asked questions about customer values and attitudes, which, when combined with demographic data, has been termed 'psychographics'. The Electric Power Research Institute (EPRI) developed software and a questionnaire that a number of utilities have used or adapted in order to learn about and to classify their customers based on psychographic data. An example of market segmentation based on psychographic information is shown by research conducted in 1992 by the Public Service Company of Colorado (PSCo) for its renewable energy programme. This market research suggested three customer segments (Henrichs 1995):[8]

- **Laissez-faire individualists** (25%). These customers feel environmental problems are a natural result of progress. In their view, jobs are more important than the environment. They believe that PSCo is more effective than government in developing renewables and think that the best solution to environmental concerns is individual conservation.
- **Suspicious inequity avoiders** (36%). These customers are extremely troubled by programme free-riders who do not contribute to renewables funds yet enjoy the benefits of less pollution. They do not believe that PSCo can develop renewables better than government, and they do not believe that renewable energy costs will go down in time. Generally, they feel abused by both private and public institutions.
- **Environmental programme boosters** (39%). These customers are more concerned with results than with free-riders. They are optimistic about the availability of renewable sources of energy. They do not believe gradual destruction of the environment is the price for economic progress, and they feel group efforts are more effective than individual efforts in dealing with environmental problems.

The key point is that supporters of green pricing for electricity hold certain attitudes that distinguish them from other consumers. According to Baugh *et al.* (1995):

> customers who support green pricing believe that collective action offers the best chance of addressing environmental problems. They are focused

8 These customer groups are based on *stated* preferences in ('hypothetical') surveys. One should not confuse these data with actual (revealed) market behaviour, especially when a trade-off between environmental goals and other preferences exists (e.g. employment, households budget constraints).

> on problems of pollution and resource conservation and are . . . willing to contribute to improve environmental externalities without worrying that those who do not contribute will also enjoy the environmental benefits. They tend to favor market solutions over governmental programs, and some express a distrust of government involvement. These customers understand the profit motive and support reasonable profits for participating utilities.[9]

What is Green Electricity Pricing/Marketing?

The key aspect of green marketing in the electricity sector is providing market-based choices for electricity consumers to purchase their electricity from environmentally preferred sources. Green marketing therefore focuses on those customers whose choices depend crucially on an ecological unique selling proposition (ecological USP). Green pricing in the electricity sector is an optional service of companies that allows customers to support higher investment in renewable energy technologies by their utility company. Participating customers pay a premium (a voluntary 'ecological fee') on their electricity bill to cover the incremental cost of the additional renewable energy, and are rewarded with the assertion of ecologically sound electricity supply. Green pricing programmes typically fall into one of the following categories (Swezey 1997):

- **Renewable energy contribution programmes.** Customers can contribute to a utility-managed fund for renewable project development. The utility company does not specify exactly which renewable energy projects (e.g. electricity based on biomass or wind energy) will be realised. There are only some general guidelines on how to invest the premiums collected from the programme.
- **Tailored renewable energy projects.** The utility identifies a particular renewable project for which it solicits contributions. After it receives some minimum number of subscriptions, i.e. a minimum number of customers willing to pay a premium for this specific source of renewable energy, the utility realises the project. Subscriptions can take the form of fee-type contributions to a fund as well as purchase of capital shares of a separate utility company producing green electricity.[10]

---

9 This section applies only to household consumers. Non-households' willingness to pay for green electricity is lower than that of households (see Table 1 on page 375).

10 A prominent recent example is that of DonauWind GmbH (Vienna). This private company offers customers capital shares; the funds are then used for the construction of wind power stations. The electricity produced is sold to the leading provincial electricity supplier at a price of around 0.70 Austrian schillings. The investment is subsidised by the Austrian Environmental Fund according to federal law; the payback period for investors is between 7 and 10 years; the internal interest rate is about 7% p.a. (the capital cannot be withdrawn within the first 15 years after purchase of capital shares).

- **Renewable electricity grid service.** The utility may acquire electricity from a number of (independent) renewable energy projects with other electricity sources for sale to customers.

Main issues when designing a green electricity pricing programme include (cf. Holt 1997b):

- How can the services of the utility company be eco-labelled? What forms of green electricity or green benefits will be sold to customers, and how can green electricity generally be defined?[11]
- How can tangibility for green electricity be created? What are the direct and indirect benefits of green electricity for customers?
- How can the programme credibility be enhanced?
- How can customers be motivated to buy indirect benefits, e.g. contribution to sustainable development with respect to possible global warming?
- Should business customers be included in green pricing programmes? Can a green electricity price for them be an additional marketing argument?
- Which legal or institutional arrangements securing the programme can be adopted?
- Which communication strategies can be developed to assure customer satisfaction?

Despite these difficult questions, which also raise a number of problems in terms of marketing transparency and ethics (Holt 1997b; Ottman 1997), green electricity marketing demonstration programmes and market surveys show that between half and three-quarters of residential customers are willing to pay a 5%–15% premium over their current electric bills for the satisfaction of purchasing green kilowatts (Holt 1997b).

Green Marketing and Eco-Labelling for Electricity

Green electricity marketing programmes have been implemented in recent years, for instance, in New Hampshire, Massachusetts and California. While they are by no

11 Green electricity can simply be defined as 'renewable energy—power from the sun, wind, plants, and moving water' (Union of Concerned Scientists 1998), which avoids such power sources as coal, oil and nuclear energy. However, contrary to the simple definition, green electricity is difficult to label because some renewable resources (e.g. hydropower) can have environmentally harmful effects (in the case of hydropower, reduced fish migration and threatened natural habitats). For environmentally friendly products, the US Federal Trade Commission (FTC) has issue standards for such terms as 'recyclable', 'recycled' and 'compostable' for green marketing and advertising purposes. FTC's 'Green Guides' were first issued in 1992 and revised again in 1996 (FTC 1996).

GO-GREEN
CUSTOMER INFORMATION
National Renewable Energy Consumers Council

System Power: Where Your Electricity Comes From
(Based on actual generation da/mo/yr through da/mo/yr)

FUEL FACTS	
Coal	50%
Natural Gas	35%
✓ Solar	1%
✓ Wind	14%
Total	100%

✓ Solar and wind energy are certified as
Environmentally Preferable
by the National Renewable Energy Consumers Council

FOR MORE INFORMATION, CALL 1-800-GO-GREEN

- Credit for this label, originally designed on the back of an envelope, belongs to Karl Rabago, Environmental Defense Fund.

Figure 1: POWER CONTENT LABEL FOR ELECTRICITY

means the only states in the US with green electricity marketing programmes, they do illustrate the scope and diverse approaches of existing programmes.

California

A number of the above questions regarding standards and eco-labels have been addressed by the new Californian legal framework on the deregulation of the electricity market (CEC 1997): Since 1 April 1998, California's electricity industry has been open to competition. Regulation requires that electricity retailers inform consumers about the energy sources of the electricity they are selling. Retailers must at least describe to consumers what sources comprise the state-wide electricity generation mix. Companies claiming to sell electricity from specific sources must disclose the source types to consumers and must be able to verify this information (standardisation of quality and product information). Electricity retailers must present this information to consumers in a format that allows consumers to compare the variety of electricity products offered to them (Wiser and Pickle 1998; cf. Salpukas 1998). This format, which could be described as a power content label for electricity (eco-labelling for electricity), may look much like a 'nutrition label' on food products (see Fig. 1).

***Green-e* programme.** A group of green electricity marketers has joined with a group of California consumers and environmental stakeholders to launch the country's first voluntary certification and verification programme for environmentally preferred electricity, i.e. electricity that is produced from at least 50% of renewable sources such as hydropower, wind energy or biomass.[12] The programme's centrepiece, the *Green-e* logo, will help consumers identify this green electricity easily. To the extent that any fossil fuel resources are used, those resources must have air emissions per kWh for SO_X, NO_X, and CO_2 less than or equal to the state-wide system electricity generation mix. In addition, nuclear energy beyond that encompassed in system power may not be included. A number of companies has been certified under the *Green-e* programme (see Box 1).

The **Renewables Portfolio Standard** (RPS) sets a minimum renewable energy requirement for a state's electricity mix, under which every electricity supplier must provide some percentage of its supply from renewable energy sources. In some cases, a tradable credit system is employed, under which electricity suppliers can buy and sell renewable generation credits in order to meet the requirement.

▢ *New Hampshire*

Unlike other retail electricity programmes in Illinois, Massachusetts or New York, where the programmes were proposed by the electric utilities, the New Hampshire Legislature initiated the deregulation process in consultation with the New Hampshire Public Utilities Commission. Under the direction of the state public utilities commission, a total of 35 competitive suppliers registered for the retail electricity pilot programme and 16,500 customers (14,765 residential, 1,728 commercial, and 16 industrial customers) became eligible to participate in the programme (Holt and Fang 1997).

Although it is too early to draw any firm conclusions from the New Hampshire retail electricity pilot programme, two important lessons can be noted. First, without a strong disclosure requirement and a public education campaign, companies may be tempted to oversell the environmental features of its power source and diminish the overall credibility of the green marketing programme. For instance, Granite State Energy, one of the competing firms in the New Hampshire electricity market, promoted its environmental commitment heavily, although it gets its power from an electric utility that relies on coal, nuclear power and oil for more than half of its total electricity generation. Despite the use of marketing and licensing tie-ins such as providing a free booklet with tips on energy conservation to project an environmentally friendly image, it is debatable whether Granite State Energy's green electricity product is as environmentally sound as the image the company is trying to project (Holt and Fang 1997).

Second, a well-crafted green marketing campaign needs to have a strong environmental perspective. A good example of a strong environmental perspective can be seen in the green marketing strategy used by the Vermont-based retail power company, Green Mountain Energy Resources (GMER). Although GMER was only one of 35 power suppliers competing for 16,800 residential customers in the state of New

12 For further detailed information on the *Green-e* programme, please refer to the programme's website (*http://www.green-e.org*).

Edison Source

Edison Source is a subsidiary of Edison International, which also owns South California Edison, a regulated public utility. Edison offers two main green electricity products, 'EarthSource 50' and 'EarthSource 100'. EarthSource 50 has 50% renewable energy content and its base price represents a 1.16¢ per kWh or 11% premium over its base power price, while EarthSource has 100% renewable energy content and its base price represents a 2.33¢ per kWh or 23% premium over its base power price.

Source: company marketing literature, US Environmental Defense Fund

Enron Energy Services

A subsidiary of the Houston, Texas-based Enron Corp., widely recognised as one of the largest natural gas/energy companies in the world, Enron Energy Services has been one of the most aggressive power companies in California. Unfortunately, Enron Energy Services was able to convince only 30,000 customers (out of a potential 10 million) to switch to its electricity products and decided temporarily to stop selling power to residential customers despite spending more than $5 million on marketing its electricity products. Enron offers one green electricity product, 'EarthSmart Power', which has 50% renewable energy content and whose cost represents a 1¢ per kWh or 10% premium over the base power price. The company has also promised its customers that it will build a new wind farm to service its green electricity customers.

Source: company marketing literature, US Environmental Defense Fund, US Natural Resources Defence Council

Green Mountain Energy Resources

Owned in part by the Vermont-based private utility Green Mountain Power Corporation, Green Mountain Energy Services offers three green electricity products: 'Water Power', '75% Renewable' and 'Wind For the Future'. Water Power has a 50% renewable energy content and its cost represents a 0.975¢ per kWh or 9% premium over its base power price. As the name implies, 75% Renewable has 75% renewable energy content and its cost represents a 1.2¢ per kWh or 11% premium over its base power price. Wind for the Future, for which the company has promised to build a new wind turbine for every 3,000 additional customers, has a 75% renewable energy content (10% of which is new power capacity) and represents a 2.1¢ per kWh or 20% premium over its base power price.

Source: company marketing literature, US Environmental Defense Fund, US Natural Resources Defence Council

Box 1: EXAMPLES OF POWER COMPANIES CERTIFIED BY THE CALIFORNIAN GREEN-E PROGRAMME

Hampshire, the company captured the public imagination with its marketing slogan, 'Choose Wisely. It's a Small Planet', and by publicising that 90% of its energy supply came from hydropower sources. They also sent a pair of beeswax candles (along with a note advising the customers to use them for a candlelit dinner on 'an occasion to slow down and enjoy life's simple pleasures') to all of its new power customers and established the *EcoCredits* programme, through which customers can earn deductions on their electricity bills for volunteering for environmental causes similar to the way consumers earn mileage awards on airline frequent-flyer programmes (Lamarre 1997).

Massachusetts

Massachusetts Electric Company (MEC)'s retail electricity pilot programme was unique in that it offered a 'green' option in the programme design. To offer power consumers the broadest possible range of service and product choices, MEC's programme asked electricity suppliers to include three service options: price, green, and other options. The price option offered customers the lowest energy price, while the green option offered customers environmentally sound energy choices. The other option offered value-added services such as energy conservation services and donations to charitable organisations. The main objective of presenting the three options was to encourage innovation and broaden the spectrum of choices available to the customer (Rothstein and Fang 1997).

Two green marketing lessons can be drawn from MEC's retail electricity programme. First, public education needs to be firmly integrated into the company's overall marketing strategy in order to build and sustain consumer environmental awareness. Because many power consumers may not be aware of and/or misinformed about the environmental dimensions of green electricity, power companies need to spend more time thinking about how to integrate public education issues, including the environmental benefits of green electricity, into their overall business strategy. For instance, Enova, one of the power marketers competing in the Massachusetts electricity market, provided its customers with an extensive amount of environmental information such as energy conservation tips, a home environmental survey, as well as an 'Earth Saver' kit containing a re-usable grocery bag and a camera to allow customers to document their environmental initiatives.

Second, it is important to give power customers a wide range of choices in terms of price and renewable energy content. By adopting a multi-tier marketing approach, a power company can give its customers a variety of choices in terms of electricity prices and equivalent environmental support. For instance, a power customer may decide to pay a premium price for electricity that has 75% renewable energy content whereas some customers would be satisfied with paying for less expensive power that has only 50% or 25% renewable energy content. A multi-tier marketing approach allows power companies to identify and analyse the environmental and price sensitivity of their customer base.

Australia. The Sustainable Energy Development (SEDA) in New South Wales, Australia, launched the Green Power Accreditation Programme in 1997. The goal of the programme is to facilitate the installation of new renewable energy projects in New South Wales by increasing consumer confidence in green power schemes developed by electricity retailers.

Canada. Ontario Hydro announced its Clear Choice green energy pilot programme in 1997, through which its business and institutional customers will be able to purchase all or a portion of their electricity from wind, solar, landfill gas and small hydro generation. Participants will receive emission reduction credits related to their support and will be entitled to use the Eco-Logo trademark that certifies Clear Choice projects as 'green'.

The Netherlands. Two major Dutch energy distributors, PNEM and EDON, have developed and are expanding green pricing schemes designed to stimulate renewable energy production. PNEM launched its programme in 1995 and attracted 7,700 customers by the end of 1996, accounting for 33,000 megawatt-hours of green electricity. All green power is currently supplied by wind energy, and PNEM has set a goal of 100,000 green users by 2000.

Sweden. The Swedish Society for the Preservation of Nature launched a green power certification programme in 1997 to evaluate environmental claims made by power suppliers. The criteria for certification requires that green electricity be produced by older hydro plants, wind power or biofuels, while nuclear, peat and the burning of waste do not meet the criteria. To date, six utilities have been certified and another 20 utilities have expressed interest in becoming certified.

United Kingdom. Eastern Electricity, one of the UK's largest suppliers and distributors of power, launched a green power scheme in 1997 by which three million of its customers can elect to pay a 5% or 10% premium under the 'EcoPower' and 'EcoPower Plus' tariffs to support the development of renewable energy generation. The money will be used to establish a fund that will help support R&D of renewable energy technologies

Box 2: INTERNATIONAL EXAMPLES OF GREEN ELECTRICITY MARKETING PROGRAMMES

Source: US Department of Energy 1998

Overview over International Experiences in Marketing Green Electricity

The overview in Box 2 gives an impression of the broad application of green electricity marketing approaches in an international context.

The Basis for Green Pricing: Consumers' Willingness to Pay for the Environment

The concept of willingness-to-pay (WTP) for valuing environmental goods is one of the core issues in environmental economics. Environmental goods such as clean air, biodiversity or environmentally sound products are normally not traded in markets. Theory of market failure explains why they cannot, in general, be traded like consumer goods because there are no prices assigned to them. In these cases, non-market valuation techniques such as 'hedonic pricing' or 'contingent valuation' are employed to clarify consumers' preferences in monetary value. By means of hedonic pricing methods, the demand for environmental goods (e.g. clean air) is derived by the market demand for complementary consumer goods, e.g. housing prices. While these methods are indirect methods, contingent valuation directly tries to investigate consumers' preferences by asking them about their willingness-to-pay for ecological goods. What is true for consumer goods (before introducing new products in the market, the product is tested in hypothetical and experimental settings), is especially applicable to public goods.

By means of product differentiation and price management, utility companies try to 'skim off' consumers' willingness to pay for ecological goals by providing them with green electricity at higher prices than 'conventional' electricity. The idea behind this strategy is to privatise ecological benefits: consumers feel a duty to contribute to ecological goals even if their single share of total electricity consumption and pollution is very small.

Ultimately, all inhabitants of the globe, whether programme participants or not, accrue the benefits of a green pricing programme. But there are also benefits only participants will receive. Specifically, only those utilities and customers choosing green pricing will financially benefit should more restrictive environmental regulations be set in the future or should fuel prices rise.

Another aspect is the 'public goods' component of electricity generation and environmental quality. That is, everyone is affected by the environmental damage caused by energy production and, conversely, everyone benefits when the system becomes cleaner. Those who choose to pay more for clean energy, in effect, subsidise those who don't (Wiser and Pickle 1997).

There are many concepts explaining consumers' WTP for ecological goals, one of them being the so-called 'warm glow' effect, which leads to a positive willingness to pay for ecological goals because consumers feel a certain social pressure to behave ecologically. Another explanation is the 'purchase of moral satisfaction' a consumer achieves when helping to improve the environment. Finally, altruism and paternalism as main motives for a general concern for others, including the environment, can explain a positive WTP.

Green pricing strategies are often founded on many ecological motives at once to overcome the free-riding argument and to attract customers in order to increase the share of environmentally sound products and services.

Examples of Consumers' Willingness to Pay for Green Electricity

> Price is not the only factor customers consider. Our lowest-priced power product is our least popular . . . People want to buy cleaner electricity (Blunden 1998).

In this section we will present a number of recent examples of consumers' willingness to pay for green electricity. There are successful programmes as well as programmes that have been abandoned due to only very marginal participation. What seems to be very important, however, is that the green electricity programme established and sold to customers has to be honest in terms of its ecological quality, and it is crucial to communicate this quality to customers. We begin with a broad description of two examples from the US, and then we present a short summary of selected green pricing programmes.

Sacramento Municipal Utility District (SMUD)

Since 1993, SMUD has operated the *Photovoltaic (PV) Pioneers* programme, under which customers can choose to pay a $4.00 flat monthly fee (for 10 years) to have a 2–4 kW, grid-connected PV system installed on their rooftops. SMUD installs, operates, maintains and owns the hardware. As of May 1997, 420 residential and 20 commercial systems had been installed. Although total installations have been limited to approximately a hundred systems per year, SMUD receives approximately a thousand new applicants annually. In June 1997, SMUD announced a *PV Pioneers II* programme, through which customers can purchase PV systems and sell excess electricity back to the utility. SMUD will 'buy down' half of the $17,000 system cost (SMUD 1995).

Looking toward the competitive retail market in California, SMUD developed a new 'green rate' that will allow its customers to obtain 100% of their electricity needs from renewable sources. SMUD already meets nearly half of its power needs with renewables. The *Renewables Energy Option* gives customers the option to buy all of their electricity from new grid-based renewable resources at a rate premium of 1.0¢ per kWh. Participating residential customers would see an average bill increase of $7.00 per month. SMUD recently announced that it will purchase electricity from a planned 8.3 MW landfill gas plant—one of the first newly constructed renewables resources to supply electricity for this programme. The *Community Solar* programme allows customers to contribute 1.0¢ per kWh for the purchase and installation of PV systems

Customer group	Additional WTP (premium) in per cent of customers' electricity bill			
	5% premium	*10% premium*	*15% premium*	*20% premium*
Residential	43%	27%	16%	7%
Business	38%	20%	10%	3%
Industrial	8%	0%	0%	0%

Table 1: PERCENTAGE OF CUSTOMERS WILLING TO PAY MORE FOR SMUD TO INVEST IN RENEWABLE SOURCES OF ENERGY

Source: SMUD 1995

on schools, churches and other community facilities. As of September 1997, more than 1,100 customers had signed up for the programme.[13]

As can be seen from Table 1, the participation rate of different customer groups varies significantly by premium levels and customer groups. Not surprisingly, the higher the premium the lower the participation rate. This pattern applies to all customer groups. Private and business customers ('small customers') are more willing to pay for green electricity than industrial customers, with the highest participation rate in the private household sector. It can be concluded from the SMUD experience that private households are more likely to participate in green pricing for electricity. One interpretation of the results is that households are a group of customers for which the price elasticity of electricity demand seems to be lower compared with other customers groups, which is not surprising because of the higher cost-awareness and competitive pressures in the non-household sectors.

◻ Massachusetts Electric Company (MEC)

A telephone survey of 403 residential customers was conducted in September 1994 asking MEC customers for their willingness to participate in green pricing programmes. A summary of customers' willingness to pay at a range of different premium levels is shown in Table 2. A follow-up mailed survey was completed with 100 of the telephone respondents who were either positive or non-committal in the telephone survey. These customers were sent a full-colour brochure describing the programme, providing information about three renewable sources of energy under consideration and about the utility's current energy mix, and specifying the premium as 1¢ per kWh, or about 10% of the monthly bill. After reading the brochure, 15% were

13 From a more general perspective, not only supply-side options (such as conventional versus renewable fuels) would have to be considered but also options on the demand side (e.g. more energy-efficient appliances such as energy-saving light bulbs). The overall optimisation of supply- and demand-side options is called Integrated Resource Planning (IRP).

Likelihood of participation	Premium level			
	Unspecified	*5%*	*10%*	*20%*
Definitely	5%	12%	6%	3%
Probably	44%	36%	26%	10%
Don't know	14%	12%	11%	11%
Probably not	21%	14%	19%	21%
Definitely not	16%	24%	37%	56%

Table 2: PERCENTAGE OF CUSTOMERS WILLING TO PARTICIPATE IN THE MEC GREEN PRICING PROGRAMME AT DIFFERENT PREMIUM LEVELS

Source: Willard & Shullman 1994

certain or almost sure they would participate, 24% very probably or probably would participate, and 35% were neutral. The proportion who were relatively certain about their commitment level was essentially unchanged from the level obtained in the telephone interview. However, the level of commitment among those who initially claimed they would probably participate decreased considerably.

Overall, a conservative estimate from this research is that 5% of MEC's residential customers may participate at a 10% premium level, and 10% of those customers will participate at a 5% surcharge.

Other Examples of Consumers' Willingness to Pay for Green Electricity[14]

Niagara Mohawk (New York) developed a *GreenChoice* programme, through which residential customers could elect to pay a $6.00 fixed monthly premium to fund the development of renewables and a tree-planting programme.

The **Kansas Electric Utility Research Programme** (KEURP) conducted phone surveys indicating that 4%–7% of customers are 'very likely' to participate in a green power programme at a contribution level of $1.00 per month.

The **Wisconsin Public Service Company** (WPS) has established a *SolarWise for Schools*™ programme, funded by voluntary customer contributions, through which

14 The following examples are taken from the website of the Electric Power Research Institute (EPRI), Palo Alto, CA (*http://www.epri.com/gg/renew/green/greenprice.html*). See also the overview of the US Department of Energy (DoE) at their website (*http://www.eren.doe.gov/greenpower/summary.html*).

PV systems will be installed on high-school rooftops in communities served by the utility. The schools will receive the power produced by the array and a student curriculum on solar energy. The utility offers monthly contribution rates of $1.00, $2.00 and $4.00; the average monthly contribution to date has been $1.70. Community and educational aspects of the programme are major selling points, as is the tax-deductible nature of the customer contribution.

Detroit Edison is offering a solar energy service to residential and small commercial customers. Customer surveys have found that 45% of respondents believe that the impact on the environment of providing electric service is an important issue. For an additional $6.59 per month, participating customers receive 100 watts of service (140 kWh per year) from a planned 28.4 kW PV facility. As of January 1996, 248 customers had signed up to participate in the *SolarCurrents*™ programme, which was enough to fund an initial project and establish a waiting list for a second project.

Traverse City Light & Power (TCL&P) is deploying a 600 kW wind turbine as a result of its green pricing programme for residential and small commercial customers. A total of 145 residential and 20 commercial customers (3.1% of the customer base) are paying a 1.58¢ per kWh premium in support of the green power, representing a 17%–25% increase in the average monthly bill.

Requirements for an Effective Green Market

In economic theory, preconditions for efficient markets include perfect competition, full information, rational behaviour of households and firms (maximising their utility and profits, respectively), absence of market failure and low barriers to entry to and exit from the market. Two of these requirements are of particular importance in the context of green pricing and green electricity marketing:

- A competitive market in which customers can choose between different suppliers and differentiated goods that best meet their preferences is one of the core elements of a functioning green market. Lack of opportunities leads to a lower demand for green products if the price of these products is above consumers' WTP for ecological goals.
- Comprehensive information actively communicated by electricity companies. In this context, certain standards controlled by independent bodies such as environmental groups are as important as eco-labelling rewarded by public authorities.

Some of the main risks to green pricing should also be briefly addressed:

- Changes in consumers' preferences can endanger the performance of green pricing programmes (e.g. tight private budgets during times of economic slumps).

- Changes in the energy context that render green electricity unprofitable (e.g. lower fossil fuel prices lead to higher necessary ecological surcharges on consumers' energy bills).

Conclusions: Lessons for Green Electricity Marketers

Marketing and education (communication and information) are essential to the success of green pricing. It is a new concept to most consumers who are, for the most part, not even thinking about customer choice of electricity. Just because large numbers of customers say they will pay more for green electricity, sponsors should not expect them to beat down the doors to get it. Marketing renewable energy is not a one-time sale. It is a process of working with customers over a long period.

One of the basic principles of efficient market operation is that consumers possess enough information to make educated purchase decisions. As a component of restructuring, requirements for uniform information disclosure that would provide consumers with objective information on the price and other important attributes of an electricity product or service, such as fuel mix and environmental emissions, are considered. Thus, the future success of green electricity marketing programmes are likely to be determined by three key issues: integrating education into the marketing strategy, utilising the ecological and strategic dimensions of information technology, and maximising customer value.

Integrating education into the marketing strategy. With many residential, business and public-sector power customers choosing their electricity providers for the first time, power companies have to contend with a relatively high degree of ignorance and misinformation about green electricity. Public education can take many forms, including a toll-free number and information brochures. However, it must ultimately explain succinctly the environmental value and message contained in the green electricity products of the respective power companies.

If a certain green electricity product is more expensive than a similar item offered by a competitor, most consumers and companies are not likely to accept the higher prices without a corresponding increase in the percentage of renewable energy content. Many consumers who do not ultimately choose green electricity products due to price or other reasons may nevertheless be sensitive about the relative greenness of the company from which they buy their electricity (Lamarre 1997; Voien 1997).

Pokorny (1994) writes about the selling of energy efficiency but, paraphrased, it applies equally well to renewable energy:

> To be successful over the long term, a utility or other marketer of renewable energy must develop a special relationship with its customers. Customers must feel that they and the supplier are working together in a kind of

partnership to achieve mutual and shared environmental or other goals. Successful companies are realistic and patient. They know they are trying to alter—in some cases dramatically—the way their customers over many years have come to think about energy and its use.

Utilising the ecological and strategic dimensions of information technology. A number of electronic commercial enterprises have been established in anticipation of a growing demand for high-quality information on green electricity products and services. Massachusetts-based Nexus Energy Software established an information clearing house for electricity products through which consumers would be able to analyse, monitor and ultimately choose their electricity through the Internet.[15]

While it remains to be seen if electricity can be sold electronically like books and compact discs, electronic commerce has the potential to be very useful in marketing green electricity as it has the potential to reach a more diverse set of technology-oriented stakeholders and to emphasise the ecological benefits of doing business on the Internet compared to traditional business practices.

Maximising customer value. 'Value' may be one of the most overused and misused terms in marketing today. Although it is often misused as a synonym for low price, the real essence of value revolves around the trade-off between the benefits customers receive from a product and the price they pay for it. More precisely, customer value equals customer-perceived benefits minus customer-perceived price. The management of this trade-off between benefits and price has long been recognised as a critical marketing component, but the concept of customer value also needs to be defined within the context of the specific market (Leszinski and Marn 1997).

In the green electricity market, value centres on the 'environmental benefits' customers perceive that they are getting from their power purchases. Customer value in the context of green electricity marketing is a dynamic concept that takes into account market price relative to the environmental benefits generated. One of the most important lessons learned in green electricity marketing in the US is that customers are willing to pay a premium for green electricity, but this willingness to pay for greater ecological value depends a great deal on how well the power companies can document and market the environmental benefits of their green electricity products.

Like the economic benefits of energy efficiency programmes, power and non-power companies are beginning to discover the market potential of green electricity. Power companies such as Green Mountain Energy Resources are learning that an effective green marketing strategy means much more than having a catchy slogan and distributing brochures with colourful trees and birds, while non-power companies such as Toyota Motor Corp. are realising that the proactive search and the purchase of green electricity can be an important component of their overall global environmental strategy.

---

15 See their website at *http://www.energyplace.com*. By logging onto Nexus's Internet website, a customer is able to search and select an energy supplier and learn more about energy efficiency, deregulation and an assortment of environmental topics.

Today, green pricing is a product for a niche market. Depending on the product concepts and how it is packaged, it may be a portfolio of green products, each of which meets the needs of a small but different market segment. Like many other innovations that have penetrated only a small share of the market, all these market segments add up. With time, green electricity may become mainstream, and the market will be transformed.

25

Exploring Organisational Recycling Market Development

The Texas–Mexico Border

John Cox, Joseph Sarkis and Wayne Wells

DEVELOPING SUSTAINABLE ECOSYSTEMS is a goal of many communities. From an organisational perspective, the involvement within, and development of, an industrial ecosystem needs to be accomplished from a proactive, competitive-advantage dimension. Thus, forging an industrial ecosystem should not be left to chance. Co-operative relationships among industries, government and academia, as well as other stakeholders, are a necessary element in this development (Leake and Kainz 1994). The tools and business conditions for forming relationships within an industrial ecosystem are beginning to appear within the environmental practices of organisations. Green marketing should play a key role in making these systems effective and efficient. Green marketing aids in the development and management of the distribution channels as well as building markets for environmentally sound materials. Even though the green marketing literature has grown in recent years (Kilbourne and Beckman 1998), research into industrial ecosystems has not been focused on, as, typically, this research and practice has been the domain of management strategists, engineers, economists and operations researchers and practitioners. A cross-disciplinary approach, which includes marketing, is required in the development, operations and maintenance of these systems. The survey presented in this chapter provides some results and a structure that would benefit the management of green market development, either for private green marketers (e.g. industry) or public green marketing developers (e.g. state governments).

This chapter focuses on the initial results of a study examining the market for recyclable materials and products. A review of environmental practices from an internal and external 'marketing' perspective sets the foundation. The work in this project is targeted at the critical and sensitive region along the Texas–Mexico border. A brief review of the results provides some insights into other inter-organisational programmes that aim to achieve similar goals. The results also point to numerous managerial implications.

Project Background

With the North American Free Trade Agreement (NAFTA) in full force, worries about environmental practices and degradation have come to the forefront of policy decisions that concern operations along the Texas–Mexico border. From an economic perspective, the most beneficial approach for developing an environmentally sustainable market system between the United States and Mexico is the formation of regional industrial ecosystems along border communities. The Texas Natural Resources Conservation Commission (TNRCC) supported an initial study on the US side of the border to evaluate some of the practices and requirements regarding recycled products, which aided in the development of a market for recycled products. The results were also used to improve the TNRCC's 'RENEW' programme (a state-operated waste exchange programme).

The RENEW programme, which is a marketing initiative by the TNRCC, has been improved by determining the potential market for recycled products along the Texas–Mexico border. Findings of a sizeable market will motivate the TNRCC to pursue and invite additional organisations to complement the recycling market needs in this region, expanding the RENEW programme. The survey also asks respondents about their familiarity with the RENEW programme, establishing a market penetration baseline for the RENEW programme in this region.

This chapter summarises some results of a survey for determination of market requirements and practices focusing on recycled materials and products. These results provide general insights into the green practices of small and medium-sized organisations from 20 industries, as classified by two-digit standard industry classification (SIC) codes. The responses indicate the existence of various industrial practices and/or needs from the perspective of recycling and a market for recycled products. To set the foundation for the survey and general industry greening practices (and their link to green marketing), a brief review of industrial ecosystems and their relationships to recycling market development is presented. Put simply, without development of green markets and efforts of green marketers, it will be difficult, if not impossible, for industry ecosystems to exist.

Industrial Ecosystems and Recycling Markets

Lowe (1993) defines industrial ecology as

> a systematic organizing framework for the many facets of environmental management. It views the industrial world as a natural system—a part of the local ecosystems and the global biosphere. Industrial ecology offers a fundamental understanding of the value of modeling the industrial system on ecosystems to achieve sustainable environmental performance.

Industrial ecology (ecosystems) has been described on three levels (Jelinski *et al.* 1992). The first is a linear materials flow system. Input into the linear materials flow system is represented by unlimited materials and resources that are transformed by the ecosystem component with unlimited waste as an output. The linear materials flow system assumes unlimited resources and capacity for formation of unlimited waste.

The second-level ecosystem is a quasi-cyclic materials flow ecosystem. The quasi-cyclic materials flow ecosystem has energy and 'limited' resources input which is transformed by ecosystem components with limited waste flow output. The third-level system has energy input with ecosystem components linked together and no waste outflows. The third-level ecosystem is an idealistic level of attainment that is sustainable as long as energy flows into the system.

An example of a quasi-cyclic materials flow ecosystem model is shown in Figure 1. The elements within this system include materials extractors and growers that supply raw materials to manufacturers and processors. The outputs of these manufacturers and processors are used by consumers. These elements produce outputs that are transformed by waste processors, eventually to be used by other ecosystem components.

A number of real-world, industrial ecosystem development programmes currently exist. Most of these systems are based on relationships that have been arranged via private business-to-business agreements. Local, state and federal governments are also involved in the formation of industrial ecosystems through support of industrial 'ecoparks' and setting up marketing relationships with various users and suppliers of waste products in close geographical proximity. For example, the President's Council on Sustainable Development in 1995 supported pilot Eco-Industrial Park (EIP) programmes in Chattanooga, Tennessee, and Brownsville, Texas. The EIPs serve as models for extracting policies and technologies; they are designed to group a variety of manufacturing and service businesses in an industrial ecosystem in which waste products of one company serve as the raw material for another.

There are currently between 20 and 25 EIPs in operation or being studied by US communities (Dwortzan 1998; Indigo Development 1998). This concept has also gained international importance (Erkman 1997). The process of developing an industrial ecosystem still requires efforts from the major groups shown in Figure 1, as well as other stakeholders that define boundaries and provide resources, such as communities and governments. Many institutions and organisations have been facilitating the development of these systems either through publishing sources or resources, brokering, or developing partnerships.

Organisational Industrial Ecosystem Elements

There are a number of elements and approaches for evaluating an industrial ecosystem. One method is to consider it from an organisational perspective by evaluating the linkages as part of a multi-organisation supply chain (see Walton *et al.* 1998 and Sarkis 1995a for a detailed discussion on environmental inter-organisational and supply chain issues). Figure 2 shows the core elements of one link in this

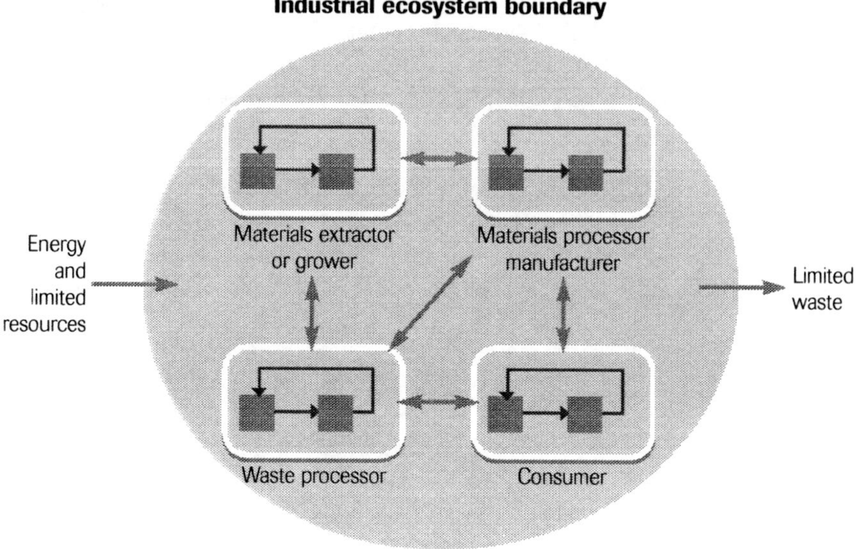

Figure 1: A GENERIC QUASI-CYCLIC INDUSTRIAL ECOSYSTEM FRAMEWORK

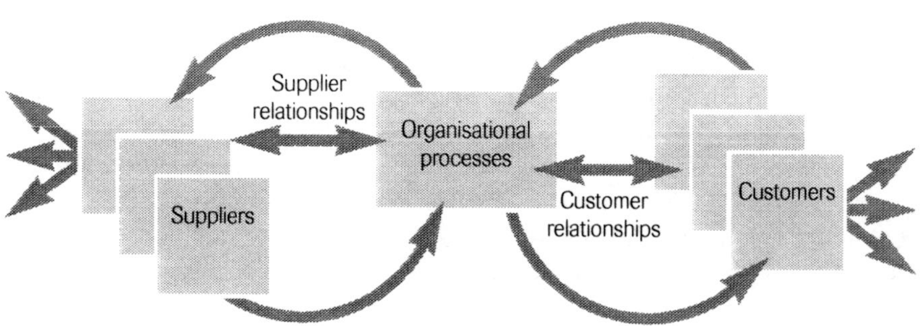

Figure 2: ELEMENTS OF AN ORGANISATIONAL SUPPLY CHAIN

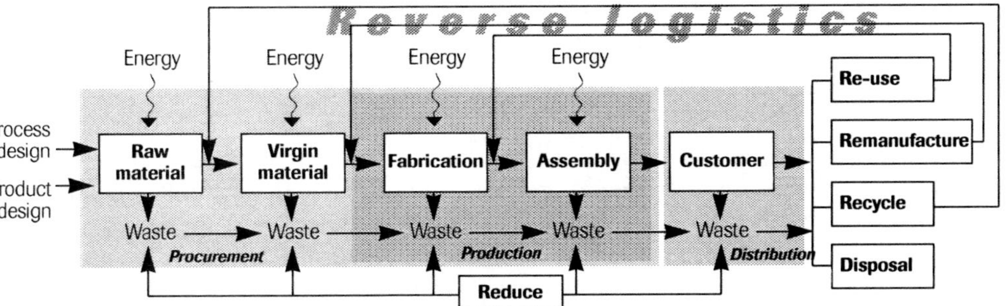

Figure 3: OPERATIONAL DEVELOPMENT LIFE-CYCLE FOR MANUFACTURING OPERATIONS

Source: Adapted from Sarkis 1995b

supply chain. The supply chain considers internal organisational processes and relationships that need managing, in addition to external linkages. The external linkages include the relationships with customers and suppliers.

The internal processes for a typical manufacturing organisation are shown in Figure 3. In this figure, each function impacts products, processes and materials within the operational system. Each function requires some form of energy with an ultimate goal of minimising energy and material usage and waste generated. This is the 'reduction' portion of the internal process and system. The outputs of this process can be environmentally managed (after product use by customer or from waste in the system) through remanufacturing, recycling, re-use or disposal. For these materials or products to return to the system, a 'reverse logistics' flow needs to exist.

Recycling Environmental Management Approaches

There are numerous strategies for the reprocessing and recycling of materials, requiring varying process and economic considerations. An example of three categories for recycling of manufactured products include (Rennie and MacLean 1989):

- Retaining all properties embodied in the original product, where the characteristics of the product are used over again in the manufacturing process and the material is completely sourced from the recycled material, i.e. glass bottles being manufactured again as glass bottles (re-usability level).
- Recycling of wastes into alternative products; the products are manufactured from recycled and raw material, i.e. making fibreglass from glass bottles.

- Processing of recycled material into products that are not usually made of the material, where all properties of the original product are all lost: i.e. glass to glassphalt.

The link between recycling, remanufacturing, and re-use can be defined by the amount of treatment required: minimal treatment of a material is more closely associated with re-use of a product, while a material that requires a large amount of treatment can be considered to have undergone recycling. Re-use involves the on-site or off-site use, with or without treatment, of a waste. Re-use generally refers to the introduction of waste directly to a process in place of the normal raw material. The axiom that one group's waste is another's treasure is most true for re-use. The necessary materials for various processes within an organisation (or even among organisations) can be researched to determine whether various wastes can be utilised in the system, with little additional processing. Remanufacturing falls within this spectrum.

Remanufacturing refers to the repair, rework or refurbishment of components and equipment to be held in inventory for either external sale or internal use. In a typical remanufacturing process, identical 'cores' (the worn-out components and equipment) are grouped into production batches, completely disassembled and thoroughly cleaned. These parts are replaced at least back to the level of the product when new. The product is assembled, finished, tested packaged and distributed in the same manner as new products. The remanufacturing process basically includes the disassembly of components, inspection and testing of the remanufacturable components, incorporation of any new improvements, and reassembly of components with newer systems.

Recycling Market Development Issues

A number of issues arise when seeking to develop a market for recycled products and materials. These issues are both opportunities and barriers to improving marketing and organisational management for a recycling environment. They include internal and external issues, with both a technological and managerial dimension. A summary of some of these includes:

Internal issues

- How can organisations alter the design of a product for increased recycling capabilities?
- What are the product and material characteristics that will increase the recycling and re-use of products/materials?
- How can internal processes and policies be altered to incorporate equipment and technology for recycling and re-usability?
- Systems need to be developed that will help trace, record and access recyclable products/materials and their sources.

- Managers need to be able to determine the financial and economic implications of implementing a recycling strategy.
- Technological planning, control and processing requirements of a recycling environment need to be evaluated.
- Special tooling and operations technology for disassembly and other in-process recycling measures need to be introduced and evaluated.
- What is the cost-effectiveness and feasibility of introducing recycled materials and products and do economic models exist to help make that decision?

External issues

- Levels of acceptance for recycled products and materials by consumers and industrial customers need to be increased.
- Reverse logistics channels need to be developed, made more efficient and more available.
- Customers and suppliers need to work with each other to aid in the design of products and processes to increase the capability for the introduction of recycled materials.
- Government and communities need to search and find compatible organisations that can use and re-use one another's material outputs.
- Education and publicity of recyclable products and markets need to be further developed.
- Types of material available and markets for these materials need to be identified and encouraged to develop.

This study does not intend to address all these issues, although many of them are examined.

The Survey and Respondents

The survey instrument contains three sections: the first section comprises organisation identification questions, i.e. size, type, contacts within organisation; the second section pertains to business recycling practices; and the final section focuses on specific materials and the organisational recycling of those materials. This chapter focuses its discussions on the responses to the second set of questions. The nine questions, of the second section, are meant to evaluate three dimensions that are important to developing a market for recycling material (and are based on the elements of Fig. 2). The first dimension includes requirements for recycled material (in-bound); the second dimension includes internal operational practices with respect to recycled material; and the third dimension includes the need for markets that will take materials for recycling (out-bound).

The target respondents from this study are small manufacturers along the Rio Grande Valley of the Texas–Mexico Border. The mean size of the companies that provided data was 203 employees and the median size was 15 employees. The large mean value was due to four companies that had between 2,000 and 10,000 employees. These organisations are all located within the United States along a four-county area in Texas. The geographical region included Cameron, Hildago, Willacy and Starr counties. Respondents were typically the executive officer or owner of the organisation.

Overall, 243 of 422 potential responses were completed. Of these responses, only 166 were usable. Approximately 77 organisations (32% of respondents) reported that no recycling was undertaken. The respondents were from 20 industries as classified by the two-digit SIC codes. The 20 industries represented in the respondent sample population are listed in Table 1. The initial number surveyed is also included in Table 1. Other than the single respondent in SIC 56, the respondents were manufacturing-oriented organisations. Results and analysis of each of the major organisational questions are now presented.

Survey Results: Recycling and Environmental Management Questions

As mentioned earlier, nine questions pertaining to organisational environmental/ recycling issues were asked of each company (see Table 2 for actual wording of questions). The overall responses and statistics for each question are presented.

Question 1: Materials Costs Comparison

The first question considered perceptions of the relative costs of virgin material versus recycled material. This question was meant to determine if costs present a barrier (whether real or imagined) to the adoption and purchase of recycled material. The results of this question were almost evenly divided among companies that believed that recycled materials were more expensive (52%) and those that believed that recycled materials were not more expensive (48%). This even breakdown may be due to some materials being viewed as more expensive and others less expensive.

A recycled product may be expensive due to the increased costs associated with inefficient reverse logistics channels, or due to the high demand but low supply of the material. The latter reason may actually be encompassed by the first in that poor delivery infrastructure will impede the supply of material, and thus increase its cost. A typical reverse logistics channel may have a number of processes within it, including collection, separation, densification, transitional processing, delivery and integration. Each of these steps adds additional cost to the recycled material. (Stock [1992] estimates that reverse logistics channels and processes may add 30% more to the cost of a recycled product.) In addition, with erratic and indeterminate amounts of 'raw material' supply for recycling purposes, there may be a need to build up necessary material quantities, requiring additional storage and inventory expenses. The erratic supplies may also be very difficult to justify some of the necessary processing

SIC code	SIC code description	Number surveyed
20	Food and related products	54
22	Textile mill products	5
23	Apparel and other textile products	19
24	Lumber and wood products	10
25	Furniture and fixtures	6
26	Paper and allied products	7
27	Printing and publishing	54
28	Chemicals and allied products	14
29	Petroleum and coal products	3
30	Rubber and miscellaneous plastics products	20
31	Leather and leather products	6
32	Stone, clay and glass products	20
33	Primary metal industries	3
34	Fabricated metal products	20
35	Industrial machinery and equipment	32
36	Electronic and other electric equipment	19
37	Transportation equipment	16
38	Instruments and related products	6
39	Miscellaneous manufacturing industries	13
56	Apparel and accessory stores	0
	Total	327

Table 1: LISTING OF SIC CODES AND DESCRIPTIONS FROM RESPONDENT SAMPLE AND NUMBER OF INITIAL FIRMS IN STUDY PER SIC CODE

1. Is the average price of new material higher than the average price of recycled material?
2. If recycled items were made easily available to you, would you continue to buy new items? If yes, is this due to: (1) process requirements? (2) customer requirements? (3) strategic considerations? (4) safety stock in case of unexpected high rejection rate?
3. Are recycled items used as inputs from suppliers?
4. Do you monitor environmental compliance of your supplier firms?
5. How does the quality of material compare with the virgin materials? Same? Better? Inferior?
6. Is there any plan for improving the efficiency of recycling of materials in the near future?
7. Are employees made aware of the environmental issues, programmes, etc. affecting the company and products?
8. Do you have (a) plan(s) for further recycling of your wastes?
9. In addition to the plans stated above, do you feel more output wastes could be recycled?

Table 2: SURVEY QUESTIONS PERTAINING TO ORGANISATIONAL ENVIRONMENTAL/RECYCLING ISSUES

technology for recycling purposes. The savings may occur in fewer processing requirements for the recycled product. Some types of product or material where recycling efficiencies have been designed in (for example, coding of plastics and disassembly standards for products) may allow for operational cost savings.

◻ Question 2: Availability and Purchase of Recycled Material

This question was divided into two parts. The first part asked: if recycled materials were easily available (that is, a satisfactory and well-priced recyclable products or materials market existed), would the company continue to purchase new material. The second part of the question asked why they would continue to purchase new material. The respondents were provided with four choices for the second part of the question. These were:

- Process requirements
- Customer requirements
- Strategic consideration
- Safety stock in case of high rejection rate for recycled material

Of those companies that replied, 58% stated that they would continue to purchase new material instead of recycled material in their processes. This result implies that a substantial fraction of the companies (42%) do not believe there is any stigma (or

For those answering 'Yes' to Question 2, the reason to continue using new material is:	
Process requirements	56%
Customer requirements	74%
Strategic considerations	26%
Safety stock	18%

Table 3: PERCENTAGES ACROSS INDUSTRIES FOR REASONS FOR CONTINUED USE OF NEW MATERIALS (QUESTION 2) FROM EACH FOR THOSE ORGANISATIONS ANSWERING 'YES'

barrier) associated with the use of recycled material if it were easily available (which should include cost and price). But, even with a possible 42% acceptance rate, the issue of whether sufficient markets for an efficient reverse logistics channel exist may still be questionable. The demand quantities and type of recycled material that would be acceptable still needs to be investigated.

Overall, Table 3 presents the reasons presented by all organisations answering 'yes' to Question 2. Respondents were allowed to select more than one answer to this question, and this is why the summation of the responses is greater than 100%. The percentages in this table represent the ratio of those that selected the reason from among the total that responded 'yes.' Customer requirement was the major driver for the continued use of new material, even when recycled material was easily available. Customers are assumed to be the immediate customers of the product sold by the organisation. There may be many reasons for this response, but one of them could be the stigma associated with the perceived quality of recycled materials by customers (who are generally larger than the small organisations defined in this sample). A change in perceptions in the use of recycled materials for products by all industries may help increase the use of recycled materials in processing. With ever-increasing pressures from the end-consumer flowing through the supply chain (Ottman 1998; Stisser 1994), increased use of recycled material will occur after the availability and cost issues are addressed. The other major reason, selected by more than half of the respondents (56%), was process-related: sometimes the recycled materials may not have the necessary qualities and attributes for incorporation into the organisations' processing and manufacturing functions; or the materials or production technology may not have the capability to incorporate recycled material into the fabrication or assembly processes (i.e. the reason is primarily technological).

Question 3: Use of Recycled Materials from Suppliers

This question pertains to the purchase of recycled materials from suppliers. It helped to identify the availability of various materials. Of those organisations that replied to Question 3, most (59%) stated that they did not use or purchase recycled materials from their suppliers, even though a large portion of the sample (41%) did purchase recycled materials. This is a good measure of the market and potential market for recycled materials by industry in this region. This value may vary for different product types, but it shows that there is demand for a recycled material market. It also points to a larger potential for additional market penetration for recycled materials.

Question 4: Monitoring Supplier Firm Environmental Compliance

This question focused less on the firm's recycling practices and more on general organisational environmental practices. Specifically, it was meant to elicit information on whether organisations monitor environmental practices of their supplier firms. The question helped to determine organisational proactivity in maintaining a long-term supply for their materials, whether or not these materials are recyclable.

Monitoring environmental compliance may be necessary for long-term supplier relationships. The reason for this monitoring is that a supplier with poor environmental practices may go out of business due to poor environmental performance. If a supplier is critical (e.g. sole source suppliers) and if there is a potential for poor environmental performance, then an organisation may want to seek other potential suppliers (as with poor financial performance). In addition, regulatory penalties (especially US Superfund-related penalties) may include customers of materials as potentially responsible parties. The monitoring of environmental performance is similar to the monitoring of a vendor's quality. Supplier certification for quality is very common and is evident on an international scale with ISO 9000 quality standards. Supplier certification for environmental management systems is also gaining international popularity, evidenced by the ISO 14000 series environmental management standards (Tibor and Feldman 1996).

The results showed that 29% of the firms monitor their suppliers' environmental compliance. Since the respondents were smaller organisations, it was surprising that almost 1 in 3 of the organisations was undertaking this type of monitoring. Typically, monitoring is carried out to ensure that the organisation is following appropriate practices; if this is not the case, the customer usually has the power and ability to require better practices. Small firms usually do not have this type of leverage. From a marketing perspective, it would be beneficial for many organisations to maintain an environmental management system, since many customers will require some form of environmental performance and compliance.

Question 5: Perceived Quality of Recycled Material versus Virgin Materials

This question was asked to determine if quality, or at least perceived quality, of recycled material was a barrier for the purchase of recycled material. Of the respondents who replied to this question, 70% felt that the quality of recycled material was equal to that of virgin material. This relatively large proportion indicated that the perceived quality of recycled material may not be a barrier for its purchase.

This result may point to availability or cost as other possible reasons for the low levels of purchase of recycled material.

Approximately 1 in 5 of the respondents (21%) specified that the quality of recycled material was not as good as that of virgin material. This was in comparison to only 1 in 10 organisations stating that recycled material was of better quality. Whether the quality comparison was perception or an actual tested process within the organisation was not determined. Perception of the quality of recycled material is important, in that it is one additional barrier to the acceptance of purchasing recycled materials. Green marketers (especially those that wish to market recycled material) should be aware of this perception. Supporting evidence (e.g. promotional tools) showing that the quality of recycled material is comparable, or providing some form of incentive if it is not, will need to be considered.

Question 6: Plans for Improving Efficiency of Recycling

The purpose of this question was to determine if initial recycling practices will improve in the future. This question evaluated the possibility of a larger market (from a supply perspective) for recycled material, assuming that an efficient system will provide a better flow and need for recycled material.

According to the responses, there was interest in improving the efficiency of organisational recycling practices (61%), and many of these organisations have improved recycling practices as part of their strategic plans. However, the time-frame for these future improvements and the reasons for seeking improved recycling efficiency was unclear.

Question 7: Employee Awareness of Environmental Issues

Similarly to Question 6, this question examined internal proactive environmental measures that organisations had introduced. The awareness of environmental issues in organisations enables continuous environmental improvement and may allow for expansion of recycling and other environmental programmes within organisations. The large number of firms that undertook this activity (81%) meant that employees were made environmentally aware in relation to corporate activities. However, it is unclear whether employees were made aware of environmental issues for reactive reasons (such as meeting government regulations) or for proactive reasons (such as implementing purchasing of recycled products mandates).

The practice of increasing employee awareness may also be related to specific functions. For example, purchasing staff would be trained to evaluate recycled material and its quality, while shop-floor employees may be made aware of minimising wastes that may not be recyclable. Employee empowerment requires employee awareness, where both are necessary for improving future recycling efficiency.

Question 8: Further Plans for Recycling of Wastes

The results of this question provided some information on whether an expansion of recycling or waste-processing equipment, technology or infrastructure will be required for this geographical location. Additional details to evaluate the specific reasons need to be ascertained and include the types of material that will be pursued for recycling, which will identify types of tool, material, etc. that can be used to aid internal recycling. Another issue that needs to be addressed is the time-horizon in which firms plan to pursue these recycling objectives.

Of those that responded to this question, 65% stated that they did plan on either initiating or expanding recycling in their organisation. This result suggested a possible need for additional sources of waste processors or customers of waste products. A market for recycled products (if material is not re-used internally) will also be needed in this area, which might also require an improvement in recycling efficiencies (see Question 6) to absorb the increased amounts of recycled materials.

Question 9: Plans for Additional Recycling of Output Wastes

This question was used to determine the potential need for additional markets for waste produced by firms. If an organisation answered 'Yes' to this question, then additional markets or infrastructure-support mechanisms may improve the chances for additional recycling activities.

A total of 69% of the respondents answered in the affirmative to Question 9. These results imply that additional markets will need to be developed for wastes in the Valley region. This result also suggests there is an immature market and infrastructure for recycling in this geographical area. Future plans will be required to develop both an external and internal recycling capacity. For industrial ecosystems to work, there must be more recyclers within the area.

Summary and Conclusion

For industrial ecosystems to function properly, there must be a development of markets and channels for products and their wastes. Manufacturing companies will require partners and networks to be developed for a variety of materials. These relationships are a concern for green marketers, who must facilitate the development of these markets by having a number of stakeholders work together. The initial step is to gather information on needs, requirements and resources to determine what gaps in the systems exist and to enable private and public organisations to pursue appropriate actions. To this end, this chapter has summarised the results of a small survey containing a series of questions put to organisations along the Texas–Mexico border on various recycling and environmental practices.

This initial study was designed to aid the TNRCC and gain insights into possible markets and opportunities for markets of recyclable products. It is difficult to generalise these findings to a larger population of organisations, but the results do provide some insights into the practices of a large group of small manufacturing organisations.

There are a number of managerial implications for green marketers derived from this study. First, it is clear that both small and large enterprises are showing an increased interest in the use of recycled material. Marketers within organisations that sell products to small manufacturers should note that some firms do have preferences for recycled materials. There seems to be an opportunity for development of marketing initiatives that tap into this market segment.

Second, there is some need to improve the image of recycled materials/products, although this need may vary among industries and between managers. Green marketers should become aware of the capabilities of the recycling systems and quality perception problems associated with their materials. The large proportion of respondents who believed that the quality of recycled materials was equal or better than that of virgin materials means there is less of a barrier in their use. Green marketers can take advantage of this finding. Customer requirements for virgin material seems to be the largest barrier for use of recycled material. Green marketers need to determine why requirements do not allow for recycled materials to be used and identify how these products can be better marketed.

Third, a high percentage of small organisations are seeking to expand their environmental practices, including internal employee awareness and expansion of recycling practices; this is a marketing opportunity for materials suppliers and manufacturers. Environmental training and service organisations may be able to target this market segment. The use of external contractors by small manufacturers will aid them in extending their limited expertise and capabilities in environmental training and their understanding of organisational environmental practices.

Finally, organisations at the end of the logistics pipeline may also see opportunities for marketing their capabilities, services and products. These types of organisation, from an industrial ecosystem perspective, are needed to help set up the reverse logistics channels that would close the loops for these manufacturers. The results show that there are a number of markets for products that could be recycled (end-of-pipe). One major issue in developing this market is the perceived cost associated with materials delivered through the reverse logistics channels. Organisations will need to make these channels more efficient: the competition for the recycled materials is not only from other recycled materials, but virgin materials that may flow through more efficient forward distribution channels.

Overall, a review of these greening and marketing practices and future plans are encouraging for the Texas–Mexico border area. Similar studies in other geographical locations will be helpful for regional governmental or private recycling marketing programmes that could be developed.

From a macro-marketing perspective, the TNRCC is seeking to build up a working industrial ecosystem in this region, and has started to evaluate the needs and requirements of small organisations along this area. Even though the focus has been on organisations on the US side of the border, the *Maquiladoras* and NAFTA have opened up these resources to Mexican industry as well. This proactive measure and initial study is part of a large TNRCC effort that includes the development of Eco-Industrial Parks in the Brownsville, Texas, area.

Bibliography

Ahrens, R., and M. Behrent (1996) 'Dialogkommunikation als strategisches Konzept: Praktische Erfahrungen mit dialogischen Kampagnen', in G. Bentele *et al.*, *Dialogorientierte Unternehmenskommunikation: Grundlagen, Praxiserfahrungen, Perspektiven* (Berlin: Vistas): 333-50.

Air Conditioning, Heating and Refrigeration News (1993) 'Greenpeace and Refrigerants: Can advocacy go too far?', *Air Conditioning, Heating and Refrigeration News*, 13 December 1993: 16.

Ajzen, I., and M. Fishbein (1980) *Understanding Attitudes and Predicting Social Behavior* (Englewood Cliffs, NJ: Prentice–Hall).

Allen, F.E. (1991) 'McDonald's to reduce waste in plan developed with environmental group', *The Wall Street Journal*, 2 February 1991.

Allenby, B.R., and D.A. Richards (1994) *The Greening of Industrial Ecosystems* (Washington, DC: National Academy Press).

Anderson, P.F. (1982) 'Marketing Strategic Planning, and the Theory of the Firm', *Journal of Marketing* 46 (Spring 1982): 15-26.

Anderson, R.C. (1995) 'The Journey from There to Here: The Eco-Odyssey of a CEO', keynote address at the *US Green Building Conference*, Big Sky, MT, 14 August 1995, *http://www.ifsia.com/ecosense/odyssey.htm*.

Ansley, M. (1991) 'Body Shop Beautiful', *Listener and TV Times* 131.2693 (4 November 1991): 24-27.

Ansoff, H.I. (1987) 'The Emerging Paradigm of Strategic Behavior', *Strategic Management Journal* 8: 501-15.

Argyris, C., and D.A. Schon (1978) *Organizational Learning* (Reading, MA: Addison-Wesley).

Arora, S., and T.N. Carson (1995) 'An Experiment in Voluntary Environmental Regulation: Participation in EPA's 33/50 Program', *Journal of Environmental Economics and Management* 28 (May 1995): 271-86.

Aspinwall, D. (1993) 'Green Cleaning', *Home Economics and Technology*, November 1993.

Associated Press (1998) 'Volvo designs first smog-eating car', *Associated Press*, 2 June 1998.

Atkinson, A.A., J.H. Waterhouse and R.B. Wells (1997) 'A Stakeholder Approach to Strategic Performance Measurement', *Sloan Management Review* 38: 25-37.

Azzone, F., M. Brophy, G. Noci, R. Welford and W. Young (1997) 'A Stakeholders' View of Environmental Reporting', *Long Range Planning* 30.5: 699-709.

Azzone, G., and R. Manzini (1994) 'Measuring Strategic Environmental Performance', *Business Strategy and the Environment* 3.1 (Spring 1994): 1-14.

B&Q (1995) *How green is my front door? B&Q's Second Environmental Report* (Eastleigh, UK: B&Q).

Balderjahn, I. (1988) 'Personality Variables and Environmental Attitudes as Predictors of Ecologically Responsible Consumption Patterns', *Journal of Business Research* 17 (August 1988): 51-56.

Banerjee, S. (1998) 'Corporate Environmentalism: Perspectives from Organizational Learning', *Management Learning* 29.2: 147-64.

Banerjee, S., C. Gulas and E. Iyer (1995) 'Shades of Green: A Multidimensional Analysis of Environmental Advertising', *Journal of Advertising* 24.2 (Summer 1995): 21-31.

Bangkok Post (1998) 'Pipeline project to go ahead', *Bangkok Post*, 1 March 1998: Home 1.

Barney, J. (1991) 'Firm Resources and Sustained Competitive Advantage', *Journal of Management* 17: 99-120.

Barrett, J. (1993) 'The Future for Environmental Managers', *ENDS Report*, November 1993: 82-87.

Baugh, K., B. Byrnes, C. Jones and M. Rahimzadeh (1995) 'Green Pricing: Removing the Guesswork', *Public Utilities Fortnightly* 133.15: 26-28.

Baumol, W.J., and W.E. Oates (1971) 'The Use of Standards and Pricing for the Protection of the Environment', *Swedish Journal of Economics* 73: 42-54.

Beard, C.M. (1996) 'Environmental Training: Emerging Products', *Journal of Industrial and Commercial Training* 28.5: 18-23.

Beard, C.M., and D. Egan (1998) 'Investing in Sustainability: An Exploration of How Firms can Adopt a Balance Sheet Approach to Achieving Sustainable Production' (Proceedings of the 1998 *International Sustainable Development Research Conference*, ERP Environment, Leeds, UK, 3–4 April 1998): 19-24.

Beard, C.M., and R. Hartmann (1997a) 'Sustainable Design: Re-thinking Future Business Products', *The Journal of Sustainable Product Design* 3 (October 1997): 18-27.

Beard, C.M., and R. Hartmann (1997b) 'Naturally Enterprising: Eco-Design, Creative Thinking and the Greening of Business Products', *European Business Review* 97.5.

Beard, C.M., and R. Hartmann (1999) 'European and Asian Telecoms: Their Role in Global Sustainable Development', *European Business Review* 99.1: 42-54.

Beaumont, J.R., L.M. Pederson and B.D. Whitaker (1993) *Managing the Environment* (Oxford: Butterworth Heinemann).

Bei, L.-T., and E.M. Simpson (1995) 'The Determinants of Consumers' Purchase Decisions for Recycled Products: An Application of Acquisition–Transaction Utility Theory', *Advances in Consumer Research* 22: 257-61.

Belbin, R.M. (1981) *Management Teams* (London: Heinemann).

Belch, G.E., and M.A. Belch (1990) *Introduction to Advertising and Promotion Management* (Homewood, IL: Irwin).

Belz, F. (1998) *Ökologische Innovationen in der Kreislaufwirtschaft: Leistungs- statt Produktverkauf* (IWÖ-Diskussionsbeitrag, 62; St Gallen, Switzerland: Universität St Gallen, Institut für Wirtschaft und Ökologie).

Bennis, W., and B. Nanus (1985) 'Organizational Learning: The Management of the Collective Self', *New Management*, Summer 1985: 7-13.

Bernoth, A. (1994) 'Anita Roddick's green halo wobbles: Body Shop', *The Sunday Times*, 28 August 1994.

Bernstein, D. (1992) *In the Company of Green: Corporate Communication for the New Environment* (Geneva: ISBA).

Berry, T. (1993) 'The Viable Human', in M.E. Zimmerman, J.B. Callicott, G. Sessions, K.J. Warren and J. Clark (eds.), *Environmental Philosophy* (Englewood Cliffs, NJ: Prentice–Hall): 171-81.

Beste, D. (1994) 'The Greenfreeze Campaign', *Akzente*, December 1994: 26-29.

Bestseller (1997) 'Focus Media Research', *Bestseller* 1.2: 64.

Biddle, D. (1993) 'Recycling for Profit: The New Green Business Frontier', *Harvard Business Review*, November/December 1993: 145-56.

BiE (Business in the Environment) (1997) *The Index of Corporate Environmental Engagement: 'Green Profile' of the FTSE-100* (London: BiE).

BiE (Business in the Environment), CIPS and KPMG (1992) *Buying into the Environment: Guidelines for Integrating the Environment into Purchasing and Supply* (London: BiE).

Black, J.S., P.C. Stern and J.T. Elworth (1985) 'Personal and Contextual Influences on Household Energy Adaptations', *Journal of Applied Psychology* 70.1: 3-21.

Black, L. (1994) 'Body Shop slips on puddle of peach', *The Independent*, 20 August 1994.

Blake, G. (1994) 'TQM and Strategic Environmental Management', in J.T. Willig (ed.), *Environmental TQM* (New York: McGraw–Hill): 1-6.

Bleicher, K. (1992) *Das Konzept Integriertes Management* (Frankfurt: Campus).

Blunden, J. (1998) '"Green" power gets two boots', *Sun Francisco Examiner*, 7 July 1998.

Body Shop (1994) *Memorandum of Response to all Business Ethics Subscribers* (press release; The Body Shop, August 1994).

Body Shop (1998a) 'The Body Shop International plc Annual Results', *http://www.the-body-shop.com/new/results98.html* (12 May 1998).

Body Shop (1998b) 'Community Trade Programme', *http://www.the-body-shop.com/trade/index.html*

Bolton, L. (1997) 'Greens follow suit', *The Financial Times*, 30 December 1997.

Bonaccorsi, A., and A. Lipparini (1994) 'Strategic Partnerships in New Product Development', *Journal of Product Innovation Management* 11: 134-45.

Bonner, S. (1997) 'It's not easy being green: Strategies and Challenges', *Apparel Industry Magazine* 58.2: 52-68.

Boo, E. (1993) *Eco-Tourism: The Potentials and Pitfalls* (2 vols.; Geneva: World Wildlife Fund).
Boone, L.E., and D.L. Kurtz (1998) *Contemporary Marketing Wired* (Fort Worth, TX: Dryden Press, 9th edn).
Booz, Allen & Hamilton (1982) *New Products Management for the 1980s* (New York: Booz, Allen & Hamilton, Inc.).
Böttcher, H., and R. Hartman (1997) 'Eco-Design: Benefit for the Environment and Profit for the Company', *Industry and Environment (UNEP IE)* 20.1-2: 48-51.
Bowie, N. (1991) 'New Directions in Corporate Social Responsibility', *Business Horizons* 34.4: 56-65.
Brand, K.-W., K. Eder and A. Poferl (1997) *Ökologische Kommunikation in Deutschland* (Opladen, Germany: Westdeutscher Verlag).
Braungart, M., and J. Engelfried (1992) 'An "Intelligent Product System" to replace "Waste Management"', *Fresenius Environmental Bulletin* 1: 613-19.
Brown, L.D. (1991) 'Bridging Organizations and Sustainable Development', *Human Relations* 44.8: 807-31.
Brown, M. (1997) 'A Green Piece of the Action', *Management Today*, May 1997: 84-88.
Brown, M., and E. Wilmanns (1997) 'Quick and Dirty Environmental Analyses for Garments: What do we need to know?', *The Journal of Sustainable Product Design* 1 (April 1997): 28-35.
Brummer, A. (1994) 'Ethical Shake Out', *The Guardian*, 31 August 1994.
Brummer, A. (1994a) 'Why Corporate Integrity Needs Outside Scrutiny', *The Guardian*, 3 September 1994.
BSI (British Standards Institute) (1996) *Environmental Management Systems: Specification with Guidance for Use* (London: BSI).
BT (British Telecom) (1997) 'Telecommunications, Technologies and Sustainable Development', *http://www.bt.com/corpinfo/enviro/agenda/doc.htm,* 6 October 1997.
Buchholz, R.A. (1993) *Principles of Environmental Management: The Greening of Business* (Englewood Cliffs, NJ: Prentice–Hall).
Bunyamanee, S. (1997) 'Fresh Approach to Marketing', *Bangkok Post* 51 (14 July 1997): 18.
Burall, P. (1991) *Green Design* (New York, McGraw–Hill).
Burkart, R. (1996) 'Verständigungsorientierte Öffentlichkeitsarbeit: Der Dialog als PR-Konzeption', in G. Bentele *et al.*, *Dialogorientierte Unternehmenskommunikation: Grundlagen, Praxiserfahrungen, Perspektiven* (Berlin: Vistas): 245-70.
Burt, D.N., and W.R. Soukup (1994) 'Purchasing's Role in New Product Development', in K. Clark and S.C. Wheelwright (eds.), *The Product Development Challenge* (Boston, MA: Harvard Business School Press): 333-45.
Business and the Environment (1994) 'Environmental group sets strategies', *Business and the Environment* 12.5 (December 1994).
Button, J. (1989) *How to be Green* (London: Century Hutchinson).
Cahan, J., and M. Schweiger (1993) 'Product Life Cycle: The Key to Integrating EHS into Corporate Decision Making and Operations', *Total Quality Environmental Management*, Winter 1993/94: 141-50.
Cairncross, F. (1992) *Costing the Earth: The Challenge for Governments, the Opportunities for Business* (Boston, MA: Harvard Business School Press).
Callenbach, E., F. Capra, L. Goldman, R. Lutz and S. Marburg (1993) *EcoManagement: The Elmwood Guide to Ecological Auditing and Sustainable Business* (San Francisco: Berrett-Koehler).
Cambridge Reports/Research International (1992) *Green Consumerism: Commitment remains strong despite economic pessimism* (Cambridge Reports).
CAN (Conservation Agriculture Network) (1998a) *Transforming Tropical Agriculture One Farm at a Time* (New York: The Rainforest Alliance).
CAN (Conservation Agriculture Network) (1998b) 'Selling Certified Products: Materials and Requirements', *www.rainforest.alliance.com* (April 1998).
CAN (Conservation Agriculture Network) (1998c) *Transforming Tropical Agriculture One Farm at a Time. ECO-OK Certification: Process and Fees* (New York: Rainforest Alliance).
CAN (Conservation Agriculture Network) (1998d) 'The Better Banana Project', *www.rainforest.alliance.com* (April 1998).
Canadian Environmental Advisory Council (1991) *A Protected Areas Vision for Canada* (Cat. no. EN 92-14/1991E; Ottawa: Minister of Supply and Services Canada).
Capra, F. (1983) *The Turning Point* (New York: Bantam).
Carew, J. (1997) 'The Supply Chain as a Catalyst for Environmental Change' (Paper presented for *Green Procurement in Government Conference,* Queen Elizabeth II Conference Centre, London, 11 July 1997).

Carlson, L., N. Kangun and S.J. Grove (1992) 'A Content Analysis of Environmental Advertising Claims', in L.N. Reid (ed.), *Proceedings of the Conference of the American Academy of Advertising* 1992.

Carlson, L., N. Kangun and S.J. Grove (1995) 'A Classification Schema for Environmental Advertising Claims: Implications for Marketers and Policy Makers', in M.J. Polonsky and A.T. Mintu-Wimsatt (eds.), *Environmental Marketing: Strategies, Practice, Theory and Research* (New York: Haworth Press): 225-38.

Carmone, F.J. (1995) 'Review: Conjoint Analysis Software', *Journal of Marketing Research*, February 1995: 113-20.

Carson, R. (1962) *Silent Spring* (Boston: Houghton–Mifflin).

Castillo, C. (1994) 'Banana Plantations Leave a Toxic Legacy', *Panoscope*, April 1994.

Caswell, J.A. 'How Green Is Thy Product?', *Rural Entrepreneur* (*http://www.halcyon.com/uconn/greenproduct.html*).

Catton, W.R., and R.E. Dunlap (1980) 'A New Ecological Paradigm for Post-Exuberant Society', *American Behavioral Scientist* 24.1: 15-48.

CEC (California Energy Commission) (1997) *Electricity Report* (Draft, Document P300-97-001; Sacramento, CA: CEC).

Chaimusik, J., and C. Nivatpumin (1996) 'Coming to Grip with Green', *Bangkok Post* 50 (8 July 1996): 18.

Charter, M. (ed.) (1992a) *Greener Marketing: A Responsible Approach to Business* (Sheffield, UK: Greenleaf Publishing).

Charter, M. (1992b) 'Greener People', in M. Charter (ed.), *Greener Marketing: A Responsible Approach to Business* (Sheffield, UK: Greenleaf Publishing): 255-84.

Charter, M. (1997) 'Managing Eco-Design' *Industry and Environment (UNEP IE)* 20.102: 29-31.

Charter, M. (1998) 'Design for Environmental Sustainability', *Foresight* (UK Office of Science and Technology), May 1998.

Chase, D. (1991) 'The Green Revolution: P&G gets top marks in AA survey', *Advertising Age* 65.5 (29 January 1991): 8-10.

Chick, A. (1997) 'The "Freeplay" Radio', *The Journal of Sustainable Product Design* 1 (April 1997): 53-56.

Chipperfield, M. (1988) 'Body baroness dispenses with tradition', *Marketing Magazine* 7.3: 51-56.

Chonko, L.B. (1995) *Ethical Decision Making in Marketing* (Sage Series in Business Ethics; Oaks, CA: Sage).

Cinq-Mars, J. (1997) 'Green Public Purchasing in OECD Countries' (Paper presented for *Green Procurement in Government Conference,* Queen Elizabeth II Conference Centre, London, 11 July 1997).

CIPS (Chartered Institute of Purchasing and Supply) (1995) *Supply Chain: The Environmental Challenge* (Lincolnshire, UK: The Chartered Institute of Purchasing and Supply).

Claus, F., and P. Wiedemann (1994): *Umweltkonflikte: Vermittlungsverfahren zu ihrer Lösung* (Taunusstein, Germany: Eberhard Blottner Verlag).

Coccossis, H. (1996) 'Tourism and Sustainability: Perspectives and Implications', in G.K. Priestley, J.A. Edwards and H. Coccossis (eds.), *Sustainable Tourism* (Wallingford, UK: CAB International).

Coddington, W. (1993) *Environmental Marketing: Positive Strategies for Reaching Green Consumers* (New York: McGraw–Hill).

Colby (1991) 'Environmental Administration during Development: Evolution of the Paradigms', *Trimestre Economico* 58.231: 589-615.

Cole, S. (1993) 'Pushing the Envelope: How Environmental Professionals are Meeting the Challenges of the 1990s', in American Chemical Society, *Environmental Science and Technology* (Environmental Buyers' Guide Edition; Washington, DC: American Chemical Society).

Commoner, B. (1974) *The Closing Circle* (New York: Bantam Books).

Confino, J., and R. Cowe (1994) 'RSPCA says Body Shop fails animal testing criteria', *The Guardian*, 27 August 1994.

Consumers' Association (1995) 'Washing Machines', *Which?*, January 1995: 48-51.

Consumers' Association (1996) 'Washday Winners', *Which?*, January 1996: 39-45.

Cook, W.J. (1998) 'Men in Blue', *US News and World Report*, 16 February 1998.

Cooper, R.G. (1988) Pre-development activities determine new product success', *Industrial Marketing Management* 17: 237-47.

Cooper, R.G. (1994) 'New Products: The Factors that Drive Success', *International Marketing Review* 11.2: 60-76.

Cooper, R.G., and E.J. Kleinschmidt (1995) 'Performance Typologies of New Product Projects', *Industrial Marketing Management* 24: 439-56.

Corder, M. (1997) 'Greenpeace no longer world's savior group', *Herald Journal* (Logan, UT), 7 October 1997: 17.

Cornwell, T.B., and C.H. Schwepker, Jr (1992) 'Attitudes and Intentions Regarding Ecologically Packaged Products: Subcultural Variations', in L.N. Reid (ed.), *Proceedings of the 1992 Conference of the American Academy of Advertising*: 119-21.

Costanzo, M., D. Archer, E. Aronson and T. Pettigrew (1986) 'Energy Conservation Behavior: The Difficult Path from Information to Action', *American Psychologist*, May 1986: 521-28.

Cottam, N. (1998) 'Energy Minimalists', *The Financial Times*, 9 June 1998.

Coulter, P.D. (1994) 'Auditing for Environmental Excellence at Union Carbide', *Environmental TQM* (New York: McGraw–Hill): 245-67.

Cowe, R. (1994a) 'Body Shop shares slip again', *The Guardian*, 31 August 1994.

Cowe, R. (1994b) 'Body Shop brands US article "recycled rubbish"', *The Guardian*, 2 September 1994.

Cowe, R. (1994c) 'Accepted wisdom takes body blow: Body Shop', *The Guardian*, 26 August 1994.

Cracco, E., and J. Rostenne (1971) 'The Socio-Ecological Product', *MSU Business Topics* 19 (Summer 1971): 28-29.

Crawford, A. (1997) 'Body Shop hands Bean first ad task', *Campaign London*, 28 March 1997: 1.

Crawshaw, A.J.E., D.I. Williams and C.M. Crawshaw (1985) 'Consumer Knowledge of Electricity Consumption', *Journal of Consumer Studies and Home Economics* 9: 283-89.

CREA (Centre for Regional and Economic Analysis) (1993) *The Contribution of Tourism to the Tasmanian Economy in 1992* (Hobart: University of Tasmania).

Crul, M.R.M. (1994) 'Milieugerichte Produktontwikkeling in de Praktijk', *Promise* (The Hague: NOTA): 151-57.

Cude, B.J. (1992) 'Making Consumer Education "Green": Issues and Approaches', in V. Haldeman, (ed.), *Proceedings of the American Council on Consumer Interests*: 187-93.

Cude, B.J. (1993) 'Does it cost more to buy "green"?', *Proceedings of the American Council on Consumer Interests* 39: 108-13.

Davis, J. (1991a) *Greening Business: Managing for Sustainable Development* (Oxford: Basil Blackwell).

Davis, J. (1991b) 'A Blueprint for Green Marketing', *Journal of Business Strategy*, July/August 1991: 14-17.

Davis, J. (1993) 'Strategies for Environmental Advertising', *Journal of Consumer Marketing* 10.2: 19-36.

Davis, J.B. (1996) 'Product Stewardship and the Coming Age of Takeback: What Your Company Can Learn from the Electronics Industry's Experience', *Cutter Information Corp.* (Arlington, MA): 38, 108.

Day, G.S. (1992) 'Marketing's Contribution to the Strategy Dialogue', *Journal of the Academy of Marketing Science* 20.4 (Fall 1992): 323-29.

Day, G.S., and R. Wensley (1983) 'Marketing Theory with a Strategic Orientation', *Journal of Marketing* 47 (Fall 1983): 79-89.

Dechant, K., and B. Altman (1994) 'Environmental Leadership: From Compliance to Competitive Advantage', *Academy of Management Executive* 8: 7-20.

Dechant, K., and V.J. Marsick (1991) 'In Search of the Learning Organization: Toward a Conceptual Model of Collective Learning', paper presented at the *Eastern Academy of Management Conference*, 1991.

Del Franco, M. (1998) 'Another Go at "Eco-Fashion"', *Catalog Age*, June 1998: 78.

Deloitte Touche Tohmatsu, IIED and SustainAbility (1993) *Coming Clean: Corporate Environmental Reporting* (London: Deloitte Touche Tohmatsu).

Dembkowski, S., and S. Hanmer-Lloyd (1994), 'The Environmental Value–Attitude–System Model: A Framework to Guide the Understanding of Environmentally-Conscious Consumer Behaviour', *Journal of Marketing Management* 10.7: 593-603.

Deni Greene Consulting (1992) *Life Cycle Analysis: A View of the Environmental Impacts of Consumer Products Using Clothes Washing Machines as an Example* (Melbourne: Australian Consumers' Association).

Dennis, M.L., E.J. Soderstrom, W.S. Koncinski, Jr, and B. Cavanaugh (1990) 'Effective Dissemination of Energy-Related Information: Applied Social Psychology and Evaluation Research', *American Psychologist* 45: 1109-17.

Dermody, J., and S. Hammer-Lloyd (1996) 'Greening New Product Development: The Pathway to Corporate Environmental Excellence', in P. McDonagh and A. Prothero (eds.), *Green Management* (London: Dryden Press): 367-87.

Der Spiegel (1996) 'Sie reißen sich um jede Tonne', *Der Spiegel* 39–96 (23 September 1996): 40-48.

Die Welt (1996) 'Germany: Foron Hausgeräte files for bankruptcy', *Die Welt*, 16 March 1996: 15.

Dienel, P. (1990) *Die Planungszelle* (Opladen, Germany: Westdeutscher Verlag).

Dillman D. (1978) *Mail and Telephone Surveys: The Total Design Method* (New York: John Wiley).

Donaldson, T., and L.E. Preston (1995) 'The Stakeholder Theory of the Corporation: Concepts, Evidence, Implications', *Academy of Management Review* 20.1: 65-91.

Dowie, M. (1995) *Losing Ground: American Environmentalism at the Close of the Twentieth Century* (Cambridge, MA: MIT Press).

Drucker, P.F. (1973) *Top Management* (London: Heinemann).

Drummond, G. (1994) 'Brave New World: Fair Trade Movement', *Supermarketing*, 9 September 1994.

Drumwright, M. (1994) 'Socially Responsible Organisational Buying: Environmental Concern as a Non-Economic Buying Criterion', *Journal of Marketing* 58.3: 1-19.

DTAG (Deutsche Telekom AG) Environmental Affairs Office (1995) *Deutsche Telekom's Environmental Management System* (Darmstadt: DTAG).

Duerksen, C. (1983) *Environmental Regulations of Plant Siting: How to Make it Work Better* (Washington, DC: The Conservative Foundation).

Durning, A.T. (1992) *How Much is Enough?* (London: Earthscan).

Dwortzan, M. (1998) 'The Greening of Industrial Parks', *MIT's Technology Review* 100.9: 18-19.

Dwyer, L.M. (1990) 'Factors Affecting the Proficient Management of Product Innovation', *International Journal of Technical Management* 5.6: 721-30.

Dyllick, T. (1989) Management von Umweltbeziehungen (Wiesbaden: Gabler).

Earl, G., R. Clift and T. Moilanen (1998) 'Reducing the Uncertainty in Environmental Investments: Integrating Stakeholder Values into Corporate Decisions', in P. James and M. Bennett (eds.), *The Green Bottom Line. Environmental Management Accounting: Current Practice and Future Trends* (Sheffield, UK: Greenleaf Publishing).

Earle, R., III (1993) 'Developing an Environmental Marketing Campaign: The FTC Guidelines and Beyond', paper presented at the *1992–93 Professional Development Series: American Marketing Association (Boston Chapter) Conference*, 22 April 1993.

EC (European Commission) (1997a) 'Working Paper on the Management of Waste from Electrical and Electronic Equipment' (DG XI, E3/FE D[97]; Director General XI Environment, Nuclear Safety and Civil Protection; Brussels: European Commission, 9 October 1997).

EC (European Commission) (1997b) *Energy for the Future: Renewable Sources of Energy* (White Paper for a Community Strategy and Action Plan; Communication from the Commission COM[97]599 final; Brussels: EC).

Eckel, L., K. Fisher and R. Grant (1992) 'Environmental Performance Measurement', *CMA Magazine* 66 (March 1992): 16-23.

Economist (1994) 'Body Shop's Battles', *The Economist*, 3 September 1994.

Economist (1998a) 'Body Shop: Capitalism and Cocoa Butter', *The Economist* 86 and 89 (16–22 May 1998).

Economist (1998b) 'When Virtue Pays a Premium', *The Economist*, 18 April 1998.

Eden, S.E. (1990) *Green Consumerism and the Response from Business and Government* (Working Paper, 542; Leeds, UK: University of Leeds, School of Geography).

EIA (Energy Information Administration) (1997a) *Electricity Prices in a Competitive Environment: Marginal Cost Pricing of Generation Services and Financial Status of Electric Utilities. A Preliminary Analysis through 2015* (Document DOE/EIA-0614; Washington, DC: EIA).

EIA (Energy Information Administration) (1997b) *Renewable Energy Annual 1997: Volume I* (Document DOE/EIA-0603[97]; Washington, DC: EIA).

EIA (Energy Information Administration) (1997c) *The Changing Structure of the Electric Power Industry: An Update* (Washington DC: EIA).

Einsmann, H. (1992) 'The Environment: An Entrepreneurial Approach', *Long Range Planning* 25.4: 22-24.

Elkington, J. (1994) 'Toward the Sustainable Corporation: Win–Win–Win Business Strategies for Sustainable Development', *California Management Review* 36.2: 90-100.

Elkington, J. (1997) *Cannibals with Forks: The Triple Bottom Line of 21st Century Business* (Oxford: Capstone).

Elkington, J., and J. Hailes (1988) *The Green Consumer Guide. From Shampoo to Champagne: High Street Shopping for a Better Environment* (London: Victor Gollancz).

Elkington, J., P. Knight and J. Hailes (1991) *The Green Business Guide* (London: Victor Gollancz).

Ellen, P.S., J.L. Wiener and C. Cobb-Walgren (1991) 'The Role of Perceived Consumer Effectiveness in Motivating Environmentally Conscious Behaviors', *Journal of Public Policy and Marketing* 10.2: 102-17.

ENDS (Environmental Data Services) (1995) 'Environmental Managers Call for Greater Support from Boardroom', *ENDS Report* 241.

ENDS (Environmental Data Services) (1996a) 'Building the "Green" Factor into Product Development', *ENDS Report* 260 (September 1996): 23-25.

ENDS (Environmental Data Services) (1996b) 'Hoover remains sole bearer of eco-label for washing machines', *ENDS Report* 261 (October 1996): 29.

Entine, J. (1994) 'Shattered Image', *Business Ethics*, September/October 1994: 23-28.

Entine, J. (1995) 'Rain-forest Chic', *Toronto Globe & Mail Report on Business*, October 1995: 41-52.

EPA (US Environmental Protection Agency) (1990) *Environmental Investments: The Cost of a Clean Environment* (Washington, DC: US Government Printing Office).

EPA (US Environmental Protection Agency) (1998) 'Program Overview of Energy Star Programs and Products', *http://www.epa.gov/energystar/*

Erkman, S. (1997) 'Industrial Ecology: An Historical View', *Journal of Cleaner Production* 5.1–2: 1-10.

Esty, D.C., and M.E. Porter (1998) 'Industrial Ecology and Competitiveness: Strategic Implications for the Firm', *Journal of Industrial Ecology* 2.1: 35-44.

Evans, R. (1990) 'The Earth's New Friends', *International Management* 7: 26-31.

Everett, P. (1997) 'Setting Outrageous Goals' (Presentation to Creative Problem-Solving Institute, Buffalo, NY, 26 June 1997).

Fair Trade Federation (1997) 'Membership Criteria', *www.fairtradefederation* (14 December 1997).

Fallon, I. (1994a) 'Body Shop', *The Sunday Times*, 4 September 1994.

Fallon, I. (1994b) 'Roddicks reel from body blows', *The Sunday Times*, 28 August 1994.

Far Eastern Economic Review (1996) 'Asian Executives Poll', *Far Eastern Economic Review* 159 (11 July 1996): 30.

Farhar, B.C., and A. Houston (1996) *Willingness to Pay for Electricity from Renewable Energy* (Golden, CO: National Renewable Energy Laboratory).

Farrand, T. (1994) 'Body Shop rebuffs magazine claims', Reuters News Service, 2 September 1994.

Fergus, J. (1991) 'Anticipating Consumer Trends', in R.A. David (ed.), *The Greening of Business* (Brookfield, VT: Gower Publishing): 51-65.

Ferguson, A. (1989) 'Soapworks Good Works', *Management Today*, May 1989: 94-100.

Ferrone, R., and C.M. O'Brien (1993) 'Environmental Assessment of Computer Workstation', *IEEE International Symposium on Electronics and Environment*, Arlington, VA: 43-48.

Fietkau, H.-J. (1994) *Leitfaden Umweltmediation* (WZB-Schriftenreihe zu Mediationsverfahren im Umweltschutz, 8; Berlin: Wissenschaftszentrum Berlin für Sozialforschung).

Fiksel, J. (1996) *Design for Environment* (New York: McGraw–Hill).

Fiksel, J. (1997) 'Competitive Advantage through Environmental Excellence', *Corporate Environmental Strategy* 4.4: 55-62.

Fiksel, J. (1998) interview section (untitled), *Journal of Sustainable Product Design* 5 (April 1998): 49-52.

Fineman, S., and K. Clarke (1996) 'Green Stakeholders: Industry Interpretations and Response', *Journal of Management Studies* 33.6: 715-30.

Finger, M. (1994) 'From Knowledge to Action? Exploring the Relationships between Environmental Experiences, Learning and Behavior', *Journal of Social Issues* 50.3: 141-60.

Fischer, K., and J. Schot (eds.) (1993) *Environmental Strategies for Industry* (Washington, DC: Island Press).

Forman, A.M., and V. Sriram (1991) 'Attitudinal Differences between Dutch and American Consumers regarding Ecological Problems and Solutions: An Empirical Investigation', *Journal of Euromarketing* 1.2: 213-32.

Fox, A. (1994) 'Body Shop leader voices principles', *Waikato Times*, 12 April 1994: 12.

Fox, W. (1990) *Toward a Transpersonal Ecology* (Boston, MA: Shambhala Publications).

Frankel, C. (1998) *In Earth's Company: Business, Environment and the Challenge of Sustainability* (Gabriola Islands: New Society Publishers).

Frantz, D. (1998) 'Chiquita still under cloud after newspaper's retreat', *New York Times*, 17 July 1998: A16.

Freeman, H.L. (1989) 'Learning to Love the Service Sector', in Deutsches Institut für Wirtschaftsforschung (DIW) (ed.), *Dienstleistungen: Neue Chancen für Wirtschaft und Gesellschaft* (Berlin: DIW).

Freeman, R.E. (1984) *Strategic Management: A Stakeholder Approach* (Boston, MA: Pitman).

Friend, G. (1998) 'On the Road to Sustainability', *Green Business Letter*, March 1998: 8.

Furubotn, E.G., and S. Pejovich (1972) 'Property Rights and Economic Theory: A Survey of Recent Literature', *Journal of Economic Literature* 10: 1137-62.

FTC (Federal Trade Commission) (US) (1996) *Guides for the Use of Environmental Marketing Claims* (Washington, DC: FTC).

Fussler, C., with P. James (1996) *Driving Eco-Innovation: A Breakthrough Discipline for Innovation and Sustainability* (London: Pitman).

Gallarotti, F.M. (1995) 'It pays to be green: The Managerial Incentive Structure and Environmentally Sound Strategies', *Columbia Journal of World Business* 30: 38-57.

GEA (Group for Efficient Appliances) (1995) *Washing Machines, Driers and Dishwashers: Final Report* (Copenhagen: Danish Energy Agency, June 1995).

Geiser, K., and M. Crul (1996) 'Greening of Small and Medium-Sized Firms: Government, Industry and NGO Initiatives', in P. Groenewegen, K. Fischer, E.G. Jenkins and J. Schot (eds.), *The Greening of Industry Resource Guide and Bibliography* (Washington, DC: Island Press): 213-44.

Geller, E.S. (1990) 'Behavior Analysis and Environmental Protection: Where have all the flowers gone?', *Journal of Applied Behavior Analysis* 23.3 (Fall 1990): 269-73.

Geller, E.S. (1992) 'It Takes More Than Information to Save Energy', *American Psychologist*, June 1992: 814-15.

GEMI (Global Environmental Management Initiative) (1992a) *Total Quality Environmental Management: The Primer* (Washington, DC: GEMI).

GEMI (Global Environmental Management Initiative) (1992b) *Environmental Self-Assessment Programme* (Washington, DC: GEMI).

Ghazi, P., and R. Tredre (1994) 'Green Queen of all she Purveys? Body Shop', *The Observer*, 28 August 1994.

Gilchrist, S. (1994) 'Body Shop dismisses magazine report', *The Times*, 2 September 1994.

Gladwin, T.N. (1993) 'The Meaning of Greening: A Plea for Organisation Theory', in K. Fischer and J. Schot (eds.), *Environmental Strategies for Industry* (Washington, DC: Island Press): 37-62.

Gladwin, T.N., and I. Walters (1976) 'Multinational Enterprise, Social Responsibility, and Pollution Control', *Journal of International Business* 7 (Fall/Winter 1976): 57-74.

Gladwin, T.N., J.J. Kennelly and T.-S. Krause (1995) 'Shifting Paradigms for Sustainable Development: Implications for Management Theory and Research', *The Academy of Management Review* 20.4: 874-907.

Gleckman, H., and R. Krut (1996) 'Neither International nor Standard: The Limits of ISO 14001 as an Instrument of Global Corporate Environmental Management', in C. Sheldon (ed.), *ISO 14001 and Beyond: Environmental Management Systems in the Real World* (Sheffield, UK: Greenleaf Publishing): 45-59.

Goggin, P.A. (1994) 'An Appraisal of Ecolabelling from a Design Perspective', *Design Studies* 15.4 (October 1994): 459-77.

Goldstein, D. (1995) *Promoting Energy Efficiency in the Utility Sector through Coordinated Regulations and Incentives* (San Francisco: Natural Resources Defense Council).

Gordon, R.E. (ed.) (1993) *1993 Conservation Directory* (Washington, DC: National Wildlife Federation).

Gouldson, A. (1993) 'Fine Tuning the Dinosaur? Environmental Product Innovation and Strategic Threat in the Automotive Industry: A Case Study of the Volkswagen Audi Group', *Business Strategy and the Environment* 2.3: 12-21.

Gray, R. (1992) 'Accounting and Environmentalism: An Exploration of the Challenge of Gently Accounting for Accountability, Transparency, and Sustainability', *Accounting, Organizations and Society* 17.5: 399-425.

Green Market Alert (1993) 'The Office Electronics Industry', *Green Market Alert* 4.4 (April 1993): 5.

Green, K., A. McMeekin and A. Irwin (1994) 'Technological Trajectories and R&D for Environmental Innovation in UK Firms', *Futures* 26.10 (December 1994): 1047-59.

Green, P.E., and V. Srinivasen (1990) 'Conjoint Analysis in Marketing: New Developments with Implications for Research and Practice', *Journal of Marketing*, October 1990: 3-19.

Greenley, G.E. (1989) 'An Understanding of Marketing Strategy', *European Journal of Marketing* 23: 45-58.

Groenewegen, P., K. Fischer, E.G. Jenkins and J. Schot (eds.) (1996) *The Greening of Industry Resource Guide and Bibliography* (Washington, DC: Island Press).

Grunig, J.E., and T. Hunt (1984) *Managing Public Relations* (Fort Worth, TX: Holt Rinehart & Winston).

Guardian (1994) 'Body Shopped?', *The Guardian*, 2 September 1994.

Haddon, M. (1993) 'Making Green Labels Stick', *New Scientist*, 20 June 1993.

Hall, S., and N. Roome (1996) 'Strategic Choices and Sustainable Strategies', in P. Groenewegen, K. Fischer, E.G. Jenkins and J. Schot (eds.), *The Greening of Industry Resource Guide and Bibliography* (Washington, DC: Island Press): 9-36.

Hamel, G., and C.K. Prahalad (1994) *Competing for the Future* (Boston, MA: Harvard Business School Press).

Hammerbacher, R. (1995) 'Der Nachbarschaftskreis des Unternehmens Riedel-de-Hän', *Der Mitarbeiter: Werkzeitschrift der Riedel-de-Hän AG*, July 1995.

Handelsblatt (1995) 'Germany: Samsung withdraws from Foron deal', *Handelsblatt*, 10 March 1995: 15.

Handelsblatt (1996) 'Germany: Foron in shock after losing Turkish buyer', *Handelsblatt*, 21 February 1996: 12.

Hanna, M.D. (1995) 'Environmentally Responsible Managerial Behavior: Is Ecocentricism a Prerequisite?', *The Academy of Management Review* 20.4: 796-99.

Hansell, S. (1998) 'Now Big Blue is at your service', *New York Times*, 18 January 1998.

Hansen, U., and I. Schoenheit (1994) *HAGE: Hautpflege und Gesundheit—ein haariges Thema* (Hannover: Fackeltrager Verlag).

Hansen, U., and U. Schrader (1997) '"Leistungs- statt Produktabsatz" für einen ökologischeren Konsum ohne Eigentum', in U. Steger (ed.), *Handbuch des integrierten Umweltmanagements* (Munich: Oldenbourg): 87-110.

Hansen, U., U. Niedergesäß and B. Rottberg (1996) 'Dialogische Kommunikationsverfahren zur Vorbeugung und Bewältigung von Umweltskandalen: Das Beispiel des Unternehmensdialogs', in G. Bentele, H. Steinmann and A. Zerfasz, *Dialogorientierte Unternehmenskommunikation: Grundlagen, Praxiserfahrungen, Perspektiven* (Berlin: Vistas): 307-20.

Hansen, U., U. Niedergesäß and B. Rottberg (1997) *HAGE II: Was künftig unter die Haut geht?* (Dokumentation eines Dialogprogramms; Hannover: Hannover University, Institut für Betriebsforschung).

Hanson, D. (1996) *Eco-Sensitive Management* (unpublished PhD thesis; Hobart: University of Tasmania).

Hanson, D., and B. Tapp (1994) 'Ecologically Sustainable Development and Process for Managers', *Proceedings of the 12th Annual Conference of the Association of Management*, Dallas, TX: 13-24.

Haq, F. (1997) 'Pepsi pullout does not jar Unocal conscience', *The Nation* (Bangkok), 6 February 1997: Local A5.

Harrison, J.S., and C.H. St John (1996) 'Managing and Partnering with External Stakeholders', *Academy of Management Executive* 10.2: 46-60.

Hart, S.L. (1997) 'Beyond Greening: Strategies for a Sustainable World', *Harvard Business Review*, January/February 1997: 66-76.

Hartman Group (1996) *Food and the Environment: A Consumer's Perspective, Phase I* (The Hartman Report; Bellvue, WA: The Hartman Group).

Hartman, C.L., and E.R. Stafford (1997) 'Market-Based Environmentalism: Developing Green Marketing Strategies and Relationships', in D.T. Leclair and M. Hartline (eds.), *American Marketing Association Winter Educators' Conference Proceedings* (Chicago: American Marketing Association): 156-63.

Hartmann, R. (1997) *Environmental Management and Controlling in European Telecommunication Enterprises* (Masters thesis; Fachhochschule Fulda, Germany).

Harvey, M.G. (1995) 'The MNC's Role and Responsibility in Deforestation of Tropical Forests', *Journal of Macromarketing*, Fall 1995: 107-27.

Hemphill, T. (1996) 'The New Era of Business Regulation', *Business Horizons* 39 (July/August 1996): 26-30.

Henderson, B.D. (1983) 'The Concept of Strategy', in K.J. Albert (ed.), *Handbook of Business Strategy* (New York: McGraw–Hill).

Henion, K.E. (1976) *Ecological Marketing* (Columbus, OH: Grid).

Henrichs, C. (1995) 'Renewable Energy Alternative Program' (Paper presented at *SolTech '95* Conference and UPVG Annual Meeting, San Antonio, TX, 11 April 1995).

Henschel, C. (1997) 'Bürgerbeteiligung in der Chemieindustrie' (unpublished speech, Hoechst AG, 31 January 1997).

Hindle, P., P. White and K. Minion (1993) 'Achieving Real Environmental Improvements Using Value–Impact Assessment', *Long Range Planning* 26.3: 36-48.

Hockerts, K. (1995) *Konzeptualisierung ökologischer Dienstleistungen: Dienstleistungskonzepte als Element einer wirtschaftsökologisch effizienten Bedürfnisbefriedigung* (IWÖ-Diskussionsbeitrag, 29; St Gallen, Switzerland: Universität St Gallen, Institut für Wirtschaft und Ökologie).

Hocking, R.W.D., and S. Power (1993) 'Environmental Performance: Quality, Measurement and Improvement', *Business Strategy and the Environment* 2.4: 19-24.

Hohmeyer, O. (ed.) (1997) *Social Costs and Sustainability: Valuation and Implementation in the Energy and Transport Sector* (Berlin: Springer Verlag).

Holland, F. (1998) 'Life thrives outside the concrete jungle', *South China Morning Post* 54.110 (22 April 1998): 19.

Holt, E. (1997a) 'Disclosure and Certification: Truth and Labeling for Electric Power', *REPP Renewable Energy Policy Project Issue Brief* 5 (January 1997).

Holt, E. (1997b) *Green Pricing Resource Guide* (Harpswell, ME: Ed Holt & Associates).
Holt, E., and J. Fang (1997) *The New Hampshire Retail Competition Pilot Program and The Role of Green Marketing* (Topical Issues Brief Series; Golden, CO: National Renewable Energy Laboratory).
Hoover (1990) *The Hoover Environmental Mission* (Merthyr Tydfil, UK: Hoover Ltd).
Hosseini, J.C., and S.N. Brenner (1992) 'The Stakeholder Theory of the Firm: A Methodology to Generate Value Matrix Weights', *Business Ethics Quarterly* 2.2: 99-119.
Hunt, C.B., and E.R. Auster (1991) 'Proactive Environmental Management: Avoiding the Toxic Trap', *Sloan Management Review*, Winter 1991: 7-18.
Hunt, S., and G. Shuttleworth (1996) *Competition and Choice in Electricity* (New York: John Wiley).
Hutchinson, C. (1992) 'Corporate Strategy and the Environment', *Long Range Planning* 25.4: 9-21.
Hutt, W.E. (1936) *Economists and the Public* (London: Jonathan Cape).
Hutton, R.B., G.A. Mauser, P. Filiatrault and O.T. Ahtola (1986) 'Effects of Cost-Related Feedback on Consumer Knowledge and Consumption Behavior: A Field Experimental Approach', *Journal of Consumer Research,* December 1986: 327-36.
IEA (International Energy Agency) (1997a) *Key Issues in Developing Renewables* (Paris: IEA).
IEA (International Energy Agency) (1997b) *Renewable Energy Policy in IEA Countries*. I. *Overview* (Paris: IEA).
Inchukul, K. (1998) 'Law alone may not be able to punish illegal waste dumpers', *Bangkok Post* 52 (30 April 1998): 19.
Indigo Development (1998) 'Eco-Industrial Parks', *http://www.indigodev.com/Ecoparks.html* (Oakland, CA).
Industrial Environmental Management (1997) 'Water Stress in Europe', *Industrial Environmental Management*, March 1997: 8.
Interface (1997) *Sustainability Report* (Atlanta, GA: Interface,).
International Ecolabeling Forum (1994) *Declaration of Representatives to International Ecolabeling Forum* (Washington, DC: International Ecolabeling Forum, 10–11 March 1994).
Intimate Brands (1998) 'Intimate Brands: Who We Are', *http://www.intimatebrands.com/who/index.asp*
Irish Times (1997) 'Energy Awareness Week' (A Commercial Supplement with *The Irish Times*), 20 September 1997.
IPD (Institute of Personnel and Development) (1997) *Innovation: Opportunity through People* (consultative document; London: IPD).
Irvine, S. (1991) 'Beyond Green Consumerism', in C. Plant and J. Plant (eds.), *Green Business: Hope or Hoax?* (Bideford, UK: Green Books): 21-29.
Irwin, A., and P.D. Hooper (1992) 'Clean Technology, Successful Innovation and the Greening of Industry: A Case Study Analysis', *Business Strategy and the Environment* 1.2: 1-12.
Iyer, E.S., and S. Gooding-Williams (1998) *A Typology of Nonprofit Organizations with Interest in Environmental Issues* (Amherst, MA: University of Massachusetts).
Jaccard, M (1995) 'Oscillating Currents', *Energy Policy* 23.7: 579-92.
Jackson, W.D., J.E. Keith and R.K. Burdick (1984) 'Purchasing Agents' Perceptions of Industrial Buying Centre Influence: A Situation Approach', *Journal of Marketing* 48: 75-83.
James, P. (1994) 'Quality and the Environment: From Total Quality Management to Sustainable Quality Management', *Greener Management International* 6 (April 1994): 62-70.
Jardine, C. (1994) 'A Blemish on its Face: Body Shop', *Daily Telegraph*, 20 August 1994.
Jelinski, L.W., T.E. Graedel, W.D. Laudise, D.W. McCall and K.N. Patel (1992) 'Industrial Ecology: Concepts and Approaches', *Proceedings of the National Academy of Sciences* 89 (February 1992): 793-97.
Jennings, P.D., and P.A. Zandbergen (1995) 'Ecologically Sustainable Organizations: An Institutional Approach', *The Academy of Management Review* 20.4: 1015-52.
Jevons, S. (1871) *The Principles of Economics* (New York, repr. 1965).
Johne, A., and P. Snelson (1988) 'Auditing Product Innovation Activities in Manufacturing Firms', *R&D Management* 11.2: 227-33.
Jones, M. (1997) 'The Role of Stakeholder Participation: Linkages to Stakeholder Impact Assessment and Social Capital in Camisea, Peru', *Greener Management International* 19: 87-98.
Journal of Sustainable Product Design (1997) 'Space and Water Saving Toilet and Washbasin Combination', *Journal of Sustainable Product Design,* July 1997.
Joss, S., and J. Durant (1995) *Public Participation in Science: The Role of Consensus Conferences in Europe* (London: Science Museum).
Joy, A., and B. Motzney (1992) 'Ecotourism and Ecotourists: Preliminary Thoughts on the New Leisure Traveller', *American Marketing Association*, Winter 1992: 457-63.
Kalisvaart, S.H., and T.J.J. van der Horst (1997) 'EcoDesign Innovation' (Presentation to DFMA-Forum; TNO Institute of Industrial Technology).

Kalke, M. (1994) 'The Foron-Story', *Akzente*, December 1994: 20-25.

Kangun, N., L. Carlson and S. Grove (1991) 'Environmental Advertising Claims: A Preliminary Investigation', *Journal of Public Policy and Marketing* 10.2: 47-58.

Karamitsos, F. (1998) 'Research and the Development of the Technologies of the Information Society Applied to the Environment', *Environmental Management and Health* 9.1: 6-9.

Kardash, W.J. (1974) 'Corporate Responsibility and the Quality of Life: Developing the Ecologically Concerned Consumer', in K.E. Henion and T.C. Kinnear (eds.), *Ecological Marketing* (Chicago: American Marketing Association).

Kärnä, A., and E. Heiskanen (1998) 'The Challenge of "Product Chain" Thinking for Product Development and Design: The Example of Electrical and Electronic Products', *Journal of Sustainable Product Design* 4 (January 1998): 26-36.

Kempton, W., and L. Montgomery (1982) 'Folk Quantification of Energy', *Energy* 7: 817-27.

Kempton, W., J.M. Darley and P.C. Stern (1992) 'Psychological Research for the New Energy Problems', *American Psychologist*, October 1992: 1213-23.

Keogh, P., and M.J. Polonsky (1998a) 'Environmental Commitment: A Basis for Environmental Entrepreneurship?', *Journal of Organizational Change Management* 11.1: 38-49.

Keogh, P., and M.J. Polonsky (1998b) 'Corporate Commitment to the Natural Environment: Issues in a Team Approach', in J. Moxen and P. Strachan, *Managing Green Teams: Environmental Change in Organisations and Networks* (Sheffield, UK: Greenleaf Publishing).

Keoleian, G., and D. Menerey (1994) 'Sustainable Development by Design: Review of Life Cycle Design and Related Approaches', *Journal of the Air and Waste Management Association* 44 (May 1994): 645-68.

Kepplinger, H.M., and U. Hartung (1995) *Störfall-Fieber: Wie ein Unfall zum Schlüsselereignis einer Unfall-Serie wird* (Freiburg: Karl Alber).

Ketola, T. (1993) 'The Seven Sisters: Snow Whites, Dwarfs or Evil Queens? A Comparison of the Official Environmental Policies of the Largest Oil Corporations in the World', *Business Strategy and the Environment* 2.3: 22-33.

Kilbourne, W.E., and S.C. Beckman (1998) 'Review and Critical Assessment of Research on Marketing and the Environment', *Journal of Marketing Management* 14.6: 513-32.

Kimmel, J.P. (1989) 'Disclosing the Environmental Impact of Human Activities: How a federal pollution control program based on individual decision making and consumer demand might accomplish the environmental goals of the 1970s in the 1990s', *University Of Pennsylvania Law Review* 138.2: 505-48.

Kirton, M.J. (1976) 'Adaptors and Innovators: A Description and Measure', *Journal of Applied Psychology* 61: 622-29.

Klafter, B. (1992) 'Pollution Prevention Benchmarking: AT&T and Intel Working Together with the Best', *Total Quality Environmental Management*, Autumn 1992: 27-34.

Kleeman, M. (1994) *Energy Use and Air Pollution in Indonesia* (Aldershot, UK: Ashgate Publishing).

Kleiner, A. (1991) 'What does it mean to be green?', *Harvard Business Review* 69.4 (July/August 1991): 38-47.

Klonoski, R.J. (1991) 'Foundational Considerations in the Corporate Social Responsibility Debate', *Business Horizons* 34.4: 9-18.

Kohli, A.K., and B.J. Jaworski (1990) 'Market Orientation: the Construct, Research Propositions, and Managerial Implications', *Journal of Marketing* 54 (April 1990): 1-18.

Kotler, P., and G. Armstrong (1996) *Principles of Marketing* (London: Prentice–Hall).

Kraisornsuthasinee, S. (1997) 'Back to the Most Basic of All: Organic Rice', *The Nation* 22 (8 December 1997): A5.

Kuusela, H., M.T. Spence and A.J. Kanto (1998) 'Expertise Effects on Prechoice Decision Processes and Final Outcomes: A Protocol Analysis', *European Journal of Marketing* 32.5/6: 559-76.

Lamarre, L. (1997) 'Utility customers go for the green', *Electric Power Research Institute Journal* 22.2 (March/April): 6-15.

Lang, D., and S. Trippolt (1997) 'Neuste Kennzahlen', *Cash Trade Magazine*, October 1997: 41-44.

Lawrence, J. (1993) 'Green products sprouting again: More focused efforts avoid controversy', *Advertising Age*, 10 May 1993: 12.

Lawrence, M. (1998) 'Nation's first cause-related marketing trends report finds consumers more responsive to strategic CRM campaigns', *http://www.celf.net/iprex/news*

Lawrence, P.R., and J.W. Lorsch (1967) *Organisation and Environment: Managing Differentiation and Integration* (Boston, MA: Harvard University Press).

Leake, M., and R. Kainz (1994) 'Environmental Enhancement: An Agile Manufacturing Perspective', *Agility Forum Paper* (Bethlehem, PA) AR94-03.

Lee, B.W., and K. Green (1994) 'Towards Commercial and Environmental Excellence: A Green Portfolio Matrix', *Business Strategy and the Environment* 3.3: 1-9.

Leszinski, R., and M. Marn (1997) 'Setting Value, Not Price', *The McKinsey Quarterly* 1: 99-110.

Levene, A. (1994) 'Greenpeace uses the market to make business "green up" its act', Reuters World Service, 13 March 1994.

Levitt, B., and J.G. March (1988) 'Organizational Learning', *Annual Review of Sociology* 14: 319-40.

Levy, D.L. (1997) 'Business and International Environmental Treaties: Ozone Depletion and Climate Change', *California Management Review* 39.2: 54-71.

Linnanen, L. (1995) 'Market Dynamics and Sustainable Organisations: HRM Implications in the Pulp and Paper Industry's Management of Environmental Issues', *Greener Management International* 10 (April 1995): 110-24.

Linnanen, L., E. Markkanen and L. Ilmola (1997) *Ympäristöosaaminen: Kestävän kehityksen haaste yritysjohdolle (Environmental Excellence: The Sustainable Business Challenge)* (Helsinki: Otaniemi Consulting Group Oy).

Lober, D.J. (1997) 'Explaining the Formation of Business–Environmentalist Collaborations: Collaborative Windows and the Paper Task Force', *Policy Sciences* 30.1: 1-24.

Löfstedt, R.E., and O. Renn (1997) 'The Brent Spar Controversy: An Example of Risk Communication Gone Wrong', *Risk Analysis* 17.2: 131-36.

Long, F.J., and M.B. Arnold (1995) *The Power of Environmental Partnerships* (Fort Worth, TX: Dryden Press).

Loser, T. (1998) 'Billa/Rewe', *Cash Trade Magazine*, June 1998: 14-16.

Lowe, E. (1993) 'Industrial Ecology: An Organising Framework for Environmental Management', *Total Quality Environmental Management* 3.1: 73-85.

Lowe, E. (1994) 'Industrial Ecology: Implications for Corporate Strategy', *Journal of Corporate Environmental Strategy* 2.1: 61-65.

Mackoy, R.D., R. Calantone and C. Droge (1995) 'Environmental Marketing: Bridging the Divide between the Consumption Culture and Environmentalism', in M.J. Polonsky and A.T. Mintu-Wimsatt (eds.) *Environmental Marketing: Strategies, Practice, Theory and Research* (New York: Haworth Press): 37-54.

MacNaghten, P., *et al.* (1995) *Public Perceptions and Sustainability in Lancashire: Indicators, Institutions, Participation* (Lancashire County Council and Lancaster University Centre for the Study of Environmental Change, UK).

Makower, J. (1998) 'Green Product Introductions for 1997', *Green Business Letter*, January 1998.

Marketing News (1991) 'American Marketing Association Environmental Policy Statement', *Marketing News*, 22 July 1991: 22.

Marketing Week (1994) 'Maverick's Fall from Grace: Body Shop', *Marketing Week*, 2 September 1994.

Marks, L. (1994) 'Chattering Evangelist: Anita Roddick', *The Independent on Sunday*, 28 August 1994.

Mastrandonas, A., and P.T. Strife (1992) 'Corporate Environmental Communications: Lessons from Investors', *The Columbia Journal of World Business*, 27.3–4 (Fall/Winter 1992): 234-40.

Matichon (1998) 'Pipeline not transparent, yet given a go', *Matichon* (Bangkok), 26 February 1998: Local 1.

Maxwell, J., S. Rothenberg, F. Briscoe and A. Marcus (1997) 'Green Schemes: Corporate Environmental Strategies and their Implementation', *California Management Review* 39.3: 118-34.

Mayer, R.N., D.L. Scammon and J. Gray-Lee (1993) 'An Audit of Claims on Product Labels', in M.J. Sheffet (ed.), *Proceedings of the 1993 Marketing and Public Policy Conference* (East Lansing, MI: Michigan State University): 109-22.

Mayer, R.N., J. Gray-Lee, D.L. Scammon and B. Cude (1996) 'The Last Audit: Environmental Marketing Claims between 1992 and 1995', paper presented to the *American Marketing Association Pre-Conference*, 3 August 1996.

Mayers, K. (1997) 'The Body Shop: The Role of Supplier Evaluation and Life Cycle Analysis (LCA) in Product Development', in M. Charter (ed.), *Managing Eco-Design: A Training Solution* (Farnham, UK: Centre for Sustainable Design/Surrey Institute of Art and Design): 24-26.

McCloskey, M. (1992) 'Twenty Years of Change in the Environmental Movement: An Insider's View', in R.E. Dunlap and A.G. Mertig (eds.), *American Environmentalism: The US Environmental Movement, 1970–1990* (Philadelphia: Taylor & Francis): 77-88.

McCrea, C.D. (1993) 'Environmental Packaging and Product Design', paper presented at the *1992–93 Professional Development Series: American Marketing Association (Boston Chapter) Conference*, 22 April 1993.

McDonagh, P., and A. Prothero (eds.) (1996) *Green Management* (London: Dryden Press).

McGee, E.C., and A.K. Bhushan (1993) 'Applying the Baldridge Quality Criteria to Environmental Performance: Lessons from Leading Organizations', *Total Quality Environmental Management*, Autumn 1993: 1-18.

Meadows, D., *et al.* (1972) *The Limits to Growth: A Report for the Club of Rome's Project on the Predicament of Mankind* (New York: Universe Books).
Meffert, H., and M. Kirchgeorg (1998) *Marktorientiertes Umweltmanagement: Konzeption, Strategie, Implementierung* (Stuttgart: Schäffer Poeschel, 3rd edn).
Meijkamp, R. (1997) *Car-Sharing: Analysis of a Service Concept between Eco-Efficiency and Sufficiency* (Wirtschaftsökologische Perspektiven, Annual Report; Bayreuth, Germany: Bayreuth Initiative for Economy and the Environment).
Meister, H.-P. (1996) 'Community Advisory Panels in den USA', in H. Hill (ed.), *Dialoge über Grenzen: Kommunikation bei Public Private Partnership* (Reihe Staatskommunikation, 3; Köln: Heymann): 17-27.
Mendleson, N., and M.J. Polonsky (1995) 'Using Strategic Alliances to Develop Credible Green Marketing', *Journal of Consumer Marketing* 12.2: 4-18.
Menon, A., and A. Menon (1997) 'Enviropreneurial Marketing Strategy: The Emergence of Corporate Environmentalism as Market Strategy', *Journal of Marketing* 61.1: 51-67.
Mesoamerica (1992) 'Environment and the Banana Industry: Costa Rican growers say they're seeing green', *Mesoamerica*, July 1992.
Midttun, A. (ed.) (1997) *European Electricity Systems in Transition. A Comparative Analysis of Policy and Regulation in Western Europe* (Amsterdam: Elsevier).
Mies, M. (1993) 'The Myth of Catching up Development', in M. Mies and V. Shiva (eds.), *Ecofeminism* (London: Zed Books): 55-69.
Miles, M., and L.S. Munilla (1995) 'The Eco-Marketing Orientation: An Emerging Business Philosophy', in M.J. Polonsky and A.T. Mintu-Wimsatt (eds.), *Environmental Marketing: Strategies, Practice, Theory and Research* (New York: Haworth Press): 23-36.
Millais, C. (1996) 'Greenpeace Solutions Campaigns: Closing the Implementation Gap', *ECOS: Journal of the British Association of Nature Conservationists* 17.2: 50-58.
Miller, C. (1993) Conflicting studies still have execs wondering what to believe', *Marketing News* 27.12: 1.
Miller, R.L., and W.F. Lewis (1991) 'A Stakeholder Approach to Marketing Management Using the Value Exchange Model', *European Journal of Marketing* 25.8: 55-68.
Milliman, J., J.A. Clair and I. Mitroff (1994) 'Environmental Groups and Business Organizations: Conflict or Cooperation?', *SAM Advanced Management Journal* 59 (Spring 1994): 41-46.
Mills, D. (1998) 'Mills on Business', *Campaign London*, 22 May 1998: 27.
Milne, G.R., E.S. Iyer and S. Gooding-Williams (1996) 'Environmental Organization Alliance Relationships within and across Nonprofit, Business, and Government Sectors, *Journal of Public Policy and Marketing* 15.2 (Fall 1996): 203-15.
Mintel (1991) *The Second Green Consumer Report* (London: Mintel).
Mintel (1995) *The Green Consumer Report* (London: Mintel).
Minton, A.P., and R.L. Rose (1997) 'The Effects of Environmental Concern on Environmentally Friendly Consumer Behavior: An Exploratory Study', *Journal of Business Research* 40: 37-48
Mintzberg, H., and J.B. Quinn (1991) *The Strategy Process* (Englewood Cliffs, NJ: Prentice–Hall).
Mobility (1998) *Car Sharing and Mobility Management* (press release; Lucerne, Switzerland: Mobility, 30 March 1998).
Mohr, L.A., D. Erodlu and P.S. Ellen (1998) 'The Development and Testing of a Measure of Skepticism toward Environmental Claims in Marketers' Communications', *The Journal of Consumer Affairs* 32 (Summer 1998): 30-55.
Moore, K.J. (1993) 'Emerging Themes in Environmental Consumer Behavior', in M.J. Sheffet (ed.), *Proceedings of the 1993 Marketing and Public Policy Conference* (East Lansing, MI: Michigan State University): 109-22.
Morgan, G. (1997) *Images of Organisation* (London: Sage).
Morris, S.A. (1997) 'Internal Effects of Stakeholder Management Devices', *Journal of Business Ethics* 16.4: 413-24.
Moskovitz, D. (1992) *Renewable Energy: Barriers and Opportunities, Walls and Bridges* (Prepared for The World Resources Institute; Gardiner, ME: The Regulatory Assistance Project, July 1992, revised September 1993).
Moxen, J., and P. Strachan (1995) 'The Formulation of Standards for Environmental Management Systems: Structural and Cultural Issues', *Greener Management International* 12 (October 1994): 32-48.
Mulhern, F.J. (1992) 'Consumer Wants and Consumer Welfare', in T.C. Allen *et al.* (eds.), *Marketing Theory and Applications: Proceedings of the 1992 American Marketing Association Winter Educators' Conference* (Chicago: AMA): 407-12.

Murphy, D.F., and J. Bendell, (1997) *In the Company of Partners: Business, Environmental Groups, and Sustainable Development Post-Rio* (Bristol, UK: The Policy Press).

Naess, A. (1993) 'The Deep Ecological Debate: Some Philosophical Aspects', in M.E. Zimmerman, J.B. Callicott, G. Sessions, K.J. Warren and J. Clark (eds.), *Environmental Philosophy* (Englewood Cliffs, NJ: Prentice–Hall): 193-212.

Naj, A.K. (1990) 'Some companies cut pollution by altering production methods', *The Wall Street Journal*, 24 December 1990: A1.

National Association of Attorneys-General (1990) *The Green Report: Findings and Preliminary Recommendations for Responsible Advertising* (Washington, DC: NAAG Publications, December 1990).

Neale, A. (1997) 'Organisational Learning in Contested Environments: Lessons from Brent Spar', *Business Strategy and the Environment* 6.2 (May 1997): 93-103.

New, S., K. Green and B. Morton (1995) 'Understanding the Dynamics of Green Supply', *Proceedings of the 1995 Business Strategy and Environment Conference* (Leeds, UK: ERP Environment): 171-76.

Newbery, D.M. (1997) 'Privatisation and Liberalisation of Network Utilities', *European Economic Review* 41: 357-83.

Newman, J.C., and K.M. Breeden (1992) 'Managing in the Environmental Era: Lessons from the Environmental Era', *The Columbia Journal of World Business* 27.3–4 (Fall/Winter 1992): 210-21.

Nicolas, R. (1998) 'New age finds a new face', *Marketing*, 21 May 1998: 15.

Nius Ecos del Atlántico (1995) 'Aliados para proteger el medio ambiente', *Nius Ecos del Atlántico* 37.

Nixon, W. (1993) 'A Breakfast among Peers: Environmental whistleblowers have some stories to tell', *E Magazine*, September/October 1993: 14-19.

NREL (National Renewable Energy Laboratory) (1998) Information Brief on Green Power Marketing (Prepared by B. Swezey and A. Houston; Golden, CO: NREL, 2nd edn).

Nuttall, N. (1998) 'Paper recycling schemes fold as price slumps', *The Times*, 26 June 1998: 13.

Nuttall, N., S. Gilchrist and S. MacCarthaigh (1994) 'Body Shop pressure builds', *The Times*, 24 August 1994.

O'Donnell, F. (1993) 'Corporate America turns on to the environment', *Nature Conservancy*, January/February 1993: 28-32.

O'Neal, C. (1993) 'Concurrent Engineering with Early Supplier Involvement: A Cross-Functional Challenge', *International Journal of Purchasing and Materials Management*, Spring 1993: 3-9.

O'Reilly, B. (1997) 'Transforming the Power Business', *Fortune*, 29 September 1997: 142-56.

Oakley, B.T. (1994) 'Total Quality Product Design: How to Integrate Environmental Criteria into the Product Realization Process', in J.T. Willig (ed.), *Environmental TQM* (New York: McGraw–Hill): 63-70.

Observer (1994) 'Green Victory: Body Shop Criticism Unfair', *The Observer*, 11 September 1994.

Oosterhuis, F., F. Rubik and G. Scholl (1996) *Product Policy in Europe: New Environmental Perspectives* (Dordrecht: Kluwer Academic Publishers).

Orlin, J., P. Swalwell and C. FitzGerald (1993) 'How to Integrate Information Strategy Planning with Environmental Management Information Systems: Part 1', *Total Quality Environmental Management*, Winter 1993–94: 193-202.

Osterhus, T.L. (1997) 'Pro-Social Consumer Influence Strategies: When and how do they work?', *Journal of Marketing* 61 (October 1997): 16-29.

Ostlund, S. (1994) 'The Limits and Possibilities in Designing the Environmentally Sustainable Firm', *Business Strategy and the Environment* 3.2: 21-33.

Ottman, J.A. (1992) *Green Marketing: Challenges and Opportunities for the New Marketing Age* (Lincolnwood, IL: NTC Business Books).

Ottman, J.A. (1995) 'A little creativity could lead to a big advantage', *Marketing News* (Chicago) 29.7 (27 March 1995): 11.

Ottman, J.A. (1997) 'Renewable Energy: Ultimate Marketing Challenge', *Marketing News*, 28 April 1997.

Ottman, J.A. (1998) *Green Marketing: Opportunity for Innovation* (Lincolnwood, IL: NTC/Contemporary Books, 2nd edn).

Oxfam (1996) 'What is Fair Trade', *www.web.net/oxfamgft* (UK and Ireland: Oxfam).

PA Consulting (1991) *Environmental Labelling of Washing Machines: A Pilot Study for the DTI/DoE* (Royston, UK: PA Consulting Group, August 1991).

Passel, P. (1990) 'Rebel economists add ecological cost to price of progress', *The New York Times*, 27 November 1990.

Paton, B. (1993) 'Environmentally Conscious Product Design through Total Quality Management', *Environmental Quality Management* 2.4: 383-97.

Payne, A. (1995) 'Relationship Marketing: A Broadened View of Marketing', in A. Payne (ed.), *Advances in Relationship Marketing* (London: Kogan Page): 29-41.

Payne, J.W., J.R. Bettman and E.J. Johnson (1993) *The Adaptive Decision Maker* (New York: Cambridge University Press).

Peattie, K. (1995) *Environmental Marketing Management: Meeting the Green Challenge* (London: Pitman).

Peattie, K., and M. Charter (1994) 'Green Marketing', in M. Baker (ed.), *The Marketing Book* (London: Butterworth–Heinemann, 3rd rev. edn).

Pegram Walters Associates (1994) *Consumer Attitudes to Packaging* (PWA 2769, prepared for INCPEN; London: Pegram Walters Associates).

Peters, T. (1991) 'Lean, Green and Clean: The Profitable Company of the Year 2000', *European Management Development Journal*, January 1991: 5-8.

Petulla, J.M. (1980) *American Environmentalism: Values, Tactics, Priorities* (Galveston, TX: A&M University Press).

Petulla, J.M. (1987) 'Environmental Management in Industry', *Journal of Professional Issues in Engineering* 113. 2: 167-83.

Piasecki, B.W. (1995) *Corporate Environmental Strategy: The Avalanche of Change since Bhopal* (New York: John Wiley).

Pickard, J. (1996) 'A Fertile Grounding', *People Management* 2.21 (24 October 1996): 34.

Pinchot, G. (1996) Conference Tapes, *Institute of Personnel and Development Annual Conference*, Harrogate, UK, 23–25 October 1996.

Plant, C., and D.H. Albert (1991) 'Green Business in a Grey World: Can it be done?', in C. Plant and J. Plant (eds.), *Green Business: Hope or Hoax?* (Bideford, UK: Green Books): 1-8.

Platt, S. (1988) *Teams: A Game to Develop Group Skills* (Aldershot, UK: Gower).

Poduska, R.A., R.H. Forbes and M.A. Bober (1992) 'The Challenge of Sustainable Development', *The Columbia Journal of World Business* 27.3–4 (Fall/Winter 1992): 287-91.

Pokorny, G. (1994) *The Marketing of Energy Efficiency in the 1990s: A 20-Year Review of Market Research and Experience* (Strategic Publications Series, 214; Cambridge, MA: Cambridge Reports/Research International).

Polonsky, M.J. (1991) 'Australia sets guidelines for green marketing', *Marketing News* 25.21: 6, 18.

Polonsky, M.J. (1995a) 'Cleaning Up Green Marketing Claims: A Practical Checklist', in M.J. Polonsky and A.T. Mintu-Wimsatt (eds.), *Environmental Marketing: Strategies, Practice, Theory and Research* (New York: Haworth Press): 199-223.

Polonsky, M.J. (1995b) 'A Stakeholder Theory Approach to Designing Environmental Marketing Strategy', *Journal of Business and Industrial Marketing* 10.3: 29-46.

Polonsky, M.J. (1996) 'Stakeholder Management and the Stakeholder Matrix: Potential Strategic Marketing Tools', *Journal of Market Focused Management* 1.3: 209-29.

Polonsky, M.J., and A.T. Mintu-Wimsatt (eds.) (1995) *Environmental Marketing: Strategies, Practice, Theory and Research* (New York: Haworth Press).

Polonsky, M.J., and J. Ottman (1997) 'Including Stakeholders in the New Product Development Process for Green Products', in J. Webber and K. Rehbein (eds.), *International Association of Business and Society 8th Annual Conference* (Milwaukee, WI: IABS): 302-308.

Polonsky, M.J., and J. Ottman (1998) 'Stakeholders' Contribution to the Green New Product Development Process', *Journal of Marketing Management* 14: 533-57.

Polonsky, M.J., H. Brooks, P. Henry and C. Schweizer (1998a) 'An Exploratory Examination of Environmentally Responsible Straight Rebuy Purchases in Large Australian Organisations', in T. Russel (ed.), *Greener Purchasing: Threats and Innovations* (Sheffield, UK: Greenleaf Publishing).

Polonsky, M.J., P. Rosenburger and J. Ottman (1998b) 'Developing Green Products: Learning from Stakeholders', *Journal of Sustainable Product Design* 5 (April 1998): 7-21.

Polonsky, M.J., H. Suchard and D. Scott (1997) 'A Stakeholder Approach to Interacting with the External Environment', in P. Reed, S. Luxton and M. Shaw (eds.), *Proceedings of the 1997 Australian and New Zealand Marketing Educators' Conference* (Melbourne: ANZMA): 495-508.

Poon, A. (1993) 'The New Tourism Revolution', *Tourism Management* 15.2: 91-92.

Porter, M.E. (1979) 'How Competitive Forces Shape Strategy', *Harvard Business Review*, March/April 1979.

Porter, M.E. (1980) *Competitive Strategy: Techniques for Analyzing Industries and Competition* (New York: Free Press).

Porter, M.E. (1981) 'The Contributions of Industrial Organization to Strategic Management', *The Academy of Management Review* 6.4: 609-20.

Porter, M.E. (1985) *Competitive Advantage: Creating and Sustaining Superior Performance* (New York: Free Press).

Porter, M.E. (1991) 'America's Green Strategy', *Scientific American*, April 1991: 168.

Porter, M.E., and C. van der Linde (1995) 'Green and Competitive: Ending the Stalemate', *Harvard Business Review* 73.5 (September/October 1995): 120-33.

Post, J., and B. Altman (1998) 'Managing the Environmental Change Process: Barriers and Opportunities', in J. Moxen and P. Strachan *Managing Green Teams: Environmental Change in Organisations and Networks* (Sheffield, UK: Greenleaf Publishing): 84-91.

Pracamthong, I. (1998) 'The Thai–Burmese Gas Pipeline: The Dark Future or the Bright Route?', *Loak See Kieaw* (Bangkok), March/April 1998: 20-30.

Prahalad, C., and G. Hamel (1990) 'The Core Competence of the Corporation', *Harvard Business Review*, May/June 1990: 79-91.

Probst, G. (1987) *Selbst-Organisation: Ordnungsprozesse in sozialen Systemen aus ganzheitlicher Sicht* (Berlin: Paul Parey).

Prothero, A. (1990) 'Green Consumerism and the Social Marketing Concept: Marketing Strategies for the 1990s', *Journal of Marketing Management* 6.2: 87-103.

PTT (1997) *PTT Spirit: Bulletin* (Bangkok: PTT, 16–31 December 1997).

Pujari, D. (1996) *Environmental New Product Development* (unpublished doctoral dissertation; Bradford, UK: University of Bradford).

Pujari, D., and G. Wright (1996) 'Developing Environmentally-Conscious Product Strategy (ECPS): A Qualitative Study of Selected Companies in Britain and Germany', *Marketing Intelligence and Planning* 14.1: 19-28.

Purser, R.E., C. Park and A. Montuori (1995) 'Limits to Anthropocentrism: Toward an Ecocentric Organization Paradigm?', *The Academy of Management Review* 20.4: 1053-89.

Racho, J. (1997) 'Green Products: Trying to Break New Ground', *Rochester Business Journal*, 22 August 1997: 15.

Rainforest Alliance (1995a) *Informativo ECO-OK: Edición No. 3* (San José, Costa Rica: Rainforest Alliance).

Rainforest Alliance (1995b) *The Canopy* (New York: Rainforest Alliance, November/December 1995).

Rainforest Alliance (1995c) 'Programa de Certificación ECO-OK', *ECO-OK Manual* (San José, Costa Rica: Rainforest Alliance, March 1995).

Rainforest Alliance (1995d) *ECO-OK and Organic Certification: Two Different Approaches* (New York: Rainforest Alliance).

Rainforest Alliance (1997) *Agricultural Certification Program: General Production Standards* (New York: Rainforest Alliance, August 1997).

Rainforest Alliance (1998a) *The Rainforest Alliance's ECO-OK Certification Program* (New York: Rainforest Alliance).

Rainforest Alliance (1998b) *The Rainforest Alliance's ECO-OK Certification Program: Program Accomplishments* (New York: Rainforest Alliance).

Rainforest Alliance (1998c) *The Rainforest Alliance's Agricultural Certification Program* (New York: Rainforest Alliance).

Rainforest Alliance (1998d) *Guiding Principles of the Conservation Agriculture Network* (New York: Rainforest Alliance).

Rainforest Alliance (1998e) *The Rainforest Alliance's Agricultural Certification Program: Better Banana Project* (New York: Rainforest Alliance).

Rainforest Alliance (1998f) 'Using the ECO-OK or Better Banana Logos: Requirements for Use of the Certification Marks', *www.rainforest.alliance.com* (April 1998).

Rainforest Alliance (1998g) 'Bridging the Gulf between Environmentalists and Industry: Crucial Credibility', *www.rainforest.alliance.com* (April 1998).

Rainforest Alliance (1998h) 'Certified Farms', *www.rainforest.alliance.com* (April 1998).

Randall, C. (1994) 'Television Journalist who Crossed Swords with the Green Giants: Profile of Jon Entine', *The Daily Telegraph*, 1 September 1994.

Rank Xerox (1995) *Environmental Performance Report* (Marlow, UK: Rank Xerox Ltd).

Redclift, M., and T. Benton (1994) *Social Theory and the Global Environment* (London: Routledge).

Reilly, W.K. (1990) 'The Green Thumb of Capitalism: The Environmental Benefits of Sustainable Growth', *Policy Review*, Fall 1990: 16-21.

Reitman, V. (1997) 'Ford is investing in Daimler-Ballard fuel-cell venture', *The Wall Street Journal*, 16 December 1997: B4.

Renn, O., and D. Levine (1991) 'Credibility and Trust in Risk Communication', in R.E. Kasperson (ed.), *Communicating Risks to the Public* (Dordrecht: Kluwer).

Rennie, C., and A. MacLean (1989) *Salvaging the Future: Waste Based Production* (Washington, DC: Institute for Local Self-Reliance).

Reuters (1994) 'Body Shop says claims "riddled with errors"', Reuters News Service, 2 September 1994.

Roberts, J.A. (1996) 'Green Consumers in the 1990s: Profile and Implications for Advertising', *Journal of Business Research* 26.2: 217-31.

Roberts, N.C., and P.J. King (1989) 'The stakeholder audit goes public', *Organizational Dynamics* 17.3: 63-79.

Roddick, A. (1991) *Body and Soul* (London: Ebury Press).

Rogers, C. (1967) *On Becoming a Person: A Therapist's View of Psychotherapy* (London: Constable).

Rogers, E.M. (1995) *Diffusion of Innovations* (New York: Free Press, 4th edn).

Rohrschneider, R. (1991) 'Public Opinion toward Environmental Groups in Western Europe: One Movement or Two?', *Social Science Quarterly* 72 (June 1991): 251-66.

Rokeach, M. (1973) *The Nature of Human Values* (New York: Free Press).

Roome, N. (1994) 'Business Strategy, R&D Management and Environmental Imperatives', *R&D Management* 24.1: 65-81.

Roper Organization (1990) *The Environment: Public Attitudes and Individual Behavior* (Report on the study commissioned by S.C. Johnson & Son, Inc.; Roper Organization).

Roper Organization (1992) *Environmental Behavior. North America: Canada, Mexico, United States* (Report on the study commissioned by S.C. Johnson & Son, Inc.; Roper Organization).

Roper-Starch (1997) *Worldwide Green Gauge Study* (Mamoroneck, NY: Roper-Starch).

Rose, C. (1998) *The Turning of the Spar* (London: Greenpeace).

Rosen, H. (1995) 'The Evil Empire: The Real Scoop on Ben & Jerry's Crunchy Capitalism', *The New Republic*, 11 September 1995: 22-25.

Rossi Umaña, J. (1994) 'El caso bananero', *La Republica* (Costa Rica), 24 December 1994.

Rothenberg, S., J. Maxwell and A. Marcys (1992) 'Issues in the Implementation of Proactive Environmental Strategies', *Business Strategy and the Environment* 1.4: 1-12.

Rothermund, H. (1996) 'Out of Sight, Not out of Mind: Societal Considerations in Offshore Development' (speech to the *Offshore Northern Seas Conference*, Stavanger, Norway, 29 August 1996).

Rothstein, S., and J. Fang (1997) *Green Marketing in the Massachusetts Electric Retail Competition Program* (Topical Issues Brief Series; Golden, CO: National Renewable Energy Laboratory).

Rothwell, R. (1992) 'Successful Industrial Innovation: Critical Factors for the 1990s', *R&D Management* 22.3: 221-39.

Rowley, T. (1997) 'Moving Beyond Dyadic Ties: A Network Theory of Stakeholders Influence', *Academy of Management Review* 22.4: 887-910.

Roy, R. (1996) Video Guide for Block 4. Video 2: *Green Product Development—The Hoover 'New Wave'* (Supplementary Material, T302 'Innovation: Design, Environment and Strategy'; Milton Keynes, UK: The Open University).

Roy, R. (1997) 'Design for Environment in Practice: Development of the Hoover New Wave Washing Machine Range', *Journal of Sustainable Product Design* 1 (April 1997): 36-43.

Roy, R., M.T. Smith and S. Potter (1998) 'Green Product Development: Factors in Competition', in T. Barker and J. Köhler (eds.), *International Competitiveness and Environmental Policies* (Cheltenham, UK: Edward Elgar): 265-75.

Royal Dutch/Shell Group (1998) *The Shell Report 1998. Profits and Principles: Does there have to be a choice?* (London: Shell International; *www.shell.com*).

Rubenstein, D.B. (1992) 'Bridging the Gap between Green Accounting and Black Ink', *Accounting, Organizations and Society* 17.5: 501-508.

Rudall Blanchard Associates (1996) *Brent Spar Abandonment BPEO* (London: Shell UK; *www.shellexpro.brentspar.com*).

Rushton, B.M. (1993) 'How Protecting the Environment Impacts R&D in the United States', *Research Technology Management* 36.3 (May/June 1993): 13-21.

Russo, M.V., and P.A. Fouts (1997) 'A Resource-Based Perspective on Corporate Environmental Performance and Profitability', *Academy of Management Journal* 40.3: 534-59.

Ryan, C., M. Hosken and D. Greene (1992) 'EcoDesign: Design and the Response to the Greening of the International Market', *Design Studies* 13.1 (January 1992): 5-22.

Ryel, R., and T. Grosse (1991) 'Marketing Ecotourism: Attracting the Elusive Ecotourist', in T. Whelan (ed.), *Nature Tourism: Managing for the Environment* (Washington, DC: Island Press): 164-85.

Sadgrove, K. (1992) *The Green Guide to Profitable Management* (Aldershot, UK: Gower).

Salpukas, A. (1998) 'California's effort to promote plan for electricity is off to a slow start', *The New York Times*, 26 February 1998: D1, D6.

Salzhauer, A.L (1995) 'Obstacles and Opportunities for a Consumer Ecolabel Environment', *Sky* 33.9.

Samuelson, C.D. (1990) 'Energy Conservation: A Social Dilemma Approach', *Social Behavior* 5: 207-30.

Sarkin, D., and J. Schelkin (1991) 'Environmental Concerns and the Business of Banking', *The Journal of Commercial Bank Lending* 73.3: 7-19.

Sarkis, J. (1995a) 'Supply Chain Management and Environmentally Conscious Design and Manufacturing', *International Journal of Environmentally Conscious Design and Manufacturing* 4.2: 43-52.

Sarkis, J. (1995b) 'Manufacturing Strategy and Environmental Consciousness', *Technovation* 15.2: 79-97.

Savage, G.T., T.W. Nix, C.J. Whitehead and J.D. Blair (1991) 'Strategies for Assessing and Managing Organizational Stakeholders', *Academy of Management Executive* 5.2: 61-75.

Scammon, D.L., and R.N. Mayer (1995) 'Agency Review of Environmental Marketing Claims: Case-by-Case Decomposition of the Issues', *Journal of Advertising* 24.2 (Summer 1995): 33-43.

Schein, E.H. (1984) 'Coming to a New Awareness of Organisational Culture', *Sloan Management Review* 25 (Winter 1984): 3-16.

Schendel, D.E., and C.W. Hofer (1979) *Strategic Management: A New View of Business Policy and Planning* (Boston, MA: Little, Brown & Co.).

Schiffman, L.G., and L.L. Kanuk (1994) *Consumer Behavior* (Englewood Cliffs, NJ: Prentice–Hall, 5th edn).

Schlegelmilch, B.B., G.M. Bohlen and A. Diamantopoulos (1996) 'The Link between Green Purchasing Decisions and Environmental Consciousness', *European Journal of Marketing* 30.5: 35-55.

Schlossberg, H. (1993) 'Report says environmental marketing claims level off', *Marketing News* 27.11 (24 May 1993): 12.

Schmidheiny, S., with the Business Council for Sustainable Development (1992) *Changing Course: A Global Perspective on Development and the Environment* (Cambridge, MA: MIT Press).

Schmidt, K. (1995) *Eco-Labelling: Environmental Protection or Trade Discrimination?* (New York: Rainforest Alliance, April 1995).

Schoell, W.F., and J.P. Guiltinan (1992) *Marketing: Contemporary Concepts and Practices* (Boston, MA: Allyn & Bacon, 5th edn).

Schönefeld, L. (1996) 'Dialogorientierte Unternehmenskommunikation im lokalen Umfeld: Das Beispiel Hoechst', in G. Bentele *et al.*, *Dialogorientierte Unternehmenskommunikation: Grundlagen, Praxiserfahrungen, Perspektiven* (Berlin: Vistas): 369-92.

Schoolman, J. (1994) 'Politically correct investing faces continued challenges', Reuters News Service, 7 September 1994.

Schoon, N. (1997) 'What Price Nature? At £20 trillion a year it is truly our most precious asset', *The Independent*, 15 May 1997: 9.

Schot, J., E. Brand and K. Fischer (1997) 'The Greening of Industry for a Sustainable Future: Building an International Research Agenda', *Business Strategy and the Environment* 6.3: 153-62.

Schrader, U. (1998) *Empirische Einsichten in die Konsumentenakzeptanz öko-effizienter Dienstleistungen* (Lehr- und Forschungsbericht, 42; Hannover: Universität Hannover, Lehrstuhl Markt und Konsum).

Schwartz, J., and T. Miller (1991) 'The Earth's Best Friends', *American Demographics* 13 (February 1991): 26-35.

Schwepker, C.H., Jr, and T.B. Cornwell (1991) 'An Examination of Ecologically Concerned Consumers and their Intention to Purchase Ecologically Packaged Products', *Journal of Public Policy and Marketing* 10.2 (Fall 1991): 77-101.

Scientific American (1997) 'Change in the Wind: Utilities are starting to offer renewable energy—for a price', *Scientific American*, October 1997: 23.

Scitovsky, T. (1976) *The Joyless Economy* (New York: Oxford University Press).

Senge, P. (1990) *The Fifth Discipline: The Art and Practice of the Learning Organization* (New York: Doubleday).

Serirat, S. (1996) *Kanthlat Pua Sengwadlom* (*Green Marketing*) (Bangkok: Pattana Suksa Publications).

Shanoff, B.S (1996) 'Environmental Survey Supports State/Local Role', *World Wastes* 39.10: 18-21.

Sharma, S., H. Vredenburg and F. Westley (1994) 'Strategic Bridging: A Role for the Multinational Corporation in Third World Development', *Journal of Applied Behavioral Science* 30.4: 458-76.

Sheldon, C. (ed.) (1997) *ISO 14001 and Beyond: Environmental Management Systems in the Real World* (Sheffield, UK: Greenleaf Publishing).

Shell Transport & Trading (1996) *Annual Report 1995* (London: Shell Transport & Trading Company plc).

Shell UK (1996) *Environmental Report 1995* (London: Shell UK).

Shelton, R.D. (1994) 'Hitting the Green Wall: Why Corporate Programs Get Stalled', *Corporate Environmental Strategy* 2.2: 5-11.

Shrivastava, P. (1983) 'A Typology of Organizational Learning Systems', *Journal of Management Studies* 20.1: 7-28.

Shrivastava, P. (1994) 'Castrated Environment: Greening Organizational Studies', *Organization Studies* 15.5: 705-26.

Shrivastava, P. (1995a) 'The Role of Corporations in Achieving Ecological Sustainability', *The Academy of Management Review* 20.4: 936-60.

Shrivastava, P. (1995b) 'Ecocentric Management for a Risk Society', *Academy of Management Review* 20.1: 118-38.

Shrum, L.J., J.A. McCarty and T.M. Lowrey (1995) 'Buyer Characteristics of the Green Consumer and their Implications for Advertising Strategy', *Journal of Advertising* 24.2 (Summer 1995): 71-82.

Silver, M. (1989) *Foundations of Economic Justice* (New York: Basil Blackwell).

Simon, F.L. (1992) 'Marketing Green Products in the Triad', *The Columbia Journal Of World Business* 27.3–4 (Fall/Winter 1992): 268-85.

Simon, H. (1989) *Price Management* (Amsterdam: North-Holland).

Sinetar, M. (1992) 'Entrepreneurs, Chaos and Creativity: Can creative people survive large company structure?', in M. Syrett and C. Hogg, *Frontiers of Leadership: An Essential Reader* (Oxford: Blackwell): 109-16.

Smith, E.T. (1991) 'Doing it for Mother Earth', *Business Week*, 23 August 1991: 44-49.

Smith, G. (1990) ' How Green Is My Valley', *Marketing and Research Today* (Netherlands) 2: 76-82.

Smith, M.T., R. Roy and S. Potter (1996) *The Commercial Impacts of Green Product Development* (Report DIG-05; Milton Keynes, UK: The Open University, Design Innovation Group, July 1996).

SMUD (Sacramento Municipal Utility District) (1995) *Achieving Municipal Power Goals in a Competitive Age* (1995 Integrated Resource Plan IV; Sacramento, CA: SMUD).

SOCMA (Synthetic Organic Chemical Manufacturers Association) (n.d.) 'Responsible Care: A Journey of Continuous Improvement', *SOCMA Bulletin*.

Speer, T. (1997) 'Growing the Green Market', *American Demographics* 19.8 (August 1997): 45-50.

Stafford, E.R., and C.L. Hartman (1996) 'Green Alliances: Strategic Relations between Businesses and Environmental Groups', *Business Horizons* 39.2: 50-59.

Stahel, W.R. (1994) 'The Utilization-Focused Service Economy: Resource-Efficiency and Product-Life Extension', in B.R. Allenby (ed.), *The Greening of Industrial Ecosystems* (Washington, DC: National Academy Press): 178-90.

Stahel, W.R. (1996) *Handbuch Abfall 1: Allegemeine Kreislauf- und Rückstandswirtschaft: Intelligente Produktionsweisen und Nutzungskonzepte* (Karlsruhe: Landesanstalt für Umweltschutz Baden-Württemberg, 2nd edn).

Stansell, J. (1997) 'Filter system recycles water in the home', *The Sunday Times*, 1 June 1997: 3, 12.

Starik, M. (1995) 'Should trees have managerial standing? Toward Stakeholder Status for Non-Human Nature', *Journal of Business Ethics* 14.3: 207-17.

Steger, U. (1993) 'The Greening of the Board Room: How European Companies are Dealing with Environmental Issues', in K. Fischer and J. Schot (eds.), *Environmental Strategies for Industry* (Washington, DC: Island Press): 147-67.

Stenross, M., and G. Sweet (1992) 'Implementing an Integrated Supply Chain: The Xerox Example', in M. Christopher, *Logistics and Supply Chain Management* (London: Pitman).

Stern, P.C. (1986) 'Blind Spots in Policy Analysis: What Economics doesn't Say about Energy Use', *Journal of Policy Analysis and Management* 5: 200-27.

Stern, P.C. (1992) 'What Psychology Knows about Energy Conservation', *American Psychologist*, October 1992: 1224-32.

Sternberg, E. (1997) 'The Defects of Stakeholder Theory', *Corporate Governance: An International Review* 5.1: 3-10.

Stevels, A.L.N. (1997) 'Moving Companies towards Sustainability through Eco-Design: Conditions For Success', *Journal Of Sustainable Product Design* 3: 47-55.

Stewart, R.B. (1993) 'Economic Competitiveness and the Law', *Yale Law Review* 102 (June 1993): 2039-2106.

Stiftung Mitarbeit (1990) *Planungszelle: Bürgergutachten* (Brennpunkt-Dokumentation, 6; Bonn: Stiftung Mitarbeit).

Stisser, P. (1994) 'A Deeper Shade of Green', *American Demographics* 16.3 (March 1994): 24-29.

Stock, J.R. (1992) *Reverse Logistics* (Oakbrook, IL: Council of Logistics Management).

Strachan, P. (1997) 'EMAS: Recent Experiences of UK Firms', *European Environment* 1: 25-33.

Strachan, P., and J. Moxen (1998) *Green Teams: Environmental Change in Organisations and Networks* (Sheffield, UK: Greenleaf Publishing).

Suamapuddhi, K. (1998) 'Pipeline vs Lifeline', *Sarakadee* (Bangkok), February 1998: 62-82.

Sullivan, M.S., and J.R. Ehrenfeld (1992) 'Reducing Life-Cycle Environmental Impacts: An Industry Survey of Emerging Tools and Programs', *Total Quality Environmental Management*, Winter 1992–93: 143-57.

Sunday Telegraph (1994) 'Body Language: Body shop staff fend off reporters', *The Sunday Telegraph*, 28 August 1994.

SustainAbility/UNEP (United Nations Environment Programme) (1997) *Engaging Stakeholders: The 1997 Benchmark Survey* (London: SustainAbility/UNEP).

Swezey, B. (1997) 'Utility Green Pricing Programs: Market Evolution or Devolution?', *Solar Today*, January/February 1997: 21-23.

TAB (Büro für Technikfolgenabschätzung beim Deutschen Bundestag) (n.d.) 'Konsensus-Konferenzen: Ein neues Element demokratischer Technologiepolitik?', *TAB-Brief* (Bonn) 10: 4.

Tannenbaum, J. (1998) Taking a Bath, *The Wall Street Journal*, 22 June 1998.

Taylor, A., III (1998) 'Iacocca and Stempel go green: An Unlikely Alliance', *Fortune*, 2 March 1998: 34-35.

Taylor, P.W. (1993) 'The Ethics of Respect for Nature', in M.E. Zimmerman, J.B. Callicott, G. Sessions, K.J. Warren and J. Clark (eds.), *Environmental Philosophy* (Englewood Cliffs, NJ: Prentice–Hall): 66-83.

Terpstra, V. (1997) *International Marketing* (Fort Worth, TX: Dryden Press, 7th edn).

Thongrung, W. (1998) 'No. 5 air conditioner drive a hit', *The Nation* 21 (17 February 1998): B3.

Thorson, E., T. Page and J. Moore (1995) 'Consumer Response to Four Categories of "Green" Television Commercials', *Advances in Consumer Research* 22: 243-50.

Throop, G.M. (1993) *Strategy in a Greening Environment: Supply and Demand Matching in US and Canadian Electricity Generators* (doctoral dissertation; Amherst, MA: University of Massachusetts).

Tibor, T., and I. Feldman (1996) *ISO 14000: A Guide to New Environmental Management Standards* (Chicago: Irwin).

Tomczak, T. and S. Reineke (1996) *Der Aufgabeorientierte Ansatz* (Thexis Beiträge, 4/96; St Gallen, Switzerland: University of St Gallen).

Tomlinson, A. (1990) 'Consumer Culture and the Aura of the Commodity', in A. Tomlinson (ed.), *Consumption, Identity and Style* (London: Routledge).

Toor, M. (1992) 'ISBA's green code delays government legislation', *Marketing*, 30 January 1992: 8.

Tran, M. (1994) 'US franchisees fear retaliation: Body Shop', *The Guardian*, 26 August 1994.

Trevino, L., and K. Nelson (1995) *Managing Business Ethics* (New York: John Wiley).

Tuppen, C. (1993) 'An Environmental Policy for British Telecommunications', *Long Range Planning* 26.5: 24-26.

Union of Concerned Scientists (1998) 'Make Your Choice Count: Buy Green Power', *http://www.ucsusa.org/energy/buy.green.html* (Union of Concerned Scientists' Energy Program).

Unnarat, S., and W. Techawongtham (1998) 'Protests have come too late, says Chuan', *Bangkok Post*, 10 February 1998: Home 1.

Valentine, P.S. (1993) 'Ecotourism and Nature Conservation: A Definition with Some Recent Developments in Micronesia', *Tourism Management*, April 1993: 108-109.

Varadarajan, P.R. (1992) 'Marketing's Contribution to Strategy: The View from a Different Looking Glass', *Journal of the Academy of Marketing Science* 20.4 (Fall 1992): 335-43.

Vaughan, D., and C. Mickle (1993) *Environmental Profiles of European Business* (London: Earthscan/Royal Institute of International Affairs).

Vidal, J. (1992) 'The Big Chill', *The Guardian*, 19 November 1992: 2.

Vidal, J., and P. Brown (1994) 'The Crucifixion of St Ethica. Body Shop: Profile of Anita Roddick', *The Guardian*, 26 August 1994.

Vigon, B.W., and M.A. Curran (1993) 'Life-Cycle Improvement Analysis: Procedure Development and Demonstration', *Proceedings of IEEE International Symposium on Electronics and Environment*, Arlington, VA, May 1993.

Voien, S. (1997) 'Pricing in Competitive Markets', *Electric Power Research Institute Journal* 22.6 (November/December 1997): 6-13.

Vollmuth, H. (1994) *Controlling-Instrumente von A–Z* (Munich: Planegg, 2nd edn).

von Weizsäcker, E.U., A.B. Lovins and L.H. Lovins (1997) *Factor Four: Doubling Wealth, Halving Resource Use* (London: Earthscan).

Wagner, C., and H. Schmeck (1998) 'Gain Mobility by New Forms of Vehicle Utilisation and Mobility Management', paper presented at the *5th International Automotive Marketing Conference* (Lausanne, Switzerland: European Society for Opinion and Marketing Research, 2–4 March 1998).

Walker, O.C., and R.W. Reukert (1987) 'Marketing's Role in the Implementation of Business Strategies: A Critical Review and Conceptual Framework', *Journal of Marketing* 51 (July 1987): 15-33.

Wallace, C.P., and E. Brown (1996) 'Can The Body Shop shape up?', *Fortune*, 15 April 1996: 118-20.

Wallace, D.R., and N.P. Suh (1993) 'Information-Based Design for Environmental Problem Solving', *Annals of CIRP* 42.1: 175-80.

Walley, N., and B. Whitehead (1994) 'It's not easy being green', *Harvard Business Review* 72.3: 46-52.

Walsh, M.W. (1995) 'Case Study: Company puts faith in freon-free fridge', *Los Angeles Times*, 10 January 1995: 4.

Walton, S.V., R.B. Handfield and S.A. Melnyk (1998) 'The Green Supply Chain: Integrating Suppliers into Environmental Management Processes', *International Journal of Purchasing and Materials Management* 34.2: 2-11.

Ward, S. (1994) 'Shares rally in Body Shop fightback', *The Independent*, 2 September 1994.

Wasik, J.F. (1996) *Green Marketing and Management: A Global Perspective* (Cambridge, MA: Blackwell).

Watson, T. (1998) 'For many, it's not easy being green', *USA Today*, 22 April 1998.

WBSCD (World Business Council for Sustainable Development) (1997) *Business and Climate Change: Case Studies in Greenhouse Gas Reduction* (Geneva: WBCSD).

WCED (World Commission on Environment and Development) (1987) *Our Common Future* ('The Brundtland Report'; Oxford: Oxford University Press).

Weaver, P., with F. Schmidt-Bleek (1999) *Factor 10: Manifesto for a Sustainable Planet* (Sheffield, UK: Greenleaf Publishing).

Wehrmeyer, W. (1995) 'Environmental Management Styles, Corporate Cultures and Change', *Greener Management International* 12 (October 1995): 81-94.

Wehrmeyer, W. (1998) 'Identifying Corporate Cultures for Strategic Health and Safety Management', in A. Hale and M. Baram (eds.), *Safety Management: The Challenge of Change* (Oxford: Pergamon): 107-16.

Wehrmeyer, W., and D. Tyteca (1998) 'Measuring Environmental Performance for Industry: From Legitimacy to Sustainability?', *The International Journal of Sustainable Development and World Ecology* 5: 111-24.

Welford, R. (1993) 'Breaking the Link between Quality and the Environment: Auditing for Sustainability and Life-Cycle Assessment', *Business Strategy and the Environment* 2.4: 25-33.

Welford, R. (1995) *Environmental Strategy and Sustainable Development* (London: Routledge).

Welford, R. (1997) *Hijacking Environmentalism: Corporate Responses to Sustainable Development* (London: Earthscan).

Western, D. (1993) 'Defining Ecotourism', in K. Lindberg and D.E. Hawkins (eds.), *Ecotourism: A Guide for Planners and Managers* (North Bennington, VA: The Ecotourism Society): 7-11.

Westley, F., and H. Vredenburg (1991) 'Strategic Bridging: The Collaboration between Environmentalists and Business in the Marketing of Green Products', *Journal of Applied Behavioral Science* 27.1: 65-90.

Westley, F., and H. Vredenburg (1996) 'Sustainability and the Corporation: Criteria for Aligning Economic Practice with Environmental Protection', *Journal of Management Inquiry* 5.2: 104-19.

Wheatley, M.J. (1994) *Leadership and the New Science: Learning about Organisation from an Orderly Universe* (San Francisco: Berrett-Koehler).

Wheeler, D., and M. Sillanpää (1997) *The Stakeholder Corporation: A Blueprint for Maximizing Stakeholder Value* (London: Pitman).

White, L. (1967) 'The Historical Roots of our Ecological Crisis', *Science* 155: 1203-1207.

Willard & Shullman, Inc. (1994) *Renewable Energy Alternative Program Research* (presentation booklet for New England Power Service Company [NEPSCO]; Westborough, MA: NEPSCO).

Wille, C. (1992) 'Making Way for the Better Banana', *Audubon Activist*, December 1992.

Wilson, D.C., and W.L. Rathje (1990) 'Modern Middens', *Natural History*, May 1990: 54-58.

Wind, Y., and T.S. Robertson (1983) 'Marketing Strategy: New Directions for Theory and Research', *Journal of Marketing* 47 (Spring 1983): 12-25.

Winter, G. (1988) *Business and the Environment: A Handbook of Industrial Ecology with 22 Checklists for Practical Use* (Hamburg: McGraw–Hill).

Winter, G. (1995) *Blueprint for Green Management: Creating your Company's Own Action Plan* (Hamburg: McGraw–Hill).

Winter, L., and G. Ledgerwood (1994) 'Motivation and Compliance in Environmental Performance for Small and Medium-Sized Businesses', *Greener Management International* 7 (July 1994): 62-71.

Wiser, R., and S. Pickle (1997) 'Green Marketing and Renewables: What Role for Public Policy in a Restructured Electric Industry?', paper presented at the 18th Annual North American Conference, *International Energy Markets, Competition and Policies* (United States Association for Energy Economics, San Francisco, 7–10 September 1997): 563-72.

Wiser, R., and S. Pickle (1998) *Selling Green Power in California: Product, Industry, and Market Trends* (LBNL-41807; Berkeley, CA: Lawrence Berkeley National Laboratory).

Wong, V., W. Turner and P. Stoneman (1995) *Market Strategies and Market Prospects for Environmentally Friendly Consumer Products* (Warwick Business School Research Bureau, Paper 165; Coventry, UK: University of Warwick, March 1995); also in *British Journal of Management* 7 (1996): 263-81.

Woodward, R. (1994) 'Rumours deal blow to Body Shop shares', Reuters News Service, 29 August 1994.

Xepapadeas, A., and A. de Zeeuw (1998) *Environmental Policy and Competitiveness: The Porter Hypothesis and the Composition of Capital* (Discussion Series, 38; Tilburg, Netherlands: Tilburg University Center for Economic Research).

Yin, R.K. (1994) *Case Study Research: Design and Methods* (Thousand Oaks, CA: Sage).

Zagor, K. (1994a) 'Fairly Green but not Pristine', *The Guardian*, 10 September 1994.

Zagor, K. (1994b) 'Fund managers stand their ground after body blow', *The Guardian*, 3 September 1994.

Zeithaml, C., and V.A. Zeithaml (1984) 'Environmental Management: Revisiting the Marketing Perspective', *Journal of Marketing* 48 (Spring 1984): 46-53.

Ziffer, K. (1989) *Ecotourism: The Uneasy Alliance* (Washington, DC: Conservation International/Ernst & Young).

Biographies

Daniel Ackerstein is currently the Marketing Co-ordinator for Walsh Environmental, an environmental engineering and consulting firm in Boulder, CO, USA. He received his Master of Environmental Management in Corporate Environmental Strategy from Duke University's Nicholas School of the Environment. His areas of work and interest include green marketing, corporate environmental strategy and policy, and pollution prevention.

Subhabrata Bobby Banerjee is Associate Professor of Management and Programme Director of the Doctor of Business Administration programme at RMIT University, Australia. He received his PhD from the University of Massachusetts at Amherst and has taught there and at the University of Wollongong, Australia. His research interests are in corporate environmentalism, sustainable development, socio-cultural influences of globalisation, critical theory, post-colonialism and issues relating to indigenous communities. He has published more than 30 articles in several international journals, conference proceedings and edited collections.

Colin Beard is a lecturer, writer and trainer specialising in environmental issues. He was instrumental in creating the UK's first DTI-funded computerised environmental awareness training package called 'Earthwise', has designed and delivered environmental awareness programmes to organisations in the UK, Europe and Asia. He has a particular interest in the use of creativity, risk and innovation in learning. He is a Fellow of the Institute of Personnel and Development.

Frank Martin Belz is a researcher at the Institute for Economy and the Environment, and also lecturer for General Management at the University of St Gallen (HSG), Switzerland. He has published several books and articles on eco-marketing and environmental management. He is due to complete a post-doctoral thesis on 'Integrative Eco-Marketing: The Successful Marketing of Ecological Products and Services' in summer 1999. He has also worked as a consultant for companies such as Migros, Switzerland's largest retailer.

John Butler is an Associate Professor of Management and Director of the South-East Asia Center at the University of Washington, USA. He has spent several years as a visiting professor in Thailand, Vietnam and Hong Kong. He is interested in strategic issues related to green and social marketing.

Martin Charter has held management positions in financial services, trade exhibitions and consultancy and, over the last ten years, has established various organisations in the business and environment field, in publishing, training, research and consultancy. He is currently Co-ordinator of The Centre for Sustainable Design (CFSD) at the Surrey Institute of Art and Design, University College, and Surrey and Hampshire Environmental Business Association, UK. At CFSD, he is responsible for the strategic direction and management of international training and the research centre, which covers environmental communications, managing eco-design and sustainable product development and design. He is a Fellow of the Royal Society of Arts (RSA) and has completed two books on environmental issues and marketing—including the first edition of *Greener Marketing* (Greenleaf Publishing, 1992)—and a number of publications on managing eco-design and electronic environmental reporting. He has edited a range of green management publications and is currently Editor of *The Journal of Sustainable Design*, as well as acting as the UK expert to the ISO sub-group on 'Design for the Environment'.

Roland Clift is Professor of Environmental Technology and Founding Director of the Centre for Environmental Strategy (CES) at the University of Surrey, UK. CES was set up in 1992 as multidisciplinary research centre concerned with long-term environmental problems. In addition to its research activities, CES runs MSc, PhD and EngD programmes, the last of these being an innovative Doctor of Engineering programme in Environmental Technology. He is a member of the Royal Commission on Environmental Pollution and the UK Eco-labelling Board. Professor Clift is a Fellow of the Royal Academy of Engineering and the Institution of Chemical Engineers.

John C. Cox is Associate Professor of Marketing and Dean of the School of Business at Campbellsville University in Campbellsville, KY, USA. Dr Cox is a veteran of over 20 years of marketing with various firms in sales and distribution. He has over 10 years of experience in university teaching, research and administration and has conducted and published research on green marketing issues relating to the Maquiladora industry along the South Texas border with Mexico.

Graham Earl is a consultant with the Environmental, Safety and Risk practice at Arthur D. Little, Cambridge, UK, where he provides expertise on environmental management, reporting, policy, strategic planning and direction. Prior to this, Dr Earl worked as an industrially based research engineer for which he was awarded an Engineering Doctorate in Environmental Technology from the University of Surrey. This doctoral research, which was sponsored by Paras Ltd and the University of Surrey's Centre for Environmental Strategy, involved extensive liaison and test case applications with leading multinational companies and resulted in the development of the Stakeholder Value Analysis Toolkit. Previous to this research, Dr Earl worked as a consultant to the petrochemical industry providing expertise on safety, quantitative risk and reliability of industrial operations.

Michael Getzner's research focus is on environmental economics and infrastructure policy. From 1991 to 1997 he was researcher at the University of Technology, Vienna. He is currently Assistant Professor at the Department of Economics, University of Klagenfurt, Austria.

Sara Gooding-Williams is currently a PhD candidate in Marketing at the Isenberg School of Management, University of Massachusetts, and a Visiting Pre-doctoral Fellow at the Institute for Health Services Research and Policy Studies at Northwestern University, Evanston, IL. She received her Bachelor's degree from Yale College and her Master's degree from the Harvard School of Public Health, and has worked extensively in the healthcare sector, most recently as Director of Research and Strategic Analysis for the health insurance division of a *Fortune 500* company.

Sonia Grabner-Kräuter is Associate Professor of Marketing and International Management at the University of Klagenfurt, Austria. She teaches in the areas of international marketing and export finance, having taught students both in Austria and in the USA. The focus of her work is on business ethics and risk management in international business. Professor Grabner-Kräutner has been published in *Schmalenbachs Zeitschrift für betriebswirtschaftliche Forschung (zfbf)*, *The Journal of Business Administration* and *The Journal of Banking and Stock Exchange*.

Dallas Hanson is a Senior Lecturer in strategic management at the University of Tasmania. His research interests include strategic management, management and the natural environment, and innovation and knowledge management. His PhD work focused on issues of environmental sustainability. He has a background in higher-level government policy-making and now consults on strategic management issues.

Cathy L. Hartman holds a PhD in Business Administration from the University of Colorado. Her stakeholder research focuses on green alliances—partnerships between NGOs/environmental groups and businesses and other social/political entities. Her work has been published in *Business Horizons*, *Long Range Planning*, *The Journal of Business Ethics* and various national conference proceedings. She is currently serving as co-editor for a special issue for *Business Strategy and the Environment* on environmental partnerships and alliances for sustainability.

Rainer Hartmann is a consultant in European economics and ecology and a lecturer for Environmental Management at the School of Construction of Sheffield Hallam University, UK. Following studies in biology, chemistry, pedagogics, European economy and international business administration, he focused on EU policies and economics, international education and research programmes, comparative international approaches to environmental management, conservation and toxicology.

Kai Hockerts has been researching eco-efficient services since 1992. In 1993 he was project manager for the first conference on eco-efficient services in Bayreuth, Germany. In 1995, he published a research paper on eco-efficient services at the Institute for Business and Ecology (University of St Gallen, Switzerland), where he is today involved in PhD research on corporate sustainability and innovation. He also works as a freelance consultant on Life-Cycle Assessment and Design for Environment at the international consultancy Ecobilan, Paris.

Leena Ilmola is the managing director of communications agency Promotiva Ltd, CARTA Corporate Advisors, Helsinki, Finland. She is a co-founder of EnviroCom, and international network specialising in environmental communications. Since the 1980s, she has worked on environmental communications issue in several industry sectors. She holds an MSc in Social Sciences from the University of Helsinki.

Easwar S. Iyer is Associate Professor of Marketing at the Isenberg School of Management, University of Massachusetts, and undertook his doctoral work at the University of Pittsburgh. He is especially interested in environmental issues as they apply to business practices and has written extensively on various aspects of that topic, such as environmental management, environmental alliances and environmental advertising. He has designed and taught a graduate-level course, 'Environmental Issues in Business' in the MBA programme. He has also presented many seminars to executive audiences.

Kate Kearins is a senior lecturer in Strategic Management at the University of Waikato, Hamilton, New Zealand. She completed the case study in this volume while on leave in the International Centre for the Environment at the University of Bath, UK, in 1998. Dr Kearins's research interests include business strategy and the natural environment, business history and the role of rhetoric and local government power and politics.

Babs Klÿn is a former student in Strategic Management at the University of Waikato, Hamilton, New Zealand, and a former assistant lecturer in Marketing at Massey University, Palmerston North, New Zealand. She is currently working as an investigator with New Zealand's Ministry of Commerce.

Suthisak Kraisornsuthasinee is a lecturer in the Department of Marketing at Thammasat University, Bangkok, Thailand. His published materials and teaching interests are focused on green marketing and issues related to environmental and social marketing in Thailand.

Hannu Kuusela is Associate Professor of Marketing at the University of Tampere, Finland. Dr Kuusela's dissertation involved analysing verbal protocols. He has since produced numerous papers using this methodology to gain insight into decision-making processes, including one that appeared in *The European Journal of Marketing*.

Katherine A. Lemon is on the Marketing faculty of the Harvard Business School. She received her PhD in Business Administration from the University of California at Berkeley. Prior to receiving her PhD, she held executive marketing positions in the high-tech and healthcare industries. Professor Lemon's research examines consumer decision-making and dynamic customer relationships. Her work has appeared in *Marketing Science*, *The Journal of Marketing Research* and *Marketing Letters*. She is on the Editorial Board of *The Journal of Service Research*. In addition to teaching and research, Professor Lemon lectures and consults on the art of Customer Relationship Management.

Lassi Linnanen is the managing director of Gaia Network Oy, an environmental management and sustainable development consultancy based in Helsinki, Finland. He holds a PhD in Economics from the University of Jyväskylä and an MSc in Engineering from the Helsinki University of Technology. He has authored three environmental management handbooks in Finnish and published several articles in international journals. He works as an advisor to a number of multinational companies and public organisations.

Elina Markkanen is a project manager at Gaia Network Oy, Finland. She holds an MSc in Engineering from the Helsinki University of Technology. Her field of specialisation includes strategic environmental management, environmental management value chain management and environmental reporting.

Alan Neale is a Senior Research Fellow in Strategic and International Management at the East London Business School, University of East London, UK. He has published widely in the fields of corporate responsibility and environmental policy, including contributions to *Environmental Futures* (Macmillan, 1999) and *Organisational Behaviour: A Critical Text* (International Thomson Business Press, forthcoming). He is also co-author of *Economics in a Business Context* (International Thomson Business Press, 3rd edn forthcoming).

Jacquelyn Ottman is president of J. Ottman Consulting, Inc., a firm she founded in 1989 to help companies and other organisations find innovative solutions to environmental product and marketing challenges. She created the *Getting to Zero*^{sm} process, an innovation process designed to help companies develop concepts for environmentally sustainable products and services. She is the author of *Green Marketing: Opportunity for Innovation* (Contemporary Books, 2nd edn, 1998).

Jacob Park is a Washington DC-based environmental policy and business analyst affiliated with the Tokyo-based United Nations University's Institute of Advanced Studies.

Ken Peattie is a Senior Lecturer in Strategic Management at the Cardiff Business School, UK, which he joined in 1986. Before moving into lecturing, he worked in marketing, strategic planning and information systems in the paper and electronics industries. His research interests are primarily centred on the greening of corporate and marketing strategies; the use of social marketing techniques in skin cancer prevention; and innovation in sales promotion. He is the author of *Environmental Marketing Management: Meeting the Green Challenge* (Pitman, 1995).

Michael Jay Polonsky is a tenured senior lecturer in Marketing at the University of Newcastle in Australia. He has published a number of works examining a range of environmental/management issues, including editing three books on the topic as well as papers in *The Journal of Macromarketing, The Journal of Business Ethics, The International Journal of Advertising, The Journal of Business and Industrial Marketing, The Journal of Marketing Management* and *Business Strategy and the Environment*. He has also been invited to speak on this topic at a range of conferences in the US, Mexico and Australia.

Devashish Pujari is a lecturer in marketing and a Vice-Chair, Doctoral Programme, at the University of Bradford Management Centre. He teaches services marketing, marketing strategy and international marketing at all levels. His main research and consultancy interests are environmental marketing management and customer satisfaction and service quality. Dr Pujari has published papers in *The European Journal of Marketing, Marketing Intelligence and Planning, Business Strategy and the Environment,* and *The Journal of Euromarketing* and has presented papers at numerous international conferences.

Robin Roy is a Senior Lecturer in Design at the Open University, UK, with a background in mechanical engineering, design and planning. Since joining the OU, he has chaired and contributed to many courses, including 'Man-Made Futures'; 'Design: Processes and Products'; 'Design: Principles and Practice'; 'Design and Innovation'; 'Managing Design' and 'Innovation: Design, Environment and Strategy'. He is head of the OU/UMIST (University of Manchester Institute of Science and Technology) Design Innovation Group, which he founded in 1979. His research interests include eco-design and sustainable technologies, the management of design and innovation, and the design evolution of bicycles and railways. He has written or edited eight books and published over 60 research papers on these and other topics.

Joseph Sarkis is currently an Associate Professor of Operations Management at Clark University, Worcester, MA, USA. Dr Sarkis has published over 80 research and practitioner articles in operations management and environmentally conscious business practices.

Alexander Schwarz-Musch is Assistant Professor of Marketing and International Management at the University of Klagenfurt, Austria, where he undertakes research in customer satisfaction and international communications. He teaches in the areas of marketing communications and brand management. The focus of his work is on marketing communications in an international environment.

Mark T. Spence is Associate Professor of Marketing at Southern Connecticut State University, USA. He has published several manuscripts on expert decision-making, including ones that appeared in *Journal of Marketing Research, Business Horizons* and *Organizational Behavior and Human Decision Processes.*

Edwin R. Stafford holds a PhD in Business Administration from Arizona State University. His research centres on green alliances and environmental stakeholder collaboration. This research has been published in *Business Horizons, Long Range Planning* and American Marketing Association conference proceedings. He is currently serving as co-editor for a special issue for *Business Strategy and the Environment* on environmental partnerships and alliances for sustainability.

John Steen holds a BSc and a PhD in the natural sciences. His research interests include published studies on Tasmanian native animals, ecological sustainability and ecotourism. At present, he is employed as an Associate Lecturer within the School of Management at the University of Tasmania, teaching in Strategic Management and International Business. Recent publications include the topics of ecotourism, innovation and the management of research and development.

Rhett H. Walker is a Senior Lecturer within the School of Management at the University of Tasmania. His research and teaching interests include customer service and the nature of service-mindedness, the marketing and management of services, competitive market positioning, tourism and eco-marketing. His research papers have been presented at a variety of academic conferences nationally and internationally, and his journal publications include *The European Journal of Marketing, The Journal of Marketing Management* and *Advances in International Marketing*. He is also the co-author of *Services Marketing Australia and New Zealand*, the leading textbook in its field.

Walter Wehrmeyer is the BG Surrey Scholar at the Centre for Environmental Strategy of the University of Surrey, working on contaminated land and the regeneration of brownfield sites. His main research interests lie within corporate environmental policy and strategy formation, risk communication and participative approaches toward urban regeneration, as well as environmental and sustainability performance indicators. He is a member of the Institute of Management and sits on the Editorial Board of *Greener Management International, The Journal of Industrial Ecology* and *Environment, Development and Sustainability*, and is co-editor of a forthcoming book on Environmental Management in Developing Countries (Greenleaf Publishing, 1999).

Wayne Wells is the Chair of the Engineering Technology Program at the University of Texas-Brownsville, TX, USA. His research has been in the area of manufacturing and mechanical engineering with some focus on environmentally benign materials development. Dr Wells has also worked for many years in industry as an Industrial and Manufacturing Engineer for the Ford Motor Company.

Rebecca Winthrop is a graduate of Swarthmore College, PA, with a double major in Political Science and Dance Politics and a concentration in Public Policy. Recent work includes posts in Costa Rica and Spain with the United Nations High Commissioner for Refugees.

Norbert Wohlgemuth's research focuses on renewable energy supply options and institutional aspects of the energy industry. Currently he is on leave from the University of Klagenfurt, Austria, to work at the UNEP Collaborating Centre on Energy and Environment in Roskilde, Denmark.

Gillian Wright is a lecturer in marketing at the University of Bradford Management Centre, UK, where she teaches customer behaviour and market research. Now specialising in research in public service and environmental management, her industrial background is in electronics and pharmaceuticals, where she worked as a market analyst. Gillian works closely with senior management in organisations such as the Institute of Water and Environmental Management. She is a member of the Chartered Institute of Marketing and the Market Research Society.

Katharina Zöller is an economic geographer and is working as a scientist and facilitator at the Akademie für Technikfolgenabschätzung (Center of Technology Assessment) in Baden-Württemberg, Germany. Her current PhD is on dialogue processes in the chemical industry in the US and Germany.

Index